Handbook of Lipid Research 2

The Fat-Soluble Vitamins

Edited by

Hector F. DeLuca
University of Wisconsin-Madison

Plenum Press · **New York and London**

Library of Congress Cataloging in Publication Data

Main entry under title:

The Fat-soluble vitamins.

  (Handbook of lipid research; v. 2)
  Includes bibliographies and index.
  1. Vitamins, Fat-soluble, I. DeLuca Hector F., 1930-   II. Series.
QP751.H33 vol. 2 [QP772.F37] 574.1'9247s [599'.01'926]
ISBN 0-306-33582-4   78-2009

QP
772
.F37
F37

© 1978 Plenum Press, New York
A Division of Plenum Publishing Corporation
227 West 17th Street, New York, N.Y. 10011

Printed in the United States of America

Contributors

H. F. DeLuca, Department of Biochemistry, College of Agricultural and Life Sciences, University of Wisconsin—Madison, Madison, Wisconsin 53706

Luigi M. De Luca, National Cancer Institute, Bethesda, Maryland 20014

Milton Leonard Scott, Department of Poultry Science and Division of Nutritional Sciences, Cornell University, Ithaca, New York 14853

J. W. Suttie, Department of Biochemistry, College of Agricultural and Life Sciences, University of Wisconsin—Madison, Madison, Wisconsin 53706

Preface

The first demonstration of the existence of a vitamin and the full recognition of this fact are often attributed to the work of McCollum, who found that a substance in butterfat and cod-liver oil was necessary for growth and health of animals fed purified diets. It became obvious that an organic substance present in microconcentrations was vital to growth and reproduction of animals. Following the coining of the word *vitamine* by Funk, McCollum named this fat-soluble substance vitamin A. We can, therefore, state that vitamin A was certainly one of the first known vitamins, yet its function and the function of the other fat-soluble vitamins had remained largely unknown until recent years. However, there has been an explosion of investigation and new information in this field, which had remained quiescent for at least two or three decades. It is now obvious that the fat-soluble vitamins function quite differently from their water-soluble counterparts. We have learned that vitamin D functions by virtue of its being converted in the kidney to a hormone that functions to regulate calcium and phosphorus metabolism. This new endocrine system is in the process of being elucidated in detail, and in addition, the medical use of these hormonal forms of vitamin D in the treatment of a variety of metabolic bone diseases has excited the medical community. The elucidation of the functional metabolism of vitamin D has also provided important tools for an in-depth study of its mechanism of action at the cellular level. Perhaps the nearest to complete solution regarding mechanism of action is vitamin K. There is now clear evidence that this vitamin is involved in the γ-carboxylation of glutamic acid residues in the clotting proteins, converting them to functional proteins in the blood-clotting process. The mechanism of the vitamin K catalyzed carboxylation is not entirely elucidated but will likely be understood further in the next decade. Lagging behind in elucidation of mechanism of action are vitamin A and vitamin E. The vitamin A problem, however, has changed dramatically in the realization that vitamin A is a multifunctional vitamin, and these functions may be satisfied by different forms of vitamin A. There is great interest in the role of vitamin A in the differentiation of epithelial tissues primarily because the vitamin A compounds are being considered for use in the prevention or delay of carcinogenesis. This added interest will undoubtedly put a great deal of investigational pressure on elucidating the functional forms of vitamin A in the differentiation process and the mechanism whereby it regulates differentiation. The most significant recent advance in the vitamin E field has been the realization that selenium, which has been known to substitute for vitamin E in prevention of a number of vitamin E deficiency diseases, probably functions because it is a structural component of glutathione peroxidase. Glutathione peroxidase destroys lipid hydroperoxides, which cause membrane damage. It is believed that vitamin E prevents membrane damage by preventing the formation of the lipid hydroperoxides. There are those, however, who still believe that vitamin E carries out some other specific function, quite apart from its antioxidant activities. Much remains to be learned with regard to the mechanism of action of this vitamin, but much progress has indeed been made.

The rapid accumulation of new information in the past decade in this field has really made necessary a review of the current status of our understanding of these important functional substances. This is likely to change in the future as new information is uncovered. It is hoped that this compilation will aid in our current understanding of the fat-soluble vitamins and will provide a springboard for new investigations and further elucidation of their functions and importance.

H. F. DeLuca

Madison

Contents

Chapter 1

Vitamin A

Luigi M. De Luca

Chapter 2

Vitamin D

H. F. DeLuca

Chapter 3

Vitamin E

Milton Leonard Scott

Chapter 4

Vitamin K

J. W. Suttie

Chapter 1

Vitamin A

Luigi M. De Luca

1.1. Historical Developments in Vitamin A Research

Stepp (1909) first described a lipid-soluble compound, present in egg yolk, which was essential for life. A compound with similar biological activity was later found in butterfat, egg yolk, and cod liver oil, and was named "fat-soluble A" by McCollum and Davies (1913, 1915). The "fat-soluble A" factor was capable of restoring and maintaining growth in rats kept on a deficient diet and of preventing xerophthalmia (McCollum and Simmonds, 1917) and night blindness (Fridericia and Holm, 1925). The active lipid was named "vitamin A" by Drummond (1920).

Steenbock et al. (1921) found a growth-promoting substance in plant extracts. The provitamin role of β-carotene became obvious after Karrer et al. (1930) elucidated the structure of β-carotene and that of retinol (Karrer et al., 1931, 1933).

In 1937, Holmes and Corbett succeeded in crystallizing vitamin A from fish liver: "The vitamin appeared in beautiful rosets or radiating clusters of pale yellow needles." The crystals were optically inactive and had a melting point of 7.5–8°C.

At about the same time, Wald (1935a,b,c, 1936a,b) isolated the chromophore from bleached retinas. The Liverpool group (Morton, 1944; Morton and Goodwin, 1944) demonstrated that the chromophore was retinal.

The first total synthesis of crystalline vitamin A was announced by Isler et al. in 1947. The first total syntheses of β-carotene were reported by Karrer and Eugster (1950), Inhoffen et al. (1950a,b,c), and Milas et al. (1950).

1.2. Nomenclature and Chemistry

The nomenclature follows the rules proposed by the IUPAC–IUB Commission on Biochemical Nomenclature (IUPAC, 1960; IUPAC–IUB, 1966). The parent compounds, retinol, retinal, and retinoic acid, and their most common derivatives are shown in Fig. 1. Four relatively new compounds are included:

Luigi M. De Luca • National Cancer Institute, Bethesda, Maryland 20014.

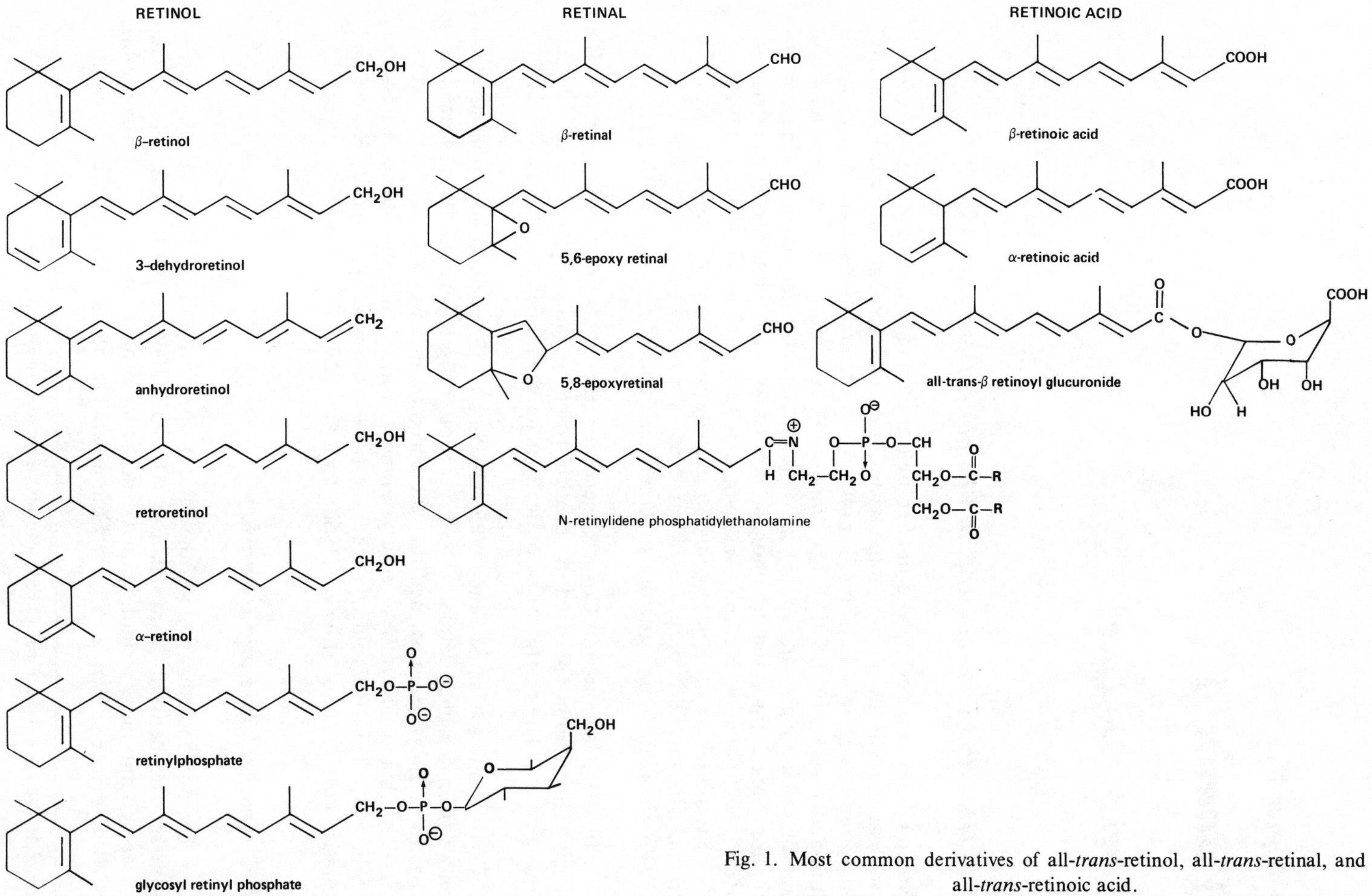

Fig. 1. Most common derivatives of all-*trans*-retinol, all-*trans*-retinal, and all-*trans*-retinoic acid.

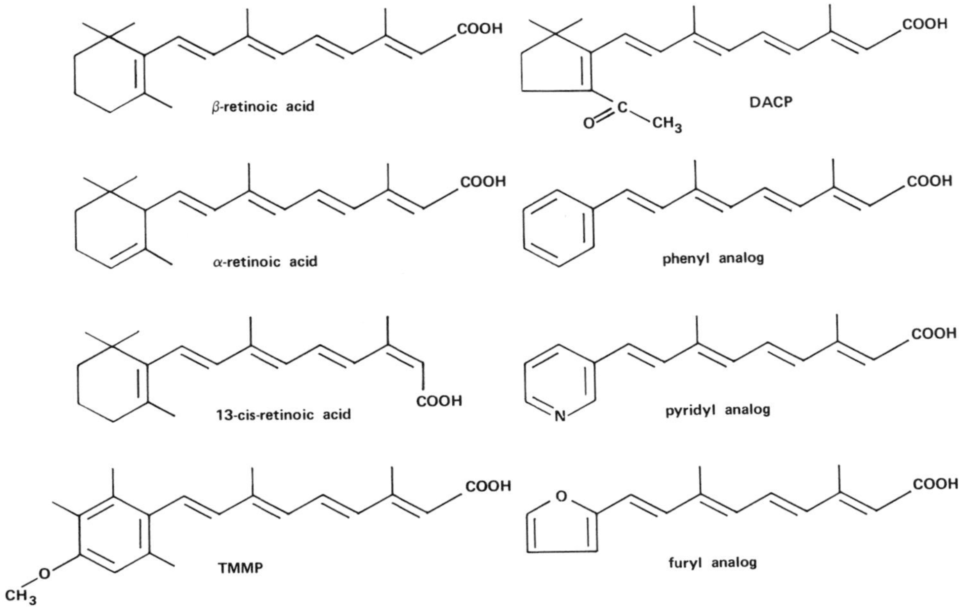

Fig. 2. Derivatives of retinoic acid. TMMP is the trimethylmethoxyphenyl analogue and DACP is the dimethylacetylcyclopentenyl analogue of retinoic acid.

retinoyl-β-glucuronide, retinyl phosphate, retinyl phosphate glycoside, and *N*-retinylidene phosphatidylethanolamine.

The chemical derivatives of retinoic acid with a modified cyclohexene ring are shown in Fig. 2. Some of these derivatives are active in reversing squamous metaplastic lesions in organ culture, although they may not be active in growth.

The synthetic chemistry of vitamin A and derivatives has been extensively reviewed (Isler *et al.*, 1967, 1970; Schwieter and Isler, 1967) and will not be presented here. A review of synthetic processes for obtaining β-carotene and other carotenoids has appeared (Mayer and Isler, 1971).

1.3. Biogenesis of Carotenoids

Carotenoids are a ubiquitous family of polyenic compounds found mostly in plants, fungi, bacteria, and some lower forms of animal life. Biosynthetic studies on this family of compounds gained momentum after the discovery of mevalonic acid (Wright *et al.*, 1956; Wolf *et al.*, 1956, 1957) and the demonstration of its role as an intermediate in the biosynthesis of cholesterol and derivatives (Tavormina *et al.*, 1956) and of carotenoids (Yokoyama *et al.*, 1962). Several common intermediates to steroid (Fig. 3) and carotenoid (Fig. 4) synthesis from mevalonic acid have been identified. Since the study of the biosynthesis and stereochemistry of squalene and its precursors has paved the way for parallel studies in the carotenoid series, the biochemical pathways of squalene synthesis will be considered first.

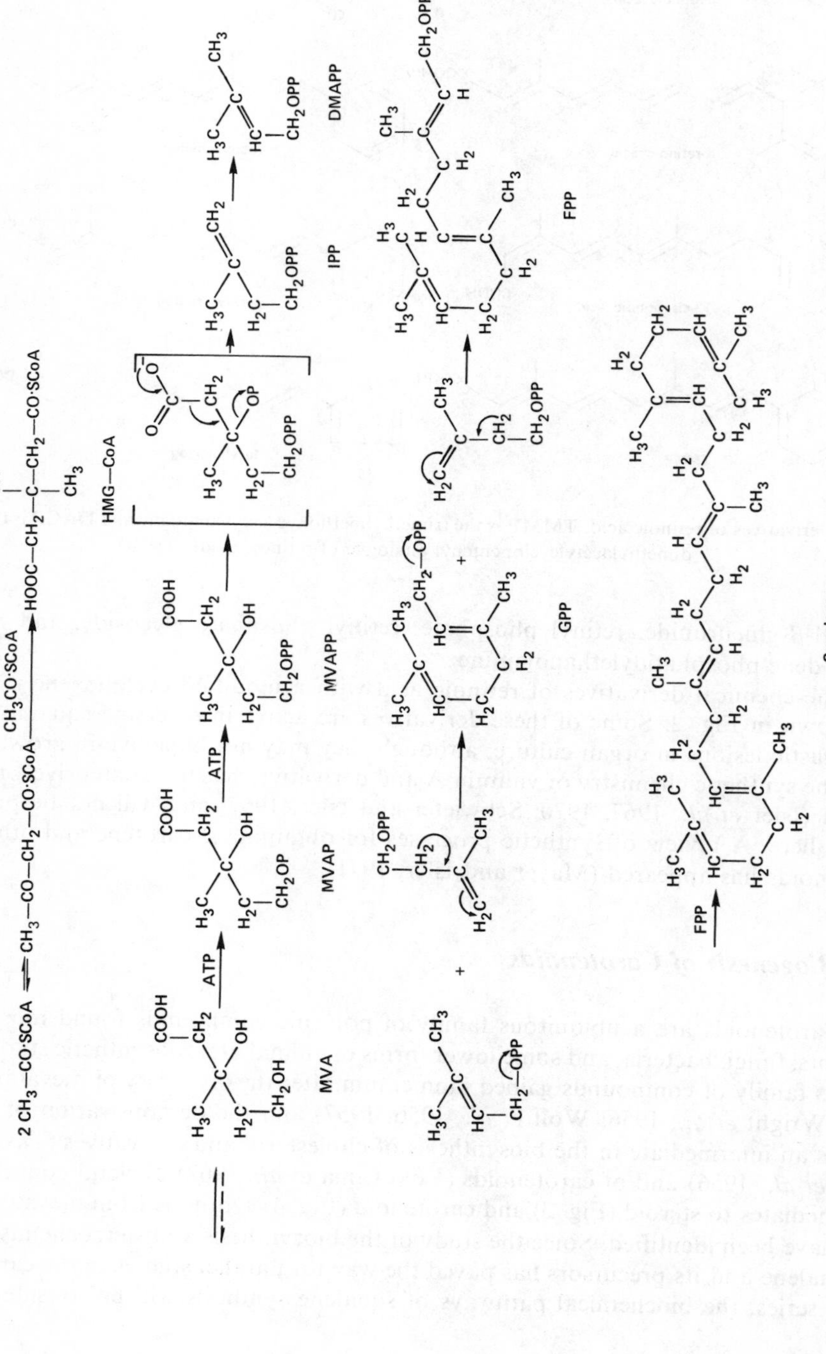

Fig. 3. The biosynthetic pathway from acetyl-CoA to squalene. From Ganguly and Murthy (1967).

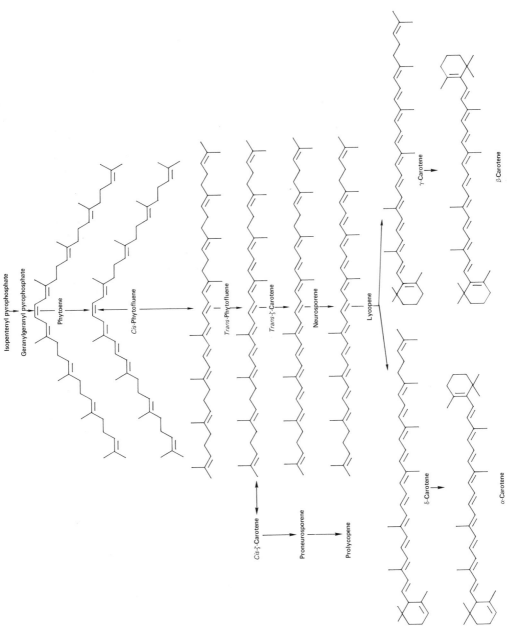

Fig. 4. The biosynthetic pathway from isopentenyl pyrophosphate to *cis*- and *trans*-carotenes. From Qureshi *et al.* (1974).

1.3.1. Squalene-Condensing System

The precursor role of squalene in the biosynthesis of cholesterol was discovered by Langdon and Bloch, who demonstrated that squalene is built up from [^{14}C]acetate by rat liver and that the resulting [^{14}C]squalene is converted to sterol when fed to mice (Langdon and Bloch, 1953*a,b*; Schwenk *et al.*, 1954). The distribution of ^{14}C in squalene biosynthesized from [Me-^{14}C]acetate was first obtained by oxidative ozonolysis (Cornforth and Popják, 1954). The derivation of the various carbon atoms from the carboxyl (C) and the methyl (M) is emphasized in Fig. 5.

A series of elegant studies followed to elucidate the pathway between mevalonic acid (MVA) and squalene (Fig. 3). Amdur *et al.* (1957) demonstrated that ATP, Mn^{2+}, and a reduced pyridine nucleotide are required by a soluble yeast extract to convert MVA into squalene. Chen (1957), working with the same yeast extract, demonstrated that the phosphorylating enzyme displayed specificity for only one of the two labeled enantiomorphs of mevalonic acid. Henning *et al.* (1959) discovered that MVA-5 P is further phosphorylated to MVA-5-PP with participation of a second molecule of ATP. Finally, Lynen *et al.* (1958) reported the isolation of isopentenyl ester obtained as the product of decarboxylation of MVA-5-PP, and Bloch *et al.* (1959) purified enzymes from yeast autolysates which catalyze the formation of mevalonic acid-5-PP and isopentenyl pyrophosphate (IPP). The phosphorylation of phosphomevalonic acid by ATP produces stoichiometric amounts of ADP. The decarboxylation of MVAPP to isopentenyl pyrophosphate and CO_2 consumes one molecule of ATP to yield ADP and Pi (Fig. 3). Both groups of investigators (Lynen *et al.*, 1958; Yuan and Bloch, 1959) confirmed the structure of biological IPP by chemical synthesis.

A key development in this series of metabolic pathways was the isolation of dimethylallyl pyrophosphate (DMAPP) as a product of isomerization of IPP (Agranoff *et al.*, 1959). Initial head-to-tail condensation of IPP and DMAPP gives geranyl pyrophosphate [GPP (C_{10})], successive head-to-tail condensation of GPP with IPP gives farnesyl pyrophosphate (FPP), and final tail-to-tail condensation of FPP gives squalene, as illustrated in Fig. 3 (Lynen *et al.*, 1958; Rilling and Bloch, 1959; Cornforth and Popják, 1959).

Up to 1959, very little was known about the stereospecific mechanisms that govern the condensing pathway from mevalonic acid to squalene. It had been demonstrated that the methyl group formed in the reaction yielding dimethylallyl pyrophosphate (DMAPP) from isopentenyl pyrophosphate is *trans* to the carbon chain (Arigoni, 1958). Also, Eberle and Arigoni (1960) established that the

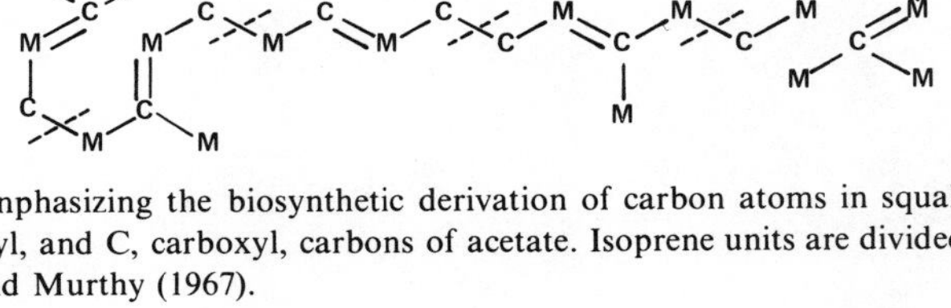

Fig. 5. Scheme enphasizing the biosynthetic derivation of carbon atoms in squalene using [2-^{14}C]-acetate. M, Methyl, and C, carboxyl, carbons of acetate. Isoprene units are divided by a dashed line. From Ganguly and Murthy (1967).

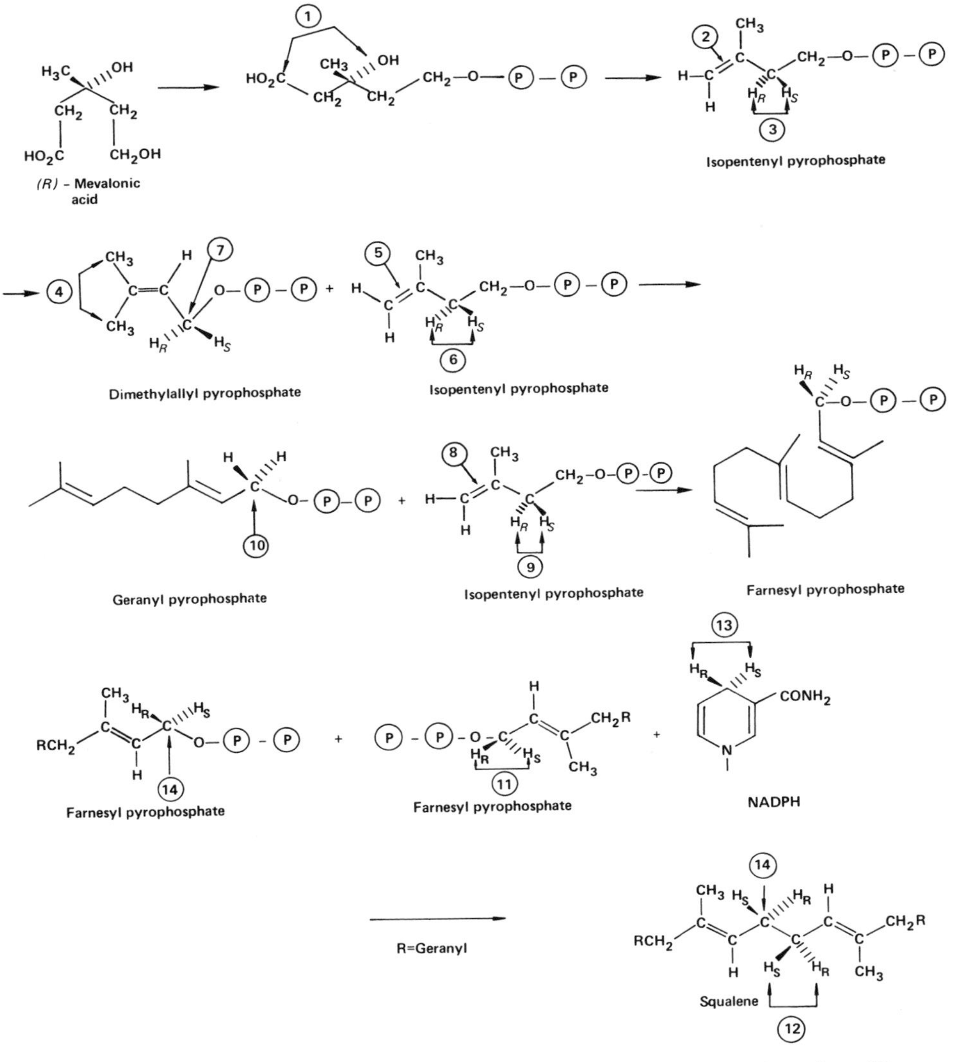

Fig. 6. Conversion of (R)-mevalonate to farnesyl pyrophosphate and squalene with points of "stereo-chemical ambiguity" indicated (Popják and Cornforth, 1966). Pyrophosphoryl groups are indicated by P-P-. From Alworth (1972).

biologically active, naturally occurring mevalonic acid has the $3R*$ configuration (Fig. 6).

Biological stereospecificity for the seven reactions leading from mevalonate to squalene was first recognized at 14 stages (Fig. 6) of the enzymatic process and brilliantly elucidated by Popják, Cornforth, and their collaborators. It was first recognized that squalene, a molecule with only *trans* geometry in its double bonds,

*The *R* and *S* convention for descriptions of absolute configuration around an asymmetrical center was introduced by Cahn *et al.* (1956).

is in fact synthesized by an asymmetrical process involving two differently derived farnesyl residues. In other words, an outwardly symmetrical substance, except for the *trans* double bonds, is not treated symmetrically by the enzyme(s) involved in the biosynthetic pathway, probably because enzyme–substrate binding imposes a particular orientation of the substrate at the physical site of reaction (Popják and Cornforth, 1966). Of the 14 points of stereochemical ambiguity, 12 were elegantly elucidated by Popják and Cornforth (1966).

This effort required the chemical synthesis of stereospecifically labeled mevalonates with a hydrogen isotope at C-4, C-2, and C-5, the three centers of chirality in mevalonic acid. It also required the syntheses of $1\text{-}^{14}C\text{-}$, $2\text{-}^{14}C\text{-}$, and $4\text{-}^{14}C$-labeled mevalonolactone, unlabeled and $3,4\text{-}^{14}C$-labeled mevalonolactone, and $(-)\text{-}R$-deuteriosuccinic acid (Cornforth and Popják, 1969).

As will be discussed, many of these stereospecifically labeled compounds were used in elucidating points of stereochemical ambiguity in the terpenoid pathway. We will discuss those advances in this field which are relevant to the understanding of stereochemical specificity in the biosynthetic pathway leading from mevalonate to terpenes. A review of the work of Popják and Cornforth has been published (Alworth, 1972).

1.3.2. Carotenoid-Condensing Pathway

Most of the early experiments on the biosynthesis of carotenoids were conducted in fungi, especially *Phycomyces blakesleeanus* and *Mucor hiemalis*; in bacteria, in particular the photosynthetic bacteria *Rhodospirillum rubrum* and *Rhodeopseudomonas spheroides*; and in carrot roots, fruit, and their homogenates (Goodwin, 1959; Shneour and Zabin, 1957). The formation of carotenoids from both acetate and mevalonate was demonstrated in microorganisms and plant tissues (Chichester *et al.*, 1959; Steele and Gurin, 1970; Grobb and Buttler, 1954). However, the first detailed demonstration that mevalonic acid is a more efficient precursor of β-carotene in *P. blakesleeanus* was reported by Yokoyama *et al.* (1962). These authors reported that the relative utilization of $[2\text{-}^{14}C]$mevalonic acid was 341 times that of $[2\text{-}^{14}C]$acetate and 31 times that of $[3\text{-}^{14}C]$hydroxymethylglutarate (HMG). Cofactor requirements for optimal incorporation of $[2\text{-}^{14}C]$MVA into β-carotene were NADH, NADP, NADPH, Mn^{2+}, and glutathione.

It soon became apparent that the biosynthesis of carotenoids occurred through a new common intermediate, phytoene (Fig. 4). Jungalwala and Porter (1967) reported the solubilization and partial purification of the enzyme system of tomato fruit plastids which catalyzes the synthesis of phytoene from geranylgeranyl pyrophosphate, in the same system. An enzyme preparation from tomato fruit plastids synthesized radioactive phytoene, phytofluene, neurosporene, and lycopene from $[4\text{-}^{14}C]$isopentenyl pyrophosphate (Suzue and Porter, 1969).

Working with a cell-free system from the R_1 mutant of *P. blakesleeanus*, Lee and Chichester (1969) demonstrated the conversion of labeled geranyl pyrophosphate to phytoene, phytofluene, β-carotene, neurosporene, and lycopene.

In 1970, Kushwaha *et al.*, using four distinct genetic types of tomato fruits, demonstrated the conversion of labeled phytoene into acyclic and cyclic carotenes (Fig. 4). The conversion of phytoene to phytofluene required NADP; that of phytoene to lycopene depended on FAD, but lack of this factor did not affect the formation of phytofluene. Mn^{2+} was required for lycopene synthesis. In the same paper, the authors demonstrated that [15,15'-^{3}H]lycopene is converted to α-, β-, γ-, and δ-carotenes by a soluble enzyme system from plastids of red tomato fruits.

Similar results on the conversion of phytoene to more unsaturated terpenes were obtained by Papastephanou *et al.* (1973), who used plastids from tomato fruits and studied the conversion of phytoene and lycopene to other carotenes. This work supported the view that phytoene is converted to more unsaturated acyclic carotenes and that lycopene is converted to the mono- and dicyclic carotenes (Fig. 4).

Additional studies by Porter's group (Qureshi *et al.*, 1974) demonstrated the conversion of *cis*- and *trans*-phytofluenes and *trans*-ζ-carotene to more unsaturated *trans* acyclic and cyclic carotenes by a soluble enzyme system from plastids of a variety of red tomato that synthesizes mainly *trans*-carotenes. Labeled *cis*- and *trans*-phytofluene and *trans*-ζ-carotene are also converted to poly-*cis*-carotenes by an enzyme from plastids of fruits of the golden jubilee tomato, which makes mostly poly-*cis*-carotenes at the expense of lycopene. All these experimental data are consistent with the pathway of carotene synthesis shown in Fig. 4.

1.3.3. Stereochemistry of the Carotenoid Pathway

Stereochemical aspects of the carotenoid pathway have been thoroughly discussed by Goodwin (1971). Only salient points of stereochemical identity or diversity between the steroid and terpene pathways will be stressed here.

By using stereospecifically labeled [2-^{14}C,4R,4-^{3}H]mevalonate and [2-^{14}C,-4S,4-^{3}H]mevalonate, Cornforth and Popják demonstrated that the hydrogen lost at C-4 of mevalonate is always in the pro-*S* configuration, whereas the pro-*R* hydrogen is retained in farnesol and squalene and derivatives (Fig. 7).

By using the same stereospecifically labeled mevalonate, Goodwin and Williams (1966) demonstrated that in plants the biosynthesis of phytoene, the carotenoid precursor (Fig. 4), proceeds by specific loss of the pro-*S* hydrogen at C-4 and retention of the pro-*R* hydrogen (Fig. 7).

A second point of stereochemical ambiguity regards the loss of two hydrogens from two molecules of geranylgeranyl pyrophosphate to generate one molecule of phytoene (Fig. 8). Popják and Cornforth (1966) had established that the pro-*S* hydrogen from C-1 of one of the two farnesyl pyrophosphates condensing to squalene is removed and replaced with H_S-hydrogen from C-4 of NADPH (step 13 of stereochemical ambiguity) (Fig. 6). To investigate the stereochemistry of the loss of two hydrogens in the condensation of geranylgeranyl pyrophosphate to phytoene, Williams *et al.* (1967) synthesized [2-^{14}C,5S,5-^{3}H]mevalonate and used it as a precursor to phytoene in enzyme preparations from tomato fruit and bean leaves.

They demonstrated that the pro-5*s* hydrogen is lost from each molecule of geranylgeranyl pyrophosphate at the condensing site (C-5 of mevalonic acid).

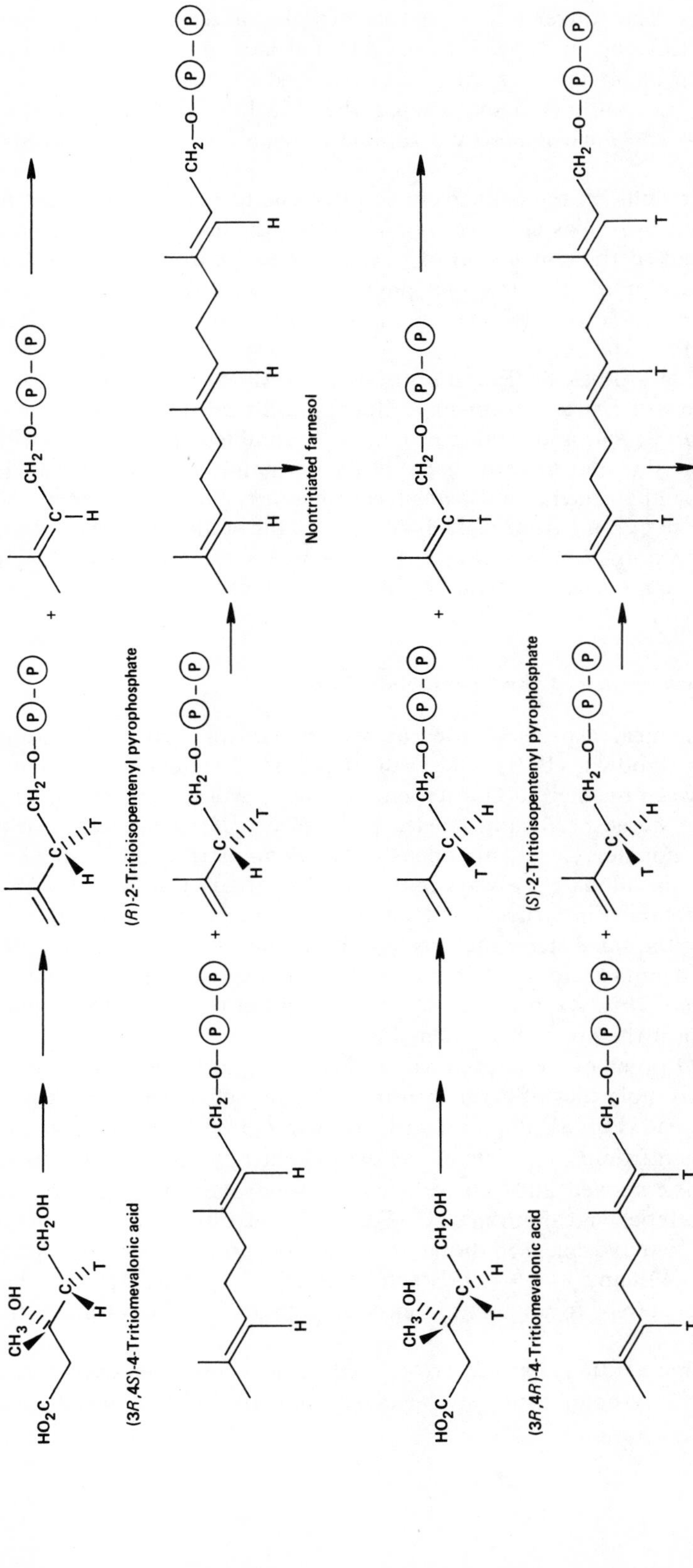

Fig. 7. Stereospecific incorporation of tritium from $(3R,4R)$-4-tritiomevalonate into farnesol. From Alworth (1972).

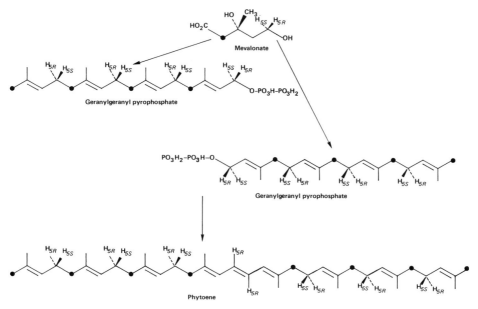

Fig. 8. Fate of the pro-5R and pro-5S hydrogens in the conversion of mevalonate into geranylgeranyl pyrophosphate and phytoene. • indicates ^{14}C; H_{5R} and H_{5S} indicate the pro-5R and pro-5S hydrogen atoms of mevalonate, respectively. From Goodwin (1971).

This is illustrated in Fig. 8. No loss of pro-(5S) 5-H occurs in the synthesis of geranylgeranyl pyrophosphate (Buggy *et al.*, 1969) (Fig. 8).

Stereochemical problems regarding the mechanism of cyclization of non-cyclic carotenoids await further elucidation, although interesting mechanisms have been proposed (Goodwin, 1971).

1.4. Conversion of β-Carotene to Retinol

Von Euler *et al.* (1928) and Moore (1930) demonstrated that β-carotene can replace vitamin A in its activity to promote growth. The provitamin is converted to vitamin A in several tissues (Thompson *et al.*, 1950; Wagner *et al.*, 1960; Olson, 1961, 1964; Zachman and Olson, 1963) including intestine and liver. A stoichiometric conversion of 1 mole of β-carotene into 2 moles of retinal is catalyzed by a soluble enzyme system from rat intestinal mucosa (Goodman and Huang, 1965) and rat liver (Olson and Hayashi, 1965). The enzyme, carotene-15,15′-oxygenase, catalyzes the reaction shown in Fig. 9.

The requirements and conditions of the reaction have been reviewed (Goodman and Olson, 1969).

The oxygenase is not present in microorganisms. It is present in mammals, and it can be found in a soluble form after centrifugation of the homogenate at 105,000*g* for 1 hr. The soluble enzyme can be purified by precipitation with ammonium sulfate between 20% and 45% saturation. Attempts to purify the enzyme by gel filtration were not successful because of instability of the enzyme. The rat

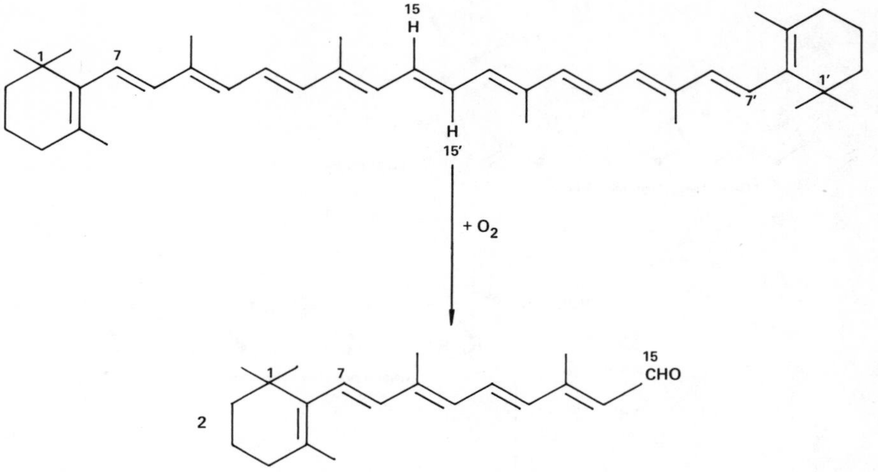

Fig. 9. Cleavage of β-carotene to retinal.

liver enzyme may be frozen and thawed repeatedly without loss of activity. The reaction is inhibited by —SH-group inhibitors and by chelators of ferrous ions such as α,α'-dipyridyl and 1,10-phenanthroline. The enzyme has no requirement for reduced pyridine nucleotides, and it retains activity after prolonged dialysis, suggesting a dioxygenase rather than a monooxygenase.

The enzyme cleaves β-carotene without loss of hydrogen from the two central carbon atoms (Goodman *et al.*, 1966). The resulting retinal is reduced to retinol by a soluble enzyme which requires a reduced pyridine nucleotide (Fidge and Goodman, 1968). Retinyl esters of long-chain fatty acids are then formed which are transported in complex with lymph chylomicrons to the circulation. Retinyl esters are transferred to the liver, where they constitute the storage form of vitamin A. Mobilization of vitamin A from the liver occurs via the retinol-binding protein.

1.5. Retinoic Acid

Retinoic acid was synthesized by Arens and Van Dorp (1946*a*), who also tested it for biological activity. When dissolved in peanut oil and given orally to rats, its activity was only one-tenth that of vitamin A. Subcutaneous administration of an aqueous solution was as active as vitamin A given in oil by mouth. However, although active in the growth test, the acid was not deposited in the liver, leading Arens and Van Dorp to the conclusion that the acid was not metabolized to vitamin A (Arens and Van Dorp, 1946*b*).

This preliminary conclusion was later substantiated by Dowling and Wald (1960), who demonstrated that treatment with retinoic acid of vitamin A deficient rats could not restore vision and synthesis of rhodopsin. Moreover, the cells of the rod outer segments were reduced to a single row of nuclei, although the rest of the retina was intact. Growth of the retinoic acid supplemented rats was normal, but no accumulation of retinoic acid in the livers was found. These studies supported the view that retinoic acid is not reduced to the aldehyde or the alcohol.

The growth-promoting activity of retinoic acid has been the subject of numerous investigations, with the essential conclusion that it equals that of retinol (Malathi *et al.*, 1963; Krishnamurthy *et al.*, 1963; Zile and DeLuca, 1968; Zachman *et al.*, 1966; Roberts and DeLuca, 1967; Fidge *et al.*, 1968; Geison and Johnson, 1970).

Another intriguing aspect of retinoic acid is its failure to maintain normal reproductive functions (Thompson, 1969; Juneja *et al.*, 1964). While retinol is necessary for the maintenance of normal reproductive function, retinoic acid fails to maintain intact testes in rats, guinea pigs, and hamsters. It also fails to maintain pregnancy in rats and guinea pigs, with resulting resorption of the fetus, and to maintain seminal vesicle size in rats. It was shown that the acid is not transferred by birds to their eggs, resulting in failure of hatching (Thompson, 1969).

At first, it was difficult to demonstrate retinoic acid in tissues. More recently, the use of radioactive precursors has allowed detection of small amounts of retinoic acid (Dunagin *et al.*, 1964, 1966; Deshmukh *et al.*, 1965; Deshmukh and Ganguly, 1967; Emerick *et al.*, 1967; Crain *et al.*, 1967).

The first unequivocal demonstration of the formation of retinoic acid from physiological doses of labeled retinyl acetate in rats was by Kleiner-Bossaller and DeLuca (1971). The labeled acid accumulated in the kidney before it appeared in blood, and its level was constant while that of other radioactive products decreased.

1.5.1. Urinary Metabolites of Retinoic Acid

Retinoic acid is absorbed from the portal vein (Fidge *et al.*, 1968) and is metabolized to polar compounds found in urine and in the bile (Zachman *et al.*, 1966). Urinary metabolites of retinoic acid have been extensively investigated by Rietz *et al.* (1974), and the structure in Fig. 10 was suggested for one of the metabolites on the basis of mass spectrometry (mol. wt. 376), from which by high-resolution mass spectroscopy a formula of $C_{22}H_{32}O_5$ was calculated. Evidence for the presence of a conjugated carbonyl function and a nonconjugated ester group came from infrared spectroscopy. No geminal methyl groups at C-1 of the cyclohexene ring were found by nuclear magnetic resonance. The ultraviolet absorption spectrum of the methylated metabolite in hexane displayed maxima at 230 nm and 314 nm, with shoulders at 302 and 329 nm and a maximum at 257 nm. This structure is in general agreement with previous work on carboxylated deriva-

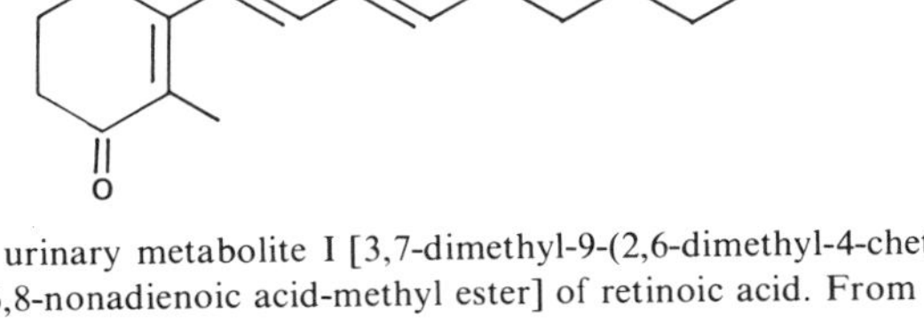

Fig. 10. Structure of urinary metabolite I [3,7-dimethyl-9-(2,6-dimethyl-4-cheto-6-carboxymethyl-1-cyclohexen-1-yl)-6,8-nonadienoic acid-methyl ester] of retinoic acid. From Rietz *et al.* (1974).

tives of vitamin A in rat urine (Wolf *et al.*, 1957). Sundaresan and Sundaresan (1973) have postulated at least six urinary metabolites of retinoic acid in rats and have suggested a metabolic attack at the hydrogen atoms of C-11 and C-12. These positions are in fact hydrogenated in metabolite I of Fig. 10. It must be kept in mind that this metabolite I and derivatives were obtained after pharmacological doses of vitamin A had been administered to rats. Thus the physiological significance of these findings is not established. The major metabolite in the bile has been identified as retinoyl-β-glucuronide (Dunagin *et al.*, 1965; Lippel and Olson, 1968) (Fig. 1).

1.5.2. Tissue Metabolism of Retinoic Acid

In a study of the metabolism of retinoic acid in the rat, Ito *et al.* (1974) examined the metabolites formed in liver, kidney, blood, small intestine, large intestine, stomach, muscle, lung, and the reproductive organs from injected [6,7,-^{14}C]retinoic acid. Retinoic acid esters and the free acid were the major polar derivatives present in all tissues. The highly polar fraction contained metabolite 8, which was present in all the tissues examined. Two more polar metabolites were found only in blood and small intestine. Of these, one was probably retinoyl-β-glucuronide. A similar spectrum of metabolites was found after injection of [11-^{14}C]retinylacetate, thus supporting the view that retinol and retinoic acid may have a common metabolic fate (Roberts and DeLuca, 1967), with the acid possibly being an intermediate toward the active metabolite.

Retinoic acid has been shown to undergo the following reactions: isomerization to 13-*cis*-retinoic acid (Zile *et al.*, 1967), esterification (Fidge *et al.*, 1968; Smith *et al.*, 1973; Lippel and Olson, 1968), metabolism of the side chain with specific loss of tritium at positions 11 and 12 (Sundaresan and Sundaresan, 1973), decarboxylation (Roberts and DeLuca, 1968*a,b*; Nelson *et al.*, 1971), and conjugation with glucuronic acid (Dunagin *et al.*, 1964, 1966; Lippel and Olson, 1968). Decarboxylation and formation of retinoyl-β-glucuronide will be presented in detail.

1.5.2.1. In Vivo Decarboxylation

[15-^{14}C]Retinoic acid given orally or intravenously to rats yielded 25–30% of its radioactivity as $^{14}CO_2$ (Roberts and DeLuca (1969*a,b*). The rate of evolution of $^{14}CO_2$ was highest in the first 6–12 hr and slowed down thereafter. [14-^{14}C]Retinoic acid was also metabolized to $^{14}CO_2$, but it represented only 18% of the administered dose (Roberts and DeLuca, 1969, 1969*a,b*). [6,7-^{14}C]Retinoic acid yielded $^{14}CO_2$ in only 0.8% of the administered dose, with 38% in urine and 64% in feces (Roberts and DeLuca, 1969*a,b*).

Much slower was the metabolism of retinyl acetate labeled in the same positions (Roberts and DeLuca, 1967). However, of the metabolized [15-^{14}C]retinol, 32% was recovered as $^{14}CO_2$, 50% in urine, and 37% in feces in the first 4 days of collection. Of [6,7-^{14}C]retinol, only 4.5% was in $^{14}CO_2$, with 48% in the urine and 47% in feces over the same period of time. The conclusion that retinol and retinoic acid might follow similar metabolic routes, with retinoic acid as an intermediate in the metabolic pathway, is reinforced by these studies.

1.5.2.2. In Vitro Decarboxylation of [15-^{14}C]Retinoic Acid

Kidney and liver slices catalyze the decarboxylation of [15-^{14}C] retinoic acid (Roberts and DeLuca, 1968a). Radioactive $^{14}CO_2$ is also generated in the same system but at a slower rate if [14-^{14}C]retinoic acid is the substrate. This is strongly inhibited by inhibitors of the Krebs cycle such as malonate and fluoroacetate, which do not affect decarboxylation of [15-^{14}C]retinoic acid to $^{14}CO_2$. On the contrary, DPPD(N,N^0-diphenyl-p-phenylene diamine), an inhibitor of free-radical reaction, has a strong inhibitory effect on both decarboxylations. This suggests that the breakdown of retinoic acid may proceed through a radical attack at the beginning and via the Krebs cycle mechanism thereafter (Roberts and DeLuca, 1969a,b).

Microsomes from rat kidney and liver carry out decarboxylation of [15-^{14}C]retinoic acid provided that Fe^{2+}, NADPH, PiPi, and O_2 are present. No $^{14}CO_2$ is detected from [14-^{14}C]retinoic acid or [6,7-^{14}C]retinoic acid in the microsomal system (Roberts and DeLuca, 1968b). At 4 min of incubation, it was found that 50% of the [15-^{14}C]retinoic acid had been decarboxylated. Other oxidative reactions did take place in this system since less than 10% of the original radioactivity was recovered as the intact compound. The reaction was blocked by electron acceptors, phenazine methosulfate, and EDTA. The lack of inhibition by carbon monoxide suggests that no hydroxylation occurred simultaneous to decarboxylation. On the other hand, DPPD completely inhibited the reaction, thus suggesting a microsomal peroxidation rather than a mechanism of drug hydroxylation (Ernster and Orrenius, 1965). However, using DPPD *in vivo*, very little inhibition of decarboxylation of [15-^{14}C]retinoic acid was detected, although slices and microsomes obtained from the treated animals showed total inhibition in an *in vitro* assay. This suggests that the *in vitro* reaction does not reflect *in vivo* metabolism. Similar results on decarboxylation of retinoic acid were obtained by Lin (1969) and Nelson *et al.* (1971). These authors employed acetone–butanol–ether-dried liver powder as well as horseradish peroxidase, and obtained 40% and 48% decarboxylation, respectively. Essentially the same requirements as found by Roberts and DeLuca (1969a,b) were valid for both reactions. Interestingly, flushing the reaction vessel with N_2 completely inhibited the reaction; the reaction took place only in phosphate buffers with a pH optimum of 6.4.

In conclusion, although the decarboxylation pathway of retinoic acid in the whole animal may be a necessary step for the synthesis of the active final form of vitamin A, more work is needed to establish the validity of this concept.

1.6. The Visual Function

Feeding fresh calf retinas (Holm, 1929) or dried pig retinas (Smith *et al.*, 1931) to vitamin A deficient rats restores normal growth and cures xerophthalmia. Wald (1933) first demonstrated the presence of vitamin A in solutions of the visual purple, in intact retinas, and in the pigment choroid layers of frogs, sheep, pigs, and cattle. Eyes from *Rana esculenta* and *Rana pipiens* contain about 4 μg of vitamin A per eye (combined pigment and choroid layers). Retinas of dark-adapted animals contain only a trace of vitamin A; instead, their chloroform extract yielded a new compound named "retinene" (Wald, 1934). This new compound did not absorb in the visible spectrum, and was "faintly yellow" due to an ascending absorption from 500 nm into the ultraviolet. The new compound gave a blue color, having an absorption maximum at 655 nm, with antimony trichloride. Light-adapted retinas did not contain free retinene but contained about 0.3 μg per retina of newly formed vitamin A. Wald related the formation of vitamin A to "bleaching" of the visual purple of dark-adapted retinas. Immediate extraction of bleached retinas by chloroform yielded retinene but not vitamin A. Keeping the bleached yellow retinas at room temperature caused fading of the yellow color, formation of vitamin A, and disappearance of retinene. The visual pigment was judged to have the characteristics of a carotenoid protein similar to the one found by Kuhn and Lederer in lobster shells (1933), with absorption maximum at 500 nm. It was termed "rhodopsin" (Wald 1935a,b, 1936a,b). Wald also studied the visual purple of freshwater fishes after extraction with 1% aqueous digitonin and found it different from rhodopsin, with an absorption maximum at 522 nm. This new pigment was termed "porphyropsin" (Wald, 1937). A phenomenon similar to rhodopsin "bleaching" was observed. The isolated dark-adapted retinas from freshwater fishes (white perch, *Monroe americana*; yellow perch, *Perca flavescens*; and pickerel, *Esox reticulatus*) are purple. If exposed to bright light, "they turn russet, then fade slowly to very pale yellow." The light process was reversible since russet retinas regenerated porphyropsin when placed in darkness. The bleached pigment generated a new product which absorbed maximally at 700 nm in the Carr Price reaction and had a broad absorption maximum at 407 nm in chloroform. This compound was termed "retinene," and the rhodopsin chromophore was also termed "retinene" (Wald, 1939).

Very elegant work by Morton and collaborators at Liverpool (Morton, 1944; Morton and Goodwin, 1944; Ball *et al.*, 1948) identified retinene of rhodopsin as retinal, and retinene of porphyropsin as 3-dehydroretinal (Fig. 11). The two vitamins retinol and 3-dehydroretinol were present in fish liver oils. They yielded mixtures of their respective aldehydes upon oxidation by manganese oxide and could be separated by chromatography (Cama *et al.*, 1952a,b). Conversion of retinal to 3-dehydroretinal is obtained chemically by means of *N*-bromosuccinimide. 3-Dehydroretinal is reduced to 3-dehydroretinol by reduction with lithium aluminum hydride (Henbest *et al.*, 1955).

The nature of the linkage between the chromophore and the protein was the object of intense effort by several groups of investigators. Lythgoe (1937) and Lythgoe and Quilliam (1938) proposed a scheme of rhodopsin degradation after exposure to light. They prepared a solution of rhodopsin displaying λ_{max} near 500 nm. Bright light caused a shift in λ_{max} to 470–480 nm, probably because of the formation of "transient orange" pigment. At room temperature, decomposition of "transient orange" gave "indicator yellow," which was deep yellow in acid and pale yellow in alkali. "Transient orange" was found to be stable when exposure of rhodopsin to light took place at very low temperatures (Broda and Goodeve, 1941). When the temperature was allowed to rise, indicator yellow and retinene were formed. Of great importance was the finding (Ball *et al.*, 1948, 1949; Collins, 1953, 1954; Collins and Morton, 1950a,b) that retinal reacted with amino groups of several compounds to give Schiff's bases with spectroscopic proper-

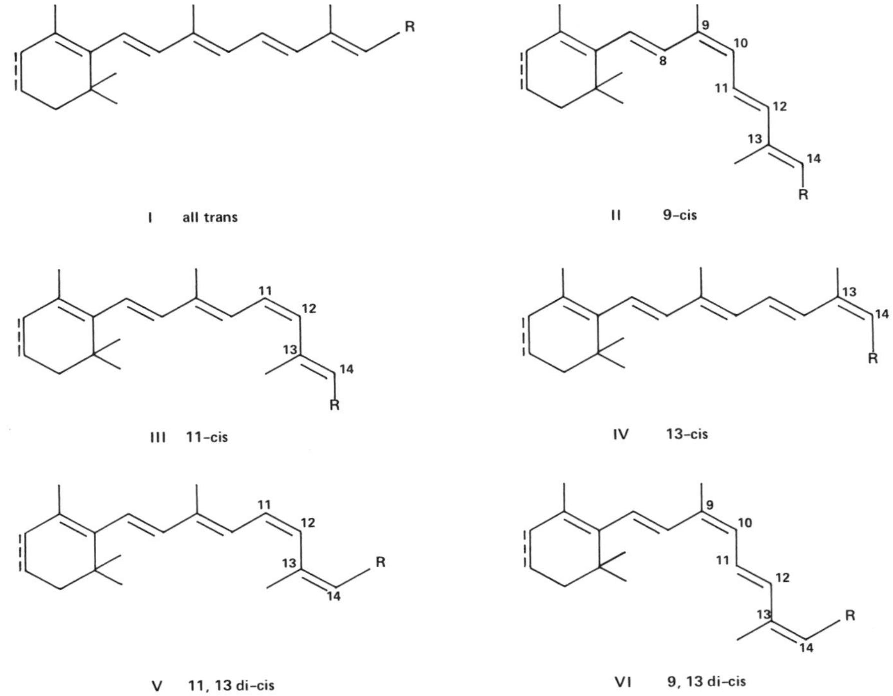

Fig. 11. Structures of the isomeric forms of retinols and 3-dehydroretinols (R is —CH$_2$OH) and retinals and 3-dehydroretinals (R is —CHO). The dashed lines indicate the additional double bond of the 3-dehydro derivatives. The formula numbers refer to structure in Table I. From Morton (1972).

ties of indicator yellow (λ_{max} 440 nm in acid and 365 nm in alkali), suggesting the possibility that indicator yellow might just be an artifact of a similar nonspecific interaction, with little significance in the visual cycle. This possibility was discarded by the studies of Collins (1953), who showed that 10 M formaldehyde, a strong reagent for all free amino groups, does not impede the formation of indicator yellow upon bleaching of rhodopsin; thus the C—N link in indicator yellow comes from rhodopsin. This link in rhodopsin was further investigated by Morton and Pitt (1955), who, upon addition of acid to bring a solution of rhodopsin quickly to pH 1, found *N*-retinylidene opsin, with spectrophotometric characteristics very similar to those of synthetic retinylidene-methylamine (Morton and Pitt, 1955). The *N*-retinylidene opsin obtained by irradiation of rhodopsin was later reduced to *N*-retinylopsin by sodium borohydride reduction, which does not act on intact rhodopsin (Bownds and Wald, 1965; Bownds, 1967; Akhtar *et al.*, 1965, 1967):

$$C_{19}H_{27}CH{=}N\text{—opsin} \longrightarrow C_{19}H_{27}CH\text{—NH—opsin}$$

$$\text{\textit{N}-retinylidene opsin} \qquad\qquad \text{\textit{N}-retinyl opsin}$$

The elegant work of Bownds (1967) and Akhtar *et al.* (1967) established that retinal and 3-dehydroretinal are linked to the ϵ-amino group of lysine. The specificity for the interaction of retinal and opsin is very high. Retinal isomers with a

Table I. Spectral Properties of cis-trans Isomers of Retinol in Ethanol[a]

Isomer	Formula	λ_{max} (nm)	ϵ_{max} (liters cm^{-1} mole^{-1})
All-*trans*	I	325	52,800
9-*cis*	II	323	42,300
11-*cis*	III	319	34,900
13-*cis*	IV	328	48,300
11,13-di-*cis*	V	311	26,200
9,13-di-*cis*	VI	324	39,500

[a]From Morton and Pitt (1957). The formula numbers refer to structures in Fig. 11.

straight side chain (all-*trans*, 13-cis, 11,13-*cis*) do not form a photosensitive pigment, whereas bent-chain isomers like 11-*cis* and 9-*cis* do (Fig. 11). The early history of the demonstration of the 11-*cis* isomer as the biologically active chromophore in rhodopsin is interesting and has been described by Morton (1972). Oroshnick and collaborators first considered and later demonstrated 11-*cis* retinal as the active candidate (Oroshnick and Mebane, 1954; Oroshnick, 1956; Oroshnick *et al.*, 1956). This work was aided by the discovery that at least five isomeric forms of retinal exist (Robeson *et al.*, 1955). Spectral properties of *cis-trans* isomers of retinol and retinal and 3-dehydroretinol and 3-dehydroretinal in ethanol are shown in Tables I, II, III, and IV. Wald (1953) had already recognized that some of his 3-dehydroretinal preparations from porphyropsin contained two *cis* isomers. One combined in darkness with opsin from freshwater fish to give porphyropsin of λ_{max} of 523 nm, while the other gave a pigment (isoporphyropsin) with λ_{max} of 507 nm. Hubbard and Wald (1952) demonstrated that 9-*cis*- and 11-*cis*-retinal form light-sensitive pigments. 11-*cis*-Retinal combines with opsin in a nonsymmetrical fashion as shown by an increase in optical rotatory dispersion and circular dichroism around 500 nm (Budowski *et al.*, 1963; Deshmukh *et al.*, 1965; Dingle and Lucy, 1965).

Moreover, an increase in the Cotton effect at 235 nm is observed during rhodopsin formation (Crescitelli *et al.*, 1966; Hubbard *et al.*, 1965; Kito and

Table II. Spectral Properties of cis-trans Isomers of Retinal in Ethanol[a]

Isomer	Formula	λ_{max} (nm)	ϵ_{max} (liters cm^{-1} mole^{-1})
All-*trans*	I	381	43,400
9-*cis*	II	373	36,100
11-*cis*	III	376.5	24,900
13-*cis*	IV	375	35,800
11,13-di-*cis*	V	373	19,900
9,13-di-*cis*	VI	368	32,400

[a]From Morton and Pitt (1957). The formula numbers refer to structures in Fig. 11.

Table III. Spectral Properties of cis-trans Isomers of 3-Dehydroretinol in Ethanol[a]

Isomer	Formula	λ_{max} (nm)			ϵ_{max} (liters cm^{-1} mole^{-1})		
		a	b	c	a	b	c
All-*trans*	I	350	286	276	41,300	20,300	15,800
9-*cis*	II	348	287	277	32,500	26,100	21,800
11-*cis*	III	344	286	278	28,100	16,100	14,000
13-*cis*	IV	352	288	277	39,000	18,400	14,000
11,13-di-*cis*	V	337	290	277	25,700	13,100	13,300
9,13-di-*cis*	VI	350	288	280	29,300	21,600	18,000

[a]From Von Planta *et al.* (1962). a, b, c represent the absorption maxima in decreasing order of magnitude. The formula numbers refer to structures in Fig. 11.

Takezaki, 1966), suggesting increased helical content or other conformational changes. As shown in Table V, the reaction of 11-*cis*-retinal with opsin to form rhodopsin is accompanied by a spectral shift for retinal from 382 nm to 498 nm, a hyperchromic effect of 70%, and inaccessibility of the chromophore by reagents like sodium borohydride, lipoxidase, and hydroxylamine (Bownds and Wald, 1965). A well-established photochemical event in rhodopsin breakdown by light is the isomerization of 11-*cis*-retinal to the all-*trans* form. The product prelumi-rhodopsin can re-form rhodopsin at very low temperatures (Yoshizawa and Wald, 1963). Prelumirhodopsin (Table V) may exist in three different forms with different half-lives (Erhardt *et al.*, 1966). Hubbard *et al.* (1965) showed that the first large changes in kinetic parameters occur in the following step in the bleaching sequence from lumirhodopsin to metarhodopsin I (Table V), possibly due to a conformational change in the opsin structure. However, their measurements for heat and entropy of activation differed considerably from those reported by Erhardt *et al.* (1966). This is probably because kinetic studies of this conversion in digitonin micelles are difficult to carry out since the spectra of the two intermediates are very similar. This discrepancy suggests that the multiform first-order analysis used to obtain the values may not be adequate in the interpretation of kinetic data (Abrahamson and Wiesenfeld, 1972). An analysis of the data as a bimolecular reaction yields results which are more in agreement with studies on

Table IV. Spectral Properties of cis-trans Isomers of 3-Dehydroretinal in Ethanol[a]

Isomer	Formula	λ_{max} (nm)			ϵ_{max} (liters cm^{-1} mole^{-1})		
		a	b	c	a	b	c
All-*trans*	I	401	314		41,500	11,100	
9-*cis*	II	391	315		34,100	19,000	
11-*cis*	III	393	321	252	24,900	14,400	12,700
13-*cis*	IV	395	314		33,300	11,600	
11,13-di-*cis*	V	386	269	261	27,200	11,000	11,000

[a]From Von Planta *et al.* (1962). a, b, c represent the absorption maxima in decreasing order of magnitude. The formula numbers refer to structures in Fig. 11.

Table V. Cycle of Rhodopsin Breakdown and Synthesis[a] and Involvement of N-Retinylidene Phosphatidylethanolamine in the All-trans to 11-cis Isomerization of Retinal by Light[b]

Compound	λ_{max} (nm)	Chemical changes		
		In chromophore	In protein conformation	By titration
Rhodopsin	498			
$hv \parallel hv$				
Prelumirhodopsin (3 forms)	543	11-*cis* → *trans*	Minor	None
Lumirhodopsin (3 forms)	497	None	Minor	None
Metarhodopsin I (3 forms?)	478	None	Extensive	None
		+ → 0	Extensive unfolding	Net proton uptake with pK 6.4
Metarhodopsin II	380	0 → +	Refolding	Unmasking of 1-2-SH
Metarhodopsin$_{465}$	465	None	Unfolding	
N-Retinylidene opsin$^+$	440	+ → 0	?	Hydrolysis of Schiff base
all-*trans*-Retinal + opsin	387			
all-*trans*-Retinylidene phosphatidylethanolamine				
$\downarrow hv$				
11-*cis*-Retinylidene phosphatidylethanolamine	459			

[a]From Olson (1967).
[b]From Shichi and Somers (1974).

binding sites (Rapp *et al.*, 1970). The conversion of metarhodopsin$_{478}$ I to meta-rhodopsin$_{380}$ II in aqueous digitonin is reversible with an entropy change of +34 to +47 units (Matthews *et al.*, 1963; Hubbard *et al.*, 1965; Erhardt *et al.*, 1966). Moreover, the complex takes up a hydrogen ion from the medium (Erhardt *et al.*, 1966). A shift of the spectral maximum from 478 nm to 380 nm is also observed, with possible conversion of a carbonium ion to a neutral Schiff base (Hubbard *et al.*, 1965).

More recently, a new mechanism has been suggested for the conversion of metarhodopsin$_{478}$ I to metarhodopsin$_{380}$ II on the basis of spectral changes, kinetic studies in rod segments and intact eyes, and studies of binding sites. This suggested molecular mechanism is a transimination of the chromophore from a protonated lipid Schiff base to a nonprotonated lysine-protein Schiff base (Poincelot and Abrahamson, 1970). Model retinyl phosphatidylethanolamine has been prepared for these studies (Fig. 1).

Ostroy *et al.* (1966) favor the view that metarhodopsin$_{380}$ II yields metarhodopsin$_{465}$, which then decays to *N*-retinylidene opsin. This compound will then give all-*trans*-retinal and opsin. Matthews *et al.* (1963) suggest that a direct decay of metarhodopsin$_{380}$ II to all-*trans*-retinal and opsin occurs with metarhodopsin$_{465}$, a secondary product. That both pathways occur is suggested by other investigations (Hagins, 1957; Ebrey, 1967; Donner and Reuter, 1969).

More recent developments (Shichi and Somers, 1974) suggest that metarhodopsin$_{465}$ III may be identical with retinylidene phosphatidylethanolamine (Fig. 1). It was first established that a phospholipid fraction stimulated the photoregeneration of rhodopsin from opsin and all-*trans*-retinal (Shichi, 1971). This regeneration process involves isomerization of all-*trans*-retinal to 11-*cis*-retinal and binding of 11-*cis*-retinal with opsin. Later studies (Shichi and Somers, 1974) on these reactions have demonstrated that all-*trans*-retinal reacts specifically with phosphatidylethanolamine to form all-*trans*-retinylidene phosphatidylethanolamine, λ_{max} 458 nm. Light alone (without enzyme) acts on this complex to form the isomeric 11-*cis*-retinal derivative, which is then available for rhodopsin synthesis. For the demonstration of the specific requirement of the phospholipid, bleached rod outer segments were used in which opsin maintained its native conformation.

About 30% of the phosphatidylethanolamine in rod outer segments is in complex with all-*trans*-retinal. All these findings are consistent with the cycle shown in Fig. 12. The suggestion by Shichi and Somers (1974) that metarhodop-

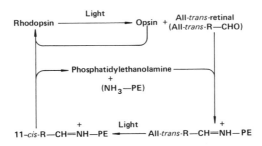

Fig. 12. Scheme of the visual cycle according to Shichi and Somers (1974).

sin III (λ_{max} 465 nm) is identical with *N*-retinylelene phosphatidylethanolamine (λ_{max} 458 nm) is very attractive. It would, in fact, explain the loss of optical activity from metarhodopsin II (Waggoner and Stryer, 1971; Yoshizawa and Horiuchi, 1972). $NRPE^+$ is optically inactive (Johnston and Zand, 1973). Whether 11-*cis*-retinal is transferred to opsin directly from $NRPE^+$ or is first released as free 11-*cis*-retinal is unclear. Yet a bigger problem concerns the mechanism by which the photochemical event triggers neurotransmission.

Some recent studies have demonstrated that phosphorylation of opsin is a light-dependent reaction (Kuhn, 1974; Miller and Paulsen, 1975; Bownds *et al.*, 1972, 1974). Light-stimulated phosphorylation is catalyzed by a kinase which is membrane bound in rod outer segments (Miller and Paulsen, 1975; Weller *et al.*, 1975). Opsin is as good a substrate as unbleached rhodopsin, suggesting that the chromophore does not take part in the phosphorylation reaction (Miller and Paulsen, 1975). However, isorhodopsin obtained by regeneration of the visual pigment with 9-*cis*-retinal prevents phosphorylation (Shichi *et al.*, 1974).

Studies on papain-digested rhodopsin from bovine retinas have demonstrated that the phosphorylation site is distinct from the retinaldehyde-binding site (Virmaux *et al.*, 1975).

Another important characteristic of the rhodopsin molecule is its high content of sulfhydryl groups. More than 90% of the total sulfhydryl groups in the rod outer segments reside in the rhodopsin molecule, with a ratio of six sulfhydryl groups per rhodopsin molecule. Of these, two are exposed to the aqueous environment (type I, De Grip *et al.*, 1975) and two are normally buried in the photoreceptor membrane (type II, De Grip *et al.*, 1975). Finally, 1.8 ± 0.4 sulfhydryl groups are buried in the photoreceptor membrane but become exposed on illumination under partially denaturing conditions or in darkness upon denaturation by sodium dodecylsulfate (type III, De Grip *et al.*, 1975). Studies of sulfhydryl groups in rhodopsin have also been conducted by fluorescence (Wu and Stryer, 1972) and spin labeling (Delmelle and Pontus, 1974).

1.7. Isomers of Retinal

Six geometric isomers of retinal have been made by stereospecific syntheses (Robeson *et al.*, 1955; Wald *et al.*, 1955; Oroshnick, 1956; Oroshnick *et al.*, 1956; Isler *et al.*, 1967). They are shown in Fig. 11. There is evidence that the cyclohexene ring and the side chain form an angle of approximately 35° (Stam and MacGillavry, 1963). The spectral properties of *cis-trans* isomers of retinol, retinal, 3-dehydroretinol, and 3-dehydroretinal are given in Tables I, II, III, and IV, respectively.

The following three isomerization reactions are catalyzed by light: (1) all-*trans* to 11-*cis* in retinochrome, a pigment present in cephalopod retina (Hara and Hara, 1965), and all-*trans* to 11-*cis* in retinylidene phosphatidylethanolamine of bovine retina outer segments (Shichi and Somers, 1974); (2) 11-*cis* to all-*trans* in rhodopsin, with loss of a proton (Hubbard and Wald, 1952); and (3) 13-*cis* to all-*trans* in bacteriorhodopsin, with loss of a proton (Oesterhelt and Stoeckenius, 1971). The reverse of this isomerization, all-*trans* to 13-*cis*, occurs in darkness (Oesterhelt *et al.*, 1973; Oesterhelt and Stoeckenius, 1973).

Table VI. *Quantum Efficiencies (γ) for the Photoisomerization of All-trans-Retinal Compared to the Average Quantum Efficiencies for That of Retinylidene Chromophores*[a]

I Reaction	II Retinal γ_{25}	III Retinal γ_{-65}	IV Chromophore γ_{20}
13-*cis* to *trans*	0.4	0.1	
11-*cis* to *trans*	0.2	0.6	0.6
9-*cis* to *trans*	0.5	0.25	0.2
all-*trans* to average *cis:*			
maximum value	0.2	0.005	
minimum value	0.06	0.002	
all-*trans* to 11-*cis*			0.3
all-*trans* to 9-*cis*			0.06

[a]From Kropf and Hubbard (1970). The γ values for the photoisomerization at 365 nm of all-*trans*-retinal and the mono-*cis* isomers in hexane at 25°C and −65°C (columns II and III) are compared to the average quantum efficiencies for the photoisomerization at 450, 500, and 550 nm of the retinylidene chromophores of rhodopsin (11-*cis*), isorhodopsin (9-*cis*), and metarhodopsin I (all-*trans*) in aqueous digitonin at 20°C.

Quantum efficiencies for these and other photoisomerization reactions of retinal were obtained by Kropf and Hubbard (1970) and are given in Table VI.

Retinal isomers and related compounds were also studied by carbon-13 magnetic resonance to investigate the planarity of these molecules (Becker *et al.*, 1975). All-*trans*-, 13-*cis*-, and 9-*cis*-retinal as well as the related all-*trans* and 9-*cis* 15-carbon aldehyde exist as planar structures in their polyene portion. However, 11-*cis*-retinal is not entirely planar along the polyenic chain but displays segmental motion about the C-12, C-13 bond (Becker *et al.*, 1975).

1.8. *Bacteriorhodopsin of Halobacterium halobium*

Recently, a new exciting area in vitamin A research has developed from studies on the purple membrane of an extremely halophilic bacterium, *Halobacterium halobium*. This bacterium does not contain chlorophyll and uses its purple pigment, bacteriorhodopsin, to carry out photophosphorylation. Oesterhelt and Stoeckenius (1971) first described bacteriorhodopsin as a compound with striking similarities to the visual pigments found in higher animals. Bacteriorhodopsin contains retinal in a protonated Schiff base linkage to ϵ-amino lysine. It has absorption maximum at 560 nm due to interaction with a tryptophanyl residue of the peptide chain (Fig. 13). The purple pigment (mol. wt. 26,000) is localized in an organized pattern in the membrane, occupying 50% of its area (Oesterhelt and Stoeckenius, 1973). The protein is stable in aqueous suspensions. Two forms of bacteriorhodopsin have been isolated containing all-*trans*-retinal and 13-*cis*-retinal (Oesterhelt, 1971*a,b*; Oesterhelt *et al.*, 1973). These forms result from the conversion of 13-*cis*- to all-*trans*-retinal by light. In the dark, a mixture of the two is formed with a 20-min half-time for the conversion of the all-*trans* to the 13-*cis* form (Fig. 14). Light, on the other hand, causes the formation of a com-

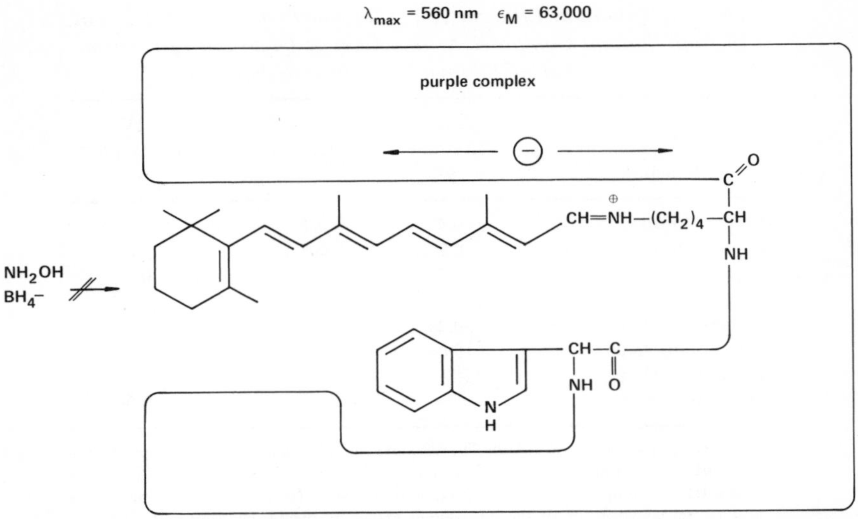

λ_{max} = 560 nm ϵ_M = 63,000

Fig. 13. Schematic representation of the binding site of all-*trans*-retinal in bacteriorhodopsin. Hydroxylamine and borohydride do not react with retinal in the purple complex. From Oesterhelt (1975).

plex with absorption maximum at 412–415 nm, which in a few milliseconds converts back to bacteriorhodopsin. Simultaneous change in the pK of bacteriorhodopsin is observed, consistent with the loss of a proton upon illumination (Fig. 14). The short-lived 412-nm pigment can be trapped by reaction with hydroxylamine to form bacteriopsin and retinaloxime (Oesterhelt *et al.*, 1973). The light-catalyzed reaction (K_1 rate constant) from bacteriorhodopsin to the 412-nm pigment occurs via simultaneous expulsion of protons. The reaction from the 412-nm pigment to bacteriorhodopsin draws a proton from the bacterial cytoplasm, with K_2 rate constant. The result is that light drives a pumping system to expel protons from the bacteria into the medium. The cycle is described by the following equation

$$[412] = [B] \left(1 + \frac{K_2}{K_1} I\right)$$

where [412] is the concentration of the 412-nm complex, [B] is the concentration of bacteriorhodopsin, I is the light intensity, and K_1 and K_2 are rate constants of the cycle as described in Fig. 14 (Oesterhelt, 1975). An increase in concentration of 412-nm complex is obtained through an unknown mechanism by treatment of the cells with ethyl ether (Oesterhelt and Hess, 1973). With this expedient, the 412-nm complex was characterized as having a broad absorption maximum at 415 nm and as depending on light but not on temperature; the quantum light yield was 0.79. Also, the formation of the 560-nm bacteriorhodopsin depends on temperature with activating energy of 47.7 kJ M^{-1}. It was also found that formation of the 412-nm complex by light changes the pK of bacteriorhodopsin and the fluorescence of the tryptophanyl residue, suggesting that protons are released on formation of the 412-nm complex and taken up again on regenera-

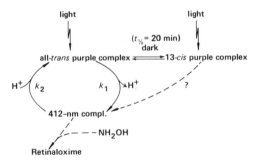

Fig. 14. Photochemical cycle in bacteriorhodospin. From Oesterhelt (1975).

tion of bacteriorhodopsin: this is the light-driven proton pump. The vectorial transfer of protons would require that the rhodopsin molecule be asymmetrical, which has been demonstrated (Blaurock and Stoeckenius, 1971). It would also create an electrochemical gradient with concomitant synthesis of ATP consistent with the hypothesis of Mitchell (1961), which proposed that reversible ATPase systems, situated in key membrane structures, are coupled with translocation of protons across the membrane.

In fact, light causes production of ATP in dark-adapted *Halobacterium halobium* (Oesterhelt, 1975). ATP steady-state concentration is a result of both oxidative phosphorylation and photophosphorylation in these bacteria. Production of ATP in the dark (oxidative phosphorylation) is stopped when air is removed by flushing with nitrogen. Carbonylcyanide-3-chlorophenylhydrazone (CCCP) will also stop production of ATP (Oesterhelt, 1975). Production of ATP by photophosphorylation can be observed in dark-adapted bacteria in which oxidative phosphorylation has been stopped by KCN or CCCP. Both phosphorylations are inhibited by dicyclohexyl carbodiimide, a compound which blocks ATP synthesis (Oesterhelt, 1975).

These and other experimental data (Oesterhelt and Stoeckenius, 1974; Oesterhelt, 1974; Racker and Stoeckenius, 1974; Kayushin and Skulachev, 1974) are consistent with the following conclusions: Light causes a photochemical cycle with release and uptake of protons; this creates an electrochemical gradient through a vectorial process due to the asymmetrical disposition of bacteriorhodopsin in the purple membrane. The energy from the electrochemical gradient is used for photophosphorylation, in agreement with Mitchell's hypothesis of chemioosmosis (West and Mitchell, 1974).

However, full comprehension of the mechanistic connection among changes in ATP concentration, inhibition of oxygen consumption by light, changes in pH of the medium, and proton transport awaits further studies.

1.9. The Reproductive Function

Evans and Bishop (1922) found that female rats kept on diets low in vitamin A suffered from prolongation of estrus and failed to enter diestrus; a vaginal smear showed the presence of cornified cells. These animals also failed to ovulate. Administration of butterfat or cod liver oil readily reversed these changes within 1–3 days (Evans, 1928).

When female rats kept on a diet low in vitamin A were mated with normal males at 100 days after placement on the diet, only 22% of the copulations resulted in the birth of living young. The large majority of the copulations did not result in implantation. These data were interpreted to mean that the spermatozoa did not reach the egg because of the keratinized surface of the vaginal epithelium (Evans, 1928).

Evans (1932) and Mason (1932, 1933) showed that diets containing adequate amounts of vitamin E but deficient in vitamin A cause degeneration of the germinal epithelium in the testes and absence of spermatozoa in the epididymal fluid. Typical pathological changes were the sloughing of cells in the ductuli efferentes and the ductus epididymis before other changes were noticed in the tubules. In contrast to the injury caused by depletion of vitamin E, which is irreversible, vitamin A can readily reverse the changes resulting from its deficiency. Mason (1932, 1933) speculated that alterations in the structure of the testes in vitamin A depleted animals might be due to some alterations of the anterior lobe of the hypophysis, which is epithelial. This suggestion is particularly attractive since the anterior lobe secretes the gonadotropins, which are glycoprotein hormones.

Dowling and Wald (1960) had demonstrated that the visual function of vitamin A is distinct from its systemic function: Rats fed on retinoic acid grew well and had all the outward appearance of normal animals; however, they went blind, since retinoic acid could not be reduced to retinal.

Thompson *et al.* (1964) maintained male and female rats on vitamin A depleted diets which were supplemented with retinoic acid. They found that the rats became blind, although they grew well. Female rats kept on a vitamin A depleted diet without retinoic acid (Evans and Bishop, 1922; Evans, 1928) had been shown not to enter diestrus. In contrast, the retinoic acid fed animals had a normal estrus cycle and became pregnant after mating, although they always resorbed their fetuses. If retinol was given to such rats during pregnancy, litters were born and were weaned successfully, at least when 1–2 µg of retinyl acetate per day was administered for the duration of pregnancy. Male rats kept on a diet containing retinoic acid as the only source of vitamin A had smaller and often edematous testes (Thompson *et al.*, 1964). Cells of the germinal epithelium were sloughed off and the lumina of the tubules were full of Sertoli cells (Howell *et al.*, 1963). Administration of vitamin A alcohol reversed these changes.

The conclusion of these studies was that depletion of retinol causes a basic lesion at a stage in spermatocytogenesis before the meiotic division of the spermatocytes to spermatids (Howell *et al.*, 1963). Similar results were obtained in the guinea pig (Howell *et al.*, 1967).

More recent studies in male rats maintained on retinoic acid were conducted to determine the amount of retinol necessary to maintain normal reproduction and vision (Coward *et al.*, 1969). Daily doses of up to 1 µg of retinyl acetate were only partially effective; 5 µg of retinyl acetate per day was effective in maintaining spermatogenesis throughout the experimental period of 28 weeks after weaning.

Measurements of electroretinogram thresholds in rats on 1 μg of retinyl acetate per day showed this amount to be sufficient to maintain normal vision, but it did not maintain spermatogenesis at 21 weeks after weaning; 2 μg of retinyl acetate was required to maintain spermatogenesis (Coward *et al.*, 1969).

Similar studies were conducted by Thompson *et al.* (1969) in the reproductive cycle of the domestic fowl. Birds reared from hatching on a vitamin A free diet died from avitaminosis A before 4 weeks of age. Birds on a diet supplemented with retinoic acid grew well and, as in the rat, showed no overt signs of depletion of vitamin A (Thompson *et al.*, 1969) except for their lack of vision.

At variance with the rat, cocks on a retinoic acid or methylretinoate diet had normal testes and the hens laid eggs at a normal rate. These eggs were fertile, but the development of the embryo did not proceed after the first 2 days of incubation (Thompson *et al.*, 1969). Administration of retinyl acetate to the hens or to the eggs allowed development of the embryo.

Thus at least three distinct functions appear to exist for vitamin A: (1) the visual function, (2) the systemic function, and (3) the reproductive function.

1.10. Vitamin A and Bone

The pioneering work of Moore *et al.* (1935) suggested that deficiency of vitamin A is the cause of blindness in calves after birth and in young dairy animals grown on a poor-quality roughage for periods of more than 3 months. Such blindness was not associated with xerophthalmia, but was due to a constriction of the optic nerve by overgrowth of the bone surrounding the optic foramen (Moore *et al.*, 1935). Administration of carotene in the diet prevented the development of nyctalopia, papillary edema, and permanent blindness (Moore, 1939). These workers also observed that vitamin A deficiency in calves is associated with a condition of syncope which could result from intracranial pressure.

Mellanby (1944), in an extensive study of vitamin A deficiency in dogs, found that a prominent change in some bony tissue is an overgrowth of the bone in areas of contact with the nerve. Under normal conditions, this is where bone resorption occurs, to allow for the accommodation of the nerve. Facial bones and the vertebrae were particularly affected and had a coarser appearance.

Histological analysis of the bones from deficient dogs showed that osteoclasts and osteoblasts were distributed in areas different from those where they are usually found. Both types of cells were abundant in deficient bone but their topology varied. It was argued that the excessive bone which narrowed the foramina of the vertebrae was of cancellous nature and accumulated because of the absence of the osteoclasts in the area near the nerve in the deficient animals. Thus Mellanby likened vitamin A to "the director of building operations" in the bone and concluded that it might be expected that in vitamin A deficiency "the working builders and demolition squad (osteoblasts and osteoclasts) would either cease to work or work in a completely disorderly way."

Results essentially in agreement with these findings were obtained in fowl (Howell and Thompson, 1967*a,b*). In these studies, avitaminosis A affected the nervous system indirectly by causing an overgrowth of the periosteum, which was considered to be the primary abnormality. Similar findings were obtained from fowl on retinoic acid in which deficiency of vitamin A was precipitated suddenly by removal of the acid from the diet. In these animals, new periosteal cancellous bone was formed (Howell and Thompson, 1967a). Because observed changes were similar in chicks and in adult fowl, it was concluded that endochondral growth is not affected by deficiency of the vitamin. It appears that excessive osteoblastic activity occurs in vitamin A deficiency.

1.11. Introduction to the Epithelial Function

Epithelial tissues of both endodermal and ectodermal origin depend on vitamin A for their function. Those of endodermal derivation undergo drastic morphological changes which are clearly visible by light microscopy. In particular, mucosal linings from animals fed a vitamin A deficient diet often degenerate and their columnar and mucus-secreting cells are replaced by a metaplastic squamous epithelium which resembles the epidermal phenotype, with keratohyaline granules and keratin production (Wolbach and Howe, 1925). This phenomenon has been reproduced *in vitro* with cultures of tracheas grown in a vitamin A free environment (Harris *et al.*, 1972). Conversely, the squamous metaplastic changes due to deficiency can be reversed by administration of the vitamin, and the normal columnar phenotype reappears after such treatment.

It is impossible to attribute these morphological changes to a specific biochemical mechanism at this stage, and morphological observations may not find an obvious explanation in the available biochemical data.

A logical explanation for the phenotypic changes which characterize some tissues under conditions of vitamin A deficiency is that the vitamin directly controls genetic expression. Alternatively, while a change in phenotype expression will have to implicate the genome, it is not necessary that this be through a direct molecular action or intervention of the vitamin at the genomic level. In fact, it may be that one of the molecular events which depends directly on the vitamin may in turn affect the genome.

On the basis of recent data (De Luca *et al.*, 1970b, 1973; Rosso *et al.*, 1975; Frot-Coutaz *et al.*, 1976), one may think of vitamin A and its phosphorylated derivatives as a building block or structural component of some specialized membranes which acquire the vitamin either at the stage of stem cell or after commitment to a specific differentiation pattern has occurred. This structurally engaged vitamin A may be in equilibrium with vitamin A in the cellular pool (which is usually a reflection of body resources) or may be a stable entity which dissociates itself from the cell after its death. The structural vitamin A may participate in catalytic processes which are involved in perpetuating the structure to which it belongs (e.g., membrane growth by glycosylation), or it may be involved in processes that synthesize products for export. Moreover, one important consequence of this is that if the turnover rate of structural vitamin A is equal to the turnover rate of the organelle which contains it, or to that of the cell itself, in vitamin A deficiency the tissues with a higher turnover would become more rapidly depleted of vitamin A than tissues with a lower turnover, provided that liver vitamin A stores are exhausted and no reutilization of vitamin A occurs.

An important consequence of the structural catalytic involvement is that, if vitamin A is not available at the time of assembly of a cell or specialized cell membrane, either the assembly may not occur or a defective cellular structure may result as a product of adaptation.

The proposed structural catalytic involvement of vitamin A and its phosphorylated derivatives in membranes may in the end explain the variety of morphological and biochemical changes resulting from vitamin A deficiency and some of the effects of excess vitamin A.

1.11.1. Excess Vitamin A

Another fascinating aspect of vitamin A research deals with the effect of excess vitamin A on different biological systems. These effects may or may not be related to the normal physiological function of the vitamin. Probably some of the effects are and some are not related to its physiological mode of action. The pharmacology and toxicology of vitamin A have been reviewed extensively (Moore, 1957).

An interesting system investigated in excess vitamin A is the organ culture of chick ectodermal explants, studied by Fell (1957). In this system, a concentration of 10 IU of retinyl acetate per milliliter of incubation medium causes the appearance of PAS-positive granules after 1 week of culture. The PAS-positive material seems to be confined to glandular structures arising among the keratinizing elements. This organ culture system contains both dermis and epidermis.

Another approach has been used in which epidermal cells have been grown in monolayers and their morphological changes have been observed in the presence and absence of vitamin A. These changes are similar to those observed in the organ culture system, with formation of PAS-positive material (Yuspa and Harris, 1974; De Luca and Yuspa, 1974).

Vitamin A excess has also been used to produce an experimental model for the study of cleft palate in the rat (Morriss, 1973). Doses of 100,000 IU of retinyl palmitate given to the mother between days 7 and 8 of gestation caused a high incidence of cleft palate in the offspring.

Thus it is clear even from this limited list of morphological events that there is a complexity of biological phenomena which depend on the vitamin. These phenomena, although morphologically different, may represent peripheral or central manifestations of the same or similar biochemical processes in different cell types. Conversely, it cannot be excluded that the vitamin may play more than one systemic biochemical role.

1.11.2. Vitamin A and Macromolecular Synthesis in Epithelial Tissues: Morphological Observations

The intestinal mucosal lining is a simple columnar epithelium of endodermal derivation. It contains two main cell types: the absorbing cell and the goblet cell. Two less-represented cells also belong to this lining: the argentaffin cell and the Paneth cell. Submucosal glands contribute to the luminar secretion.

Only the goblet cell responds to vitamin A deficiency in a morphologically detectable way. This cell is very rich in mucus and it can be considered an exocrine unicellular gland which discharges its contents into the lumen of the intestine. This cell is essential for maintenance of the moist mucus environment of the surface of several epithelia, including the respiratory and intestinal tracts. Thus an impairment of its capacity to synthesize mucus may have disastrous consequences for the epithelium. A sharp decrease in the number of mucus-secreting goblet cells of the intestinal crypts is caused by vitamin A deficiency, although in a series of random measurements, no difference in epithelial height was found (De Luca *et al.*, 1969, 1970*a*). A count of goblet cells per crypt in 100 crypts showed a difference between normal (17.3 ± 1.3) and vitamin A deficient rat intestine (11.0 ± 1.9). The difference was statistically significant at the $P \leqslant 0.01$ level. In this work animals were pair-fed in an attempt to minimize nutritional differences. Generally, the deficiency does not appear to affect the tissue, except in the reduction of PAS-staining goblet cells. This point was further investigated by electron microscopy. No obvious reduction of any cellular elements, e.g., endoplasmic reticulum, was found in vitamin A deficient intestine (De Luca *et al.*, 1969).

Studies conducted in tadpoles have demonstrated that excess vitamin A causes mucous metaplasia and a remarkable increase in the amount of mucus produced by the intestinal cells (Weissman, 1961). These and other morphological observations have prompted studies on the protein-, nucleic acid-, and glycoprotein-synthesizing capacity of various epithelial tissues.

1.11.2.1. *Protein and Nucleic Acid Synthesis*

A biochemical investigation of the protein-synthesizing machinery was conducted in cell-free systems involving free and membrane-bound polyribosomes prepared by sucrose density gradient.

Interestingly, protein synthesis on membrane-bound polyribosomes was affected by vitamin A deficiency. This effect was localized in the pH 5 enzyme fraction rather than the particulate fraction, as demonstrated by crossing-over experiments. The pH 5 enzyme fraction contains both amino acid activating enzymes and transfer RNAs. Vitamin A deficient polyribosomes are organized in aggregates larger than the normal polysomes. It was shown that, by ligation of the bile duct and the pancreatic duct, this difference in polysome size disappeared (De Luca *et al.*, 1969). This finding suggests that some hydrolase, usually active in normal intestine, may not be active in vitamin A deficiency. This might be due to atrophy of pancreatic glands, to keratinization of pancreatic ducts, or to both, since both are known to occur in vitamin A deficiency, as described by Wolbach and Howe (1925).

An investigation of the concentration of RNA in the intestinal epithelium showed a 30% decrease in the amount of the total RNA per gram of wet tissue. When incubations were equalized for RNA bound to the particulate fraction in the study of protein synthesis, a difference was still found in the amount of labeled leucine incorporated into protein. This reduction was due to the pH 5 fraction responsible for the charging of tRNAs.

The rates of charging for leucyl-tRNA were measured using pH 5 enzyme from normal and vitamin A deficient rat intestine. The deficient preparation had a slower charging rate. This was obviated by the addition of purified tRNA prepared from either normal or vitamin A deficient rat intestine (De Luca *et al.*, 1971*a*). Leucyl-tRNA synthetase was not affected by vitamin A deficiency, but a decreased amount of tRNA was responsible for the decreased rate of charging. Investigation of the charging of isoaccepting species of tRNA for L-leucine and for L-threonine by separation on Freon columns confirmed the finding that an overall decrease (30%) of tRNA available for charging occurs in vitamin A deficiency (De Luca *et al.*, 1971*a*).

RNA synthesis was also investigated in normal and vitamin A deficient intestinal mucosa by *in vivo* incorporation of $[6\text{-}^{14}C]$orotic acid and $[G\text{-}^{3}H]$uridine in three different experiments. A 30% decrease was constantly observed in the total amount of label incorporated into ribosomal and tRNAs. The differences were not merely a reflection of the change in precursor pool sizes since vitamin A deficiency decreased neither the pool nor the specific radioactivities of uridine, uridine monophosphate, and uridine diphosphate. These observations were made at a relatively early stage of vitamin A deficiency (5% weight loss), when no difference in DNA content is found between normal and vitamin A deficient preparations. It was concluded that a true difference in rate of incorporation of uridine into RNA of normal and vitamin A deficient rat intestinal mucosa occurred, resulting in decreased RNA.

Results from several laboratories have demonstrated an influence of vitamin A on the synthesis of RNA. Thus Kaufman *et al.* (1972) showed that the electrophoretic pattern of RNA species synthesized *in vitro* by tracheal epithelial cultures from hamsters deficient in vitamin A differed from that of RNA synthesized in normal, pairfed control hamsters. There was less labeled RNA of low electrophoretic mobility in the epithelial cells deficient in vitamin A. This electrophoretic change was reversed to normal 2 weeks after administration of vitamin A to the deficient hamsters.

In a separate study (Sporn *et al.*, 1975), the RNA and DNA content of mouse epidermal cells, cultured in a chemically defined medium, in response to different doses of β-retinoic acid and other vitamin A analogues showed proportional increase in RNA and DNA at concentrations of β-retinoic acid between 10^{-11} and 10^{-7}M. Similarly, both the dimethylacetyl cyclopentenyl (DACP) (Fig. 2) and trimethyl methoxyphenyl (TMMP) (Fig. 2) analogues were active in increasing the amounts of RNA and DNA present in the epidermal cultures.

In conclusion, the results obtained in the tracheal organ culture system and in the epidermal cell culture system suggest that vitamin A is directly or indirectly involved in the metabolism of RNA and possibly DNA.

Zile and DeLuca (1970) reported that 50 μg of retinol, given by intravenous injection, increases the incorporation of $[5\text{-}^{3}H]$orotic acid into colon RNA at 2, 4, and 8 hr after the injection of vitamin. The studies were conducted 1 hr after the injection of the label. These workers could not find any stimulation in duodenal and ileal mucosal RNA. Liver showed a decrease in incorporation into total RNA at least 2 and 4 hr after administration of retinol. Prostate tissue showed a threefold increase at 4 hr of treatment. When colon, liver, and small intestinal mucosal nuclear RNA were analyzed for incorporation of $[5\text{-}^{3}H]$orotic acid, a consistent increase was observed in the retinol-treated animals at 2 hr after injection of retinol. These results are in good agreement with those obtained by Zachman (1967), who showed a two- to three-fold stimulation of total RNA synthesis in colon due to administration of 20 μg of retinol to vitamin A deficient rats.

Johnson *et al.* (1969) reported a 2.5-fold stimulation in rapidly labeled nuclear RNA of liver and intestinal mucosa 15 min after the administration of potassium retinoate to vitamin A deficient rats.

In conclusion, this body of work tends to support the concept that the effect of vitamin A on maintenance of normal epithelia involves the synthesis of nucleic acids. While this is probably the case, the important questions of how the vitamin causes such changes in nucleic acid metabolism and whether these changes are the consequence of a direct molecular involvement of vitamin A in nucleic acid synthesis remain to be answered.

1.11.2.2. Glycoprotein Synthesis

Another approach has been to start from the morphological observation that deficiency of the vitamin causes a decrease in PAS-staining material of most if not all epithelial tissues. This finding was amply substantiated in the early work of Wolbach and Howe (1925). More recent studies, mostly on the respiratory mucosa, have confirmed these data (Harris *et al.*, 1972). PAS of the vitamin A deficient intestinal mucosa revealed fewer stainable goblet cells. While this means a reduction in mucus, it does not necessarily signify the lack of goblet cells, as goblet cells devoid of mucus might be present but would not be stained. However, an electron microscopic study did not reveal such cells (De Luca *et al.*, 1969). It may be useful here to remark that deficiency does not cause squamous metaplasia and keratinization of the intestinal lining, at variance with the respiratory tract and other epithelia. Thus biochemical observations of the intestinal mucosa are not complicated by the presence of new proteins such as keratins.

A factor difficult to control in these biochemical studies on the epithelium of the small intestinal mucosa is the amount of mesenchymal tissue which is scraped in the preparation of the epithelium. A weaker epithelium may cause more of the mesenchymal fraction to be included in the epithelial scraping. Quantitation of the mesenchymal component is very difficult, and results must be considered with caution in a comparative study. Incorporation of D-[1-^{14}C]glucosamine into total microsomal material did not appear to be affected by the deficiency status over a period of 1–8 hr after the intraperitoneal injection of the label. Incorporation of the label into free polysome constituents increased constantly for 8 hr with no remarkable differences between deficient and normal epithelia. The kinetics of incorporation into the membrane fraction was completely different, with a maximum at 1 hr or earlier and a fast drop thereafter. Incorporation of [1-^{14}C]glucosamine into intestinal glycopeptides was studied 2 hr after the injection of label into normal and vitamin A deficient rats (De Luca *et al.*, 1970a). No difference was detected in total radioactivity or weight, but upon chromatography on DEAE-Sephadex A50, peak III showed a 70% decrease in the incorporated label in deficiency. This peak was eluted with 0.4 M LiCl and could be separated into one major higher molecular weight component and two minor smaller components by gel filtration on Sepharose 4B.

Administration of retinyl acetate restored the incorporation of [^{14}C]glucosamine into peak III to normal levels within 18 hr after the injection of vitamin A

Table VII. *Effect of Administration of Retinyl Acetate (500 µg i.p. and 500 µg in Food)*
on the Number of Goblet Cells and the Incorporation of [^{14}C]*Glucosamine into*
Peak III of Vitamin A Deficient Rat Intestinal Epithelium[a]

Hours after vitamin A administration	[^{14}C]Glucosamine incorporation into peak III (dpm)	Hours after vitamin A administration	Number of goblet[b] cells per crypt of Lieberkühn
0	425	0	13.88 ± 2.04
4	630	4	14.35 ± 2.08
18	1,297	10	15.11 ± 2.19
26	12,750	18	19.58 ± 2.41
		23	19.81 ± 2.34

[a] From De Luca *et al.* (1970a) .
[b] Statistical significance between 0 and 4 hr is at the level of $p < 0.01$; between 4 and 10 hr and between 10 and 18 hr, $p < 0.01$; between 18 and 23 hr, $p < 0.2$.

(Table VII). This table also shows that the number of goblet cells increased to normal levels within 18 hr after the injection of the vitamin.

Of the three ^{14}C-labeled glycopeptide peaks obtained by chromatography on Sepharose 4B, only the high molecular weight compound contained fucose and showed a dependence on retinol for its synthesis. For this reason, the vitamin A dependent glycopeptide was termed "fucose-glycopeptide" (FG). This compound was prepared in large amounts and its sugar composition was studied. Eighty percent of its dry weight was constituted by L-fucose 11.8%, D-glucosamine 22.5%, D-galactosamine 11.1%, galactose 25.5%, and N-acetylneuraminic acid 10.9%, in the molar ratio of 1.0:1.73:0.87:1.97:0.49, respectively (De Luca *et al.*, 1970a, 1971b; De Luca and Wolf, 1972). Amino acids constituted 14.72% of the dry weight with threonine and glutamic acid the most abundant (De Luca *et al.*, 1970a). The glycopeptide had an S value of 3.6 in distilled water and it aggregated into two components at 6.2 S and 7.2 S when studied in 0.1 M phosphate buffer, pH 7.5. This finding would suggest a change from a rod shape at pH 6 to a spherical shape at pH 7.5.

1.11.2.3. Localization of FG in Goblet Cells

Demonstration of the specific localization of fucose-glycopeptide in the goblet cell was performed by indirect immunofluorescence using an antiserum prepared in chickens against the vitamin A dependent fucose-glycopeptide, which was shown to be chromatographically and electrophoretically homogeneous. Figure 15 illustrates that the glycopeptide is specifically localized in the goblet cell (De Luca *et al.*, 1971b).

This first suggestion of a role of vitamin A in glycoprotein synthesis was followed by the isolation and characterization of the goblet cell glycoprotein, which showed the identical polysaccharide composition as the glycopeptide (Kleinman and Wolf, 1974). Further studies demonstrated similar effects of vitamin A deficiency in the epithelium of respiratory tract glycoproteins (Bonanni *et al.*, 1973; Bonanni and De Luca, 1974). In this tissue, a cross-reactive glycopeptide was

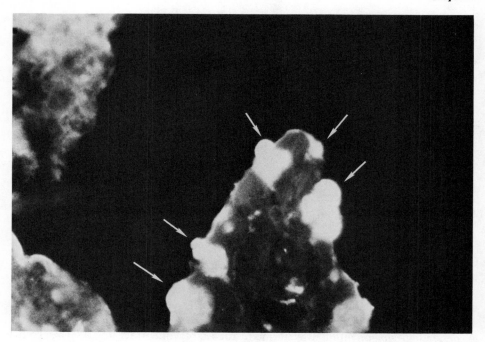

Fig. 15. Localization of the vitamin A dependent fucose-glycopeptides in goblet cells by indirect immunofluorescence. Goblet cells are indicated by arrows. From De Luca *et al.* (1971*b*).

shown to be visibly decreased in vitamin A deficient respiratory mucosa by im-munofluorescence (De Luca *et al.*, 1972).

1.11.2.4. Biosynthetic Studies on [^{14}C]Mannose Incorporation into Liver Mannolipids and Mannoproteins

A more profound effect of vitamin A deficiency was found in hamster liver. Studies on the incorporation of [^{14}C]mannose into liver glycolipids and glycopro-teins were conducted in normal and vitamin A deficient hamsters (De Luca *et al.*, 1975). A 95% drop in the incorporation of [^{14}C]mannose into total chloroform–methanol (2:1, v/v) extractable lipids was found in severe vitamin A deficiency. To exclude the possibility that weight loss might account for the dramatic differ-ence, studies were also conducted in animals of equal weight and appetite (De Luca *et al.*, 1975). Both retinol and retinoic acid were active in causing a two- to threefold increase in the incorporation of [^{14}C]mannose in both mannolipids and manno-proteins. A detailed study of the [^{14}C]glycopeptides after proteolytic digestion of the material insoluble in chloroform–methanol (2:1, v/v) was conducted to find out whether specific molecules are particularly affected by deficiency. Eighty-five percent of the affected labeled glycopeptides was not retained by DEAE-Sephadex equilibrated with 0.05 M LiCl. This fraction from normal hamsters, termed "N-1," contained 2,521,900 ^{14}C dpm compared to D-1 (from deficient hamsters), which had only 426,000 dpm. Fractions N-1 and D-1 were further separated from lower molecular weight material by successive chromatography on Sephadex G25 superfine. The higher molecular weight fraction thus obtained from N-1 contained 2,431,500 ^{14}C dpm and that obtained from D-1 contained 108,400 dpm— a decrease of more than 95% in the amount of [^{14}C]mannose incorporated in the glycopeptide from deficient livers.

Analysis of the [^{14}C]monosaccharide from acid hydrolysis was performed by gas–liquid chromatography of the boronic acid derivative of the hexitols (Eisenberg, 1972). Most (90%) of the radioactivity was found as [^{14}C]mannose. The amount of mannose present in N-1 was 0.94 mg per gram of wet normal liver tissue and only 0.19 mg per gram of vitamin A deficient liver tissue. Notwithstanding this dramatic decrease in covalently bound mannose, the specific radioactivity in peak 1A was still smaller in vitamin A deficiency (from 187 in N-1 to 48 in D-1). This may be due to the fact that the incorporation of labeled mannose reflected only the last 20 min of life (hence the time of most severe deficiency), while the weight of 1A represented accumulation from several days, when deficiency was less severe.

These results supported the idea that vitamin A is involved in mannose transfer. However, another possibility had to be considered: that the vitamin controls the availability of mannose to synthetic pools via a specific monosaccharide transport function across the membrane. This second possibility was suggested by the fact that the synthesis of dolichyl phosphate [^{14}C]mannose was also greatly (90%) decreased in severe vitamin A deficiency (De Luca *et al.*, 1975). Such transport mechanisms should have drastic effects on pool sizes of mannose, and, in turn, on the biosynthesis of dolichyl phosphate mannose and retinyl phosphate mannose as well as other mannoconjugates.

This possibility was checked by measuring liver monosaccharide pool sizes and their specific radioactivities 20 min after the injection of [^{14}C]mannose into normal and vitamin A deficient hamsters under the same conditions as for the biosynthetic study. It was shown that the total water-soluble pool of mannose and glucose was similar in the two biological situations; however, the amount of radioactive mannose was greater (4.9×10^5 dpm/g liver) in deficient than in normal hamster livers (1.6×10^5 dpm/g liver). The same trend was also observed for glucose, resulting in an increased specific radioactivity of the mannose and glucose pools in vitamin A deficiency. Thus, if corrected for this increase, the observed differences in glycolipid and glycoprotein synthesis would be even greater (De Luca *et al.*, 1975).

These findings supported the hypothesis of a direct precursor involvement of retinyl phosphate mannose in mannose transfer to different glycoconjugates.

1.11.2.5. *Polyprenyl Glycosylphosphate*

Some glycosyltransferases are known to reside in the membrane, although their substrates, the sugar nucleotides, are water soluble. Moreover, the glycosylated proteins are synthesized on polysomes attached to membranes inside the channels of the endoplasmic reticulum, and the sugar nucleotides must cross such membranes before they become available for glycosylation of either structural or secreted glycoproteins.

A mechanism which obviates this problem is the conversion of a water-soluble sugar nucleotide into an activated lipid-soluble compound. Scher *et al.* (1968) and Lahav *et al.* (1969) reported that a C_{55}-polyisoprenol, undecaprenyl-phosphate, functions as an intermediate between guanosine diphosphate mannose and the final polymeric product, mannan, in *Micrococcus lysodeikticus*, according to the reactions

GDP-mannose $\longrightarrow$ undecaprenyl-phosphate-mannose $\longrightarrow$ mannan

Reactions involving undecaprenol were also discovered in *Salmonella typhimurium* (Wright *et al.*, 1967) for the biosynthesis of O-antigen and in *Staphylococcus aureus* and *laurentii* for the biosynthesis of the cell wall polysaccharide (Higashi *et al.*, 1970). In 1969, Caccam *et al.* reported that microsomal membranes from mammalian tissues synthesize a mannolipid with the characteristics of bacterial polyprenylphosphate–sugar compound. This compound was characteristically stable to alkali and was identified as dolichyl mannosyl phosphate (Richards and Hemming, 1972).

Retinol, a tetraprenoid derivative with an additional double bond per isoprene unit, was tested as a carrier of monosaccharides, as this might explain the effects of vitamin A deficiency and excess on glycoprotein synthesis. However, since retinol and its derivatives are present only in very small amounts in target tissues, the carrier function might be restricted to specific monosaccharides.

1.11.2.6. Synthesis of Retinyl Phosphate Mannose

It was observed that retinol, when added in the presence of ATP in a membrane system from vitamin A deficient rats, enhanced the incorporation of [^{14}C]mannose from GDP-[^{14}C]mannose into a mannolipid by at least twofold (De Luca *et al.*, 1970*b*). The reaction showed a consistent absolute requirement for ATP. Labeled retinol was incorporated in the mannolipid fraction both *in vivo* and *in vitro* (De Luca *et al.*, 1970*b*, 1973). Mild acid hydrolysis of the mannolipid fraction, obtained by incubating vitamin A deficient rat liver membranes with retinol, gave a biphasic curve, with 40% of the mannolipid hydrolyzed in 30 sec. A mannolipid obtained from an incubation without retinol showed a constant rate of hydrolysis with the same slope as the slow-hydrolyzing component in the incubation with retinol, but lacked the fast-hydrolyzing component. This suggested the presence of two compounds in the mannolipid preparation, although the two products could not be separated by chromatography on DEAE-cellulose or in three solvent systems on silica gel.

Wright *et al.* (1967) had shown that catalytic hydrogenation of allylic isoprenyl ester results in formation of the corresponding hydrocarbon, with breakage of the ester bond. Starting with [^{3}H]retinol and GDP-[^{14}C]mannose and vitamin A deficient rat liver membranes, a doubly labeled mannolipid (De Luca *et al.*, 1973) was obtained after purification by DEAE-cellulose and silicic acid. The mannolipid fraction was then hydrogenated, and the resulting filtered mixture was chromatographed on deactivated alumina, after removal of water-soluble compounds. [^{14}C]Mannose was identified in the water phase by paper chromatography. The lipid phase contained 100% of the [^{3}H]retinol-derived radioactivity, as expected for a product of hydrogenolysis, whereas 60% of the [^{14}C] mannolipid without any [^{3}H]retinol behaved as an α-dihydropolyprenyl phosphate mannose.

The mixture of mannolipids was subjected to milk alkaline hydrolysis. It yielded mannose phosphate, which gave mannose by alkaline phosphatase. This also supported the idea of an allylic phosphate, since dolichyl phosphate mannose is stable under these conditions. Most of this work was conducted with vitamin A deficient enzyme preparations and with added retinol, so it lent itself to the criticism that normal membranes behave differently and may not use retinol as a substrate for retinyl phosphate mannose. This preliminary work was put on firmer grounds by other important developments. The separation of the two mannolipids was achieved by thin-layer chromatography in chloroform–methanol–/water (60:25:4, v/v/v), as first described by Tkacz *et al.* (1974). This solvent system separates polyprenylphosphate sugar derivatives by polyprenol chain length. Dolichol (Fig. 16) derivatives have an R_f of 0.5 and retinol derivatives an R_f of 0.2–0.3. By using this solvent system for the analysis of mannolipids obtained from a normal rat liver postmitochondrial membrane fraction, two mannolipids were found: one at R_f 0.25 and one at 0.5 (Rosso *et al.*, 1975).

A second most important development was the chemical synthesis of retinyl phosphate. This compound, added *in vitro* to normal liver membranes, specifically stimulated the synthesis of retinyl phosphate mannose (Fig. 17). Retinyl phosphate and retinyl phosphate mannose were separated by DEAE-cellulose chromatography and ammonium acetate gradient elution. Retinyl phosphate mannose eluted with 0.010 M ammonium acetate and retinyl phosphate with 0.040 M ammonium acetate (Fig. 18). Labeling experiments with [^{3}H]retinyl phosphate and GDP-[^{14}C]mannose were also conducted, demonstrating the specific incorporation of [^{3}H]retinyl phosphate in retinyl phosphate mannose (Rosso *et al.*, 1975).

Retinyl phosphate mannose has typical solubility characteristics. Using the method described by Silverman-Jones *et al.* (1976), 70% of this compound can be extracted in the upper phase, whereas 100% of the dolichyl phosphate mannose is found in the lower phase of the extraction of membranes. These different solubility characteristics allow complete separation of the two mannolipids (Rosso *et al.*, 1975; Silverman-Jones *et al.*, 1976). Another technical point has emerged from recent studies on chemical and biological retinyl phosphate (Frot-

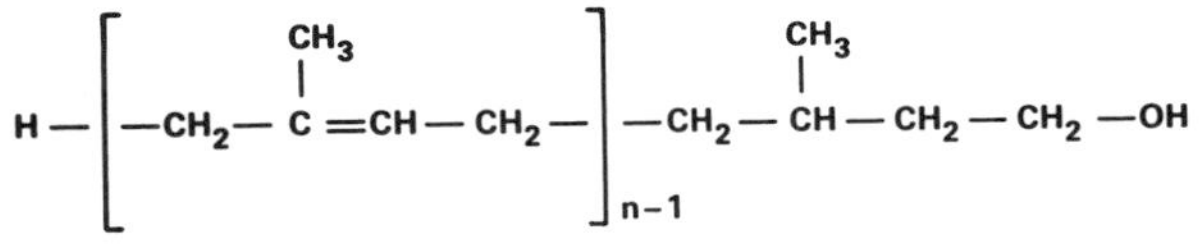

$$H - \left[-CH_2 - \underset{\underset{CH_3}{|}}{C} = CH - CH_2 - \right]_{n-1} -CH_2 - \underset{\underset{CH_3}{|}}{CH} - CH_2 - CH_2 - OH$$

No. residues (n): Mammalian tissue   16 → 21
Yeast   14 → 18

No. of internal <u>trans</u> residues: 2

Esterification: 60%

Fig. 16. Structure of dolichols.

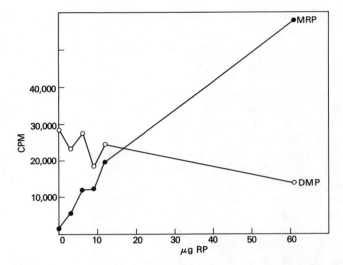

Fig. 17. Effect of the addition of retinyl phosphate on the synthesis of retinyl phosphate mannose (MRP) and dolichyl phosphate mannose (DMP) by normal rat liver membranes.

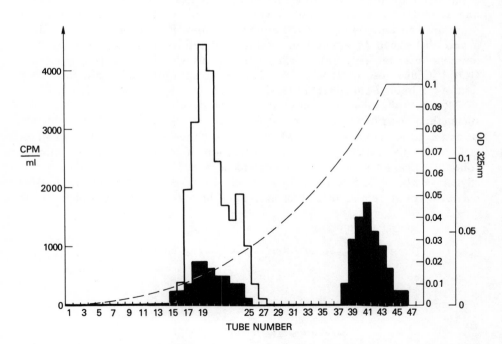

Fig. 18. Separation of retinyl phosphate from retinyl phosphate [^{14}C]mannose by gradient elution from DEAE-cellulose. The black blocks indicate absorption at 325 nm. The shape of the gradient (from 0 to 0.1 M ammonium acetate) is shown by the dashed curve. From Frot-Coutaz *et al.* (1976).

Coutaz *et al.*, 1976): chromatography on unbuffered silicic acid yields anhydro-retinol from retinyl phosphate. This is obviated by treating silicic acid with ammonia (Rosso *et al.*, 1975). Sephadex LH-20 chromatography has the same effect, yielding anhydroretinol from phosphorylated derivatives of retinol (De Luca *et al.*, 1977).

In conclusion, it has been demonstrated that:

1. Retinyl phosphate mannose is synthesized by normal rat liver membranes, although in smaller amounts than dolichyl phosphate mannose, in the absence of added chemically synthesized retinyl phosphate.
2. Retinyl phosphate functions as a substrate for the mannosyltransferase, with a hundredfold stimulation of retinyl phosphate mannose synthesis above endogenous levels at 0.27 mM in rat liver membranes.
3. The syntheses of the two mannolipids have two distinct pH optima, with pH 6 favoring the synthesis of dolichyl phosphate mannose and pH 7.0 favoring of retinyl phosphate mannose.
4. The two mannolipids can be separated by solvent extraction.
5. Membranes from vitamin A deficient livers do not synthesize retinyl phosphate mannose unless retinyl phosphate is supplied exogenously.

1.11.2.7. Retinyl Phosphate: Biological and Chemical Synthesis

The isolation and characterization of retinyl phosphate were achieved from a cell suspension culture system of intestinal epithelial cells. The culture system allows the isolation of [*carbinol*-^{14}C]retinyl phosphate starting from [*carbinol*-^{14}C]retinol. These are separated by chromatography on DEAE-cellulose and silicic acid (Frot-Coutaz *et al.*, 1976). The synthesis, followed by radioactive labeling, of retinyl phosphate proceeds linearly for 1 hr in these culture conditions. The absorption spectrum of the purified [^{14}C]retinyl phosphate is identical to that of the chemically synthesized compound (Fig. 19). They exhibit identical behavior on DEAE-cellulose, with elution at 0.06–0.07 M acetate for both chemical and biological retinyl phosphate. The biologically generated compound, as the chemical one, is labile to 0.1 N KOH at 37°C for 15 min. The hydrolysis product is identical with chemical anhydroretinol by chromatography on TLC in toluene–chloroform–methanol (50:12.5:12.5, v/v/v). It also has the same absorption spectrum, with three maxima at 490, 470, and 450 nm. Since both chemical and biological retinyl phosphate yield anhydroretinol, the reaction mechanism shown in Fig. 20 is proposed in an alkaline environment.

The concentration of retinyl phosphate in hamster intestinal epithelium is calculated at about 0.035 μg per small intestinal lining or per gram of wet weight. Biological [^{14}C]retinyl phosphate is active in reversal of squamous metaplasia in the tracheal organ culture system (Frot-Coutaz *et al.*, 1976) developed by Sporn *et al.* (1975).

In conclusion, intestinal cells synthesize retinyl phosphate. Although a detailed study has not been made of the ability of liver cells to synthesize retinyl phosphate, a compound with chromatographic properties of chemical retinyl phosphate was isolated from hamster liver after injection of [*carbinol*-^{14}C]retinol (Barr and De Luca, 1974, and unpublished).

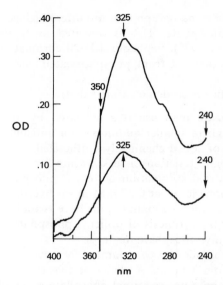

Fig. 19. Absorption spectra of synthetic and biosynthetic retinyl phosphate. The upper curve represents the synthetic compound and the lower represents the biosynthetic compound. From Frot-Coutaz *et al.* (1976).

RETINYL—PHOSPHATE

ANHYDRORETINOL

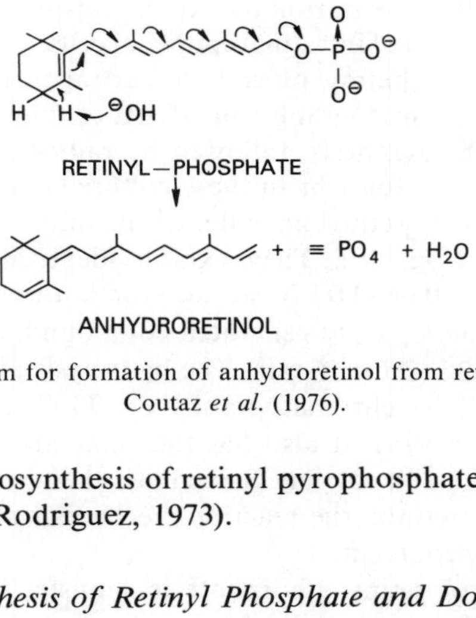

Fig. 20. Proposed mechanism for formation of anhydroretinol from retinyl phosphate. From Frot-Coutaz *et al.* (1976).

A report on the biosynthesis of retinyl pyrophosphate by rat thyroid has also appeared (Gaede and Rodriguez, 1973).

1.11.2.8. *In Vivo Synthesis of Retinyl Phosphate and Dolichyl Phosphate Derivatives*

Pennock *et al.* (1960) and Burgos-Gonzales *et al.* (1963) have shown that dolichol, the polyisoprenoid alcohol of eukaryotes, contains 16–22 isoprene units (Fig. 16). The dolichols have very characteristic mass spectra (Gough and Hemming, 1970) and can be estimated gravimetrically after purification (Burgos-Gonzales and Morton, 1962). The elegant work of Leloir's group and Hemming's group has established unequivocally that dolichyl phosphate functions as an acceptor of D-mannose from GDP-mannose (Richards and Hemming, 1972) and of D-glucose from UDP-glucose (Behrens and Leloir, 1970).

However, the problem of demonstrating the *in vivo* existence of dolichyl phosphate glycoside derivatives by specific labeling is a difficult one. Mammalian tissues utilize most [2-^{14}C]mevalonate to make sterols and ubiquinones through the pathway shown in Fig. 3. The dolichols contain mostly *cis* double bonds which are formed with retention of the hydrogen at 4S-4-H (Hemming, 1970). This hydrogen is lost to NAD in the biosynthesis of *trans* double bonds (Fig. 6) as in cholesterol and ubiquinone (Cornforth *et al.*, 1966). Thus the use of both [*carbinol*-^{14}C]retinol and [4S-4-^{3}H]mevalonic acid in the same experiment allowed differential labeling of [^{14}C]retinyl phosphate and [^{3}H]dolichyl phosphate derivatives.

These compounds were isolated by silicic acid, DEAE-cellulose, and thin-layer chromatography which demonstrated the synthesis of both retinyl phosphate and dolichyl phosphate derivatives *in vivo*. These two classes of compounds were separated by thin-layer chromatography in chloroform–methanol–water (60:25:4, v/v/v) (Barr and De Luca, 1974).

A differential characteristic of these two families of compounds is that retinyl phosphate and derivatives are extremely labile to weak alkali (0.1 N NaOH, 20 minutes at 37°C). These conditions liberated 45% of [^{14}C]retinol-derived radioactivity in the aqueous phase, while the remaining lipid-soluble material migrated in the area of retinol and anhydroretinol. The [4S-4-^{3}H]MVA-labeled lipids remained in the organic phase after alkaline treatment, and 80% behaved as dolichyl phosphate mannose (Barr and De Luca, 1974). The stability of the dolichyl phosphate to relatively mild alkaline conditions is due to its α-saturation (Fig. 16). Conversely, retinyl phosphate derivatives are particularly labile because of the five conjugated double-bond system vicinal to the phosphate linkage (Fig. 1). Thus even mild alkaline treatment, which is usually performed to remove unwanted material, should be carefully avoided in the isolation of these compounds.

In vivo experiments employing [*carbinol*-^{14}C]retinol and [4S-4-^{3}H]MVA have also been reported by Martin and Thorne (1974). These authors found that both radioactive products were incorporated in the mannolipid. In vitamin A depleted and partially hepatectomized animals, the mannolipid contained approximately 3 nmol of dolichol and 0.6 nmol of retinol when obtained from the liver. A very similar chromatographic pattern was obtained with epithelial cells from rat intestine.

1.11.2.9. Retinyl Phosphate Galactose

Helting and Peterson (1972) and Peterson *et al.* (1974) have reported that mastocytoma membranes synthesize two galactolipids, one with optimal pH at 8.4, possibly involving dolichyl phosphate as the carrier polyprenol, and the other with pH optimum 6.3, involving retinol as the carrier. The incorporation of [^{3}H]retinol displayed the same pH optimum as the incorporation of UDP-[^{14}C]galactose (pH 6.3). [^{3}H]Retinol was not incorporated into the second compound, which is suggested to be dolichyl phosphate galactose. Moreover, Peterson *et al.* (1974a) have shown that vitamin A deficiency greatly decreases the synthesis of retinyl phosphate galactose, which is restored to normal levels by the addition of retinol in the cell-free system.

By employing radioactive UDP-[^{14}C]galactose and [^{3}H]retinol as substrates for the synthetase reaction at pH 6.3, a double-labeled acidic galactolipid was obtained. This compound could be purified through DEAE-cellulose and silica gel. Mild acid hydrolysis yielded galactose, and alkaline hydrolysis yielded galactose-1-phosphate. Retinol-binding protein, the specific retinol carrier in blood, formed a complex with biosynthetic retinyl phosphate galactose, a further proof of the structure of this compound. Moreover, retinyl phosphate galactose, when incubated in the presence of UDP, produced UDP galactose according to the proposed reaction

retinyl phosphate + UDP galactose $\rightleftharpoons$ retinyl phosphate galactose + UDP

Recent brief reports (Yogeeswaran *et al.*, 1975) have suggested the participation of retinol in the biosynthesis of retinyl phosphate galactose by fibroblasts in culture. The metabolic fate and specificity in transfer reactions of retinyl phosphate galactose are not known as yet.

Since retinol is a micronutrient, one would not expect it to function as a general carrier, but rather as one which takes part in very specific functions.

1.11.2.10. Conclusions

A molecular involvement of retinol has been established in the following biochemical pathway:

retinol $\xrightarrow{\ 1\ }$ retinyl phosphate $\xrightarrow{\ 2\ }$ retinyl phosphate glycoside

Reaction 1 has been demonstrated *in vitro* in cultures of hamster intestinal cells (Frot-Coutaz *et al.*, 1976) and *in vivo* in rat liver (Barr and De Luca 1974). Reaction 2 has been demonstrated *in vitro* with or without exogenous retinyl phosphate (De Luca *et al.*, 1970*b*, 1973; Rosso *et al.*, 1975; Peterson *et al.*, 1974*a,b*).

Although the physiological role of retinyl phosphate glycoside awaits elucidation, it is reasonable to think of retinyl phosphate as a carrier of only some specific monosaccharide moieties. Dolichyl phosphate mannose has been shown to provide only α-mannose by inversion from dolichyl phosphate-β-mannose to endogenous or exogenous acceptors (Levy *et al.*, 1974; Adamany and Spiro, 1975). Similar specificity may characterize retinyl phosphate derivatives.

Two different specific synthetases seem to exist for retinyl phosphate mannose and dolichyl phosphate mannose (Rosso *et al.*, 1975). Moreover, retinyl phosphate mannose synthesis can be enhanced by retinyl phosphate (up to 0.1 mM) without affecting dolichyl phosphate mannose synthesis in hamster and rat liver membranes from normal animals. Thus all available evidence supports the concept that vitamin A acts in the phosphorylated form in the membrane as a specific carrier of some monosaccharidic moieties.

If this is a general physiological function of vitamin A, the question of how and whether retinoic acid satisfies it must be resolved since this compound can replace retinol in maintaining growth.

1.12. Retinyl Glycosides

Rodriguez *et al.* (1972) have reported that whole homogenates from rat thyroid synthesize retinyl glucoside, galactoside, and mannoside when incubated in the presence of all-*trans*-retinol and the sugar nucleotide. The [^{14}C]retinyl glycosides were separated from retinol by chromatography on a cellulose column and elution with *n*-butanol–water (78:17, v/v). No glycosidic derivatives of retinol were formed when retinol was incubated in the absence of the sugar nucleotide. Gaede and Rodriguez (1973) also reported that the reaction is reversible:

$$\text{retinol} + \text{UDP-glucose} \rightleftharpoons \text{retinyl glucoside} + \text{UDP}$$

1.13. Retinol-Binding Proteins

The diet contains vitamin A either as the preformed vitamin or as a provitamin compound such as β-carotene, which is split by a dioxygenase to yield two molecules of retinaldehyde (Fig. 9). Retinal is then reduced to retinol (Fidge and Goodman, 1968), which is esterified with long-chain fatty acids, complexed with lymph chylomicrons, and transported via the lymphatics (Huang and Goodman, 1965; Goodman *et al.*, 1966). Liver can remove the retinyl esters from circulation (Goodman *et al.*, 1965) and store them for dispensation to target organs. These are apparently presented with a complex of retinol bound to a specific binding protein (RBP). The molecular weight of RBP is approximately 21,000 and its complex with retinol would soon be lost to the urine if it were not complexed to another protein, the thyroxin-binding protein, prealbumin (PA). THE RBP—PA complex (1:1) (Kanai *et al.*, 1968; Peterson, 1971*a*) has a molecular weight of approximately 80,000, and can be dissociated in its component proteins at high ionic strength. One molecule of retinol is present per mole of RBP, and the complex retinol–RBP–PA circulates in the blood as the holoprotein. The normal level of circulating RBP in plasma is about 40–50 μg/ml and that of prealbumin is about 200–300 μg/ml (Smith *et al.*, 1970; Smith and Goodman, 1971). Prealbumin has been found to be one of three human plasma proteins which carry thyroxin (Ingbar, 1963; Oppenheimer, 1968). Physicochemical studies have established that prealbumin is tetrameric, with four very similar subunits (Blake *et al.*, 1971; Morgan *et al.*, 1971; Branch *et al.*, 1971; Rask *et al.*, 1971; Gonzales and Offord, 1971) and a molecular weight of about 54,000 (Branch *et al.*, 1971; Blake *et al.*, 1971). The approximate association constant for thyroxin has been determined as 10^7 or 10^8 M^{-1} (Raz and Goodman, 1969; Nilsson and Peterson, 1971; Oppenheimer and Surks, 1964; Woeber and Ingbar, 1968; Pages *et al.*, 1973). Studies have been conducted on the interdependence of binding between RBP and thyroxin, prealbumin, and retinol by polarization of fluorescence (van Jaarsveld *et al.*, 1973) and by equilibrium dialysis (Raz and Goodman, 1969). The absorbance and fluorescence of retinol were used to measure the interaction of RBP with PA. It was found that retinol is not necessary for such interaction, but that the interaction of retinol with RBP is stabilized by the formation of RBP-PA complex. Sucrose density gradient experiments showed that both forms of RBP bind to PA (van Jaarsveld *et al.*, 1973). The interaction of RBP and PA is best at physiological pH and declines at lower or higher pH. In general, these results agree with previous data obtained by gel filtration analysis (Raz *et al.*, 1970), which showed that the RBP-PA complex is stable at pH 5.8–7.5, with considerable dissociation at pH 10.3. Ionic strength is also a vital factor in the stability of the complex RBP-PA and can be used to dissociate RBP from PA (Peterson, 1971*b*). At their isoelectric point (near pH 4–4.5) (Peterson and Berggard, 1971; van Jaarsveld *et al.*, 1973) the binding is independent of KC1 concentration.

That conformational changes occur upon binding of RBP and PA is suggested by the fact that the circular dichroism (CD) spectrum of the proteins of the complex is not equal to the sum of the two CD spectra of the proteins (Heller and Horwitz, 1973).

Although there is high specificity for the binding of retinol to RBP, isomeric derivatives of retinol and retinal, as well as retinoic acid and retinyl acetate, bind to apo-RBP with varying affinity, whereas unrelated compounds, such as cholesterol or phytol, display very little binding (Goodman and Raz, 1972; Heller and Horwitz, 1973; Horwitz and Heller, 1973). Studies with polarization of fluorescence and velocity ultracentrifugation have demonstrated that PA has four binding sites for RBP, with an apparent association constant of approximately 1.2×10^6. The physiologically occurring 1:1 complex of RBP and PA may be related to the relative concentration of the two proteins in plasma, with the molar concentration of RBP slightly less than half that of PA (Smith and Goodman, 1971).

1.13.1. Regulation of RBP Metabolism by the Vitamin A Status in the Rat

Rat RBP was isolated and characterized by Muto and Goodman (1972). It has α_1-mobility on electrophoresis and a molecular weight of approximately 20,000. The fluorescent and ultraviolet absorption spectra of the human and rat protein are almost identical, due to a high content of aromatic amino acids. However, they appear to be immunologically distinct (Muto and Goodman, 1972). Rat RBP is found in circulation as a complex with prealbumin (PA), a protein with an apparent molecular weight of 45,000–50,000. A radioimmunoassay has been developed with highly labeled [^{125}I]RBP. This has permitted the study of variations in RBP concentration in liver and plasma under various conditions. Holo- and apo-RBP have identical immunoreactivity in the radioimmunoassay. Deficiency of vitamin A causes a sharp drop in the vitamin A and RBP content of rat plasma, especially between days 15 and 25 on the deficient diet. There is a delay of 3 days in RBP drop compared to vitamin A drop (Muto and Goodman, 1972). Because of vitamin A deficiency, the vitamin completely disappears from the serum, but RBP is present even in severe deficiency, after 40 days on the deficient diet, at levels of about 12–15 μg/ml, representing about 25–30% of the level of normally circulating RBP. While this is usually in the holoform, only apo-RBP is detected in severely deficient rats. Inanition is not the cause of these effects, since pair-fed animals fail to show any difference from *ad libitum* fed controls. Moreover, the low levels of RBP from plasma cannot be restored to normal levels by administration of retinoic acid. Apo-RBP circulating in vitamin A deficient plasma is in the form of RBP-PA complex. Oral administration of vitamin A to depleted rats is effective in raising their serum level of vitamin A to normal within 5 hr. Serum RBP levels also rise from 14 to 56 μg/ml.

Normal rats have low levels of RBP in their livers: about $0.15 \mu g$ of RBP per milligram of liver protein. Vitamin A deficiency causes a fourfold increase in immunoreactive RBP in liver. Administration of retinoic acid does not decrease the levels of immunoreactive RBP to normal values. Most (83%) of administered radioactive [^{14}C]retinoic acid in plasma is associated with lipoproteins of density greater than 1.21. Gel filtration and polyacrylamide gel electrophoresis of the labeled serum established that the main association of retinoic acid in blood is with serum albumin (Smith *et al.*, 1973) and not with RBP.

Independently, Peterson *et al.* (1974*b*) have found that rat RBP associates with PA with a constant of $8 \times 10^6 \ M^{-1}$, similar to its human and monkey counterparts. Studies with actinomycin D suggest that newly synthesized RBP requires retinol for its release from the hepatocyte. Studies in the cynomolgus monkey (Vahlquist and Peterson, 1972) and in the chicken (Abe *et al.*, 1975) have demonstrated that the vitamin is carried by retinol-binding protein in these species as well. Thus a total of five species, man, monkey, rat, chicken, and sheep (Glover *et al*, 1974), have been shown to use the RBP-PA complex for transport of vitamin A.

Unlike in man and monkey, PA seems to be the major thyroxin-binding protein in the rat (Muto *et al.*, 1972) and in the chicken (Abe *et al.*, 1975). A structural similarity for PA of all species is its tetrameric organization. However, the four prealbumins are immunologically different (Abe *et al.*, 1975).

A study of the effects of hypervitaminosis A in the rat on plasma RBP metabolism and on vitamin A transport has appeared (Mallia *et al.*, 1975). Rats fed 7.3 mg/day of vitamin A grew less well than control rats, but no other signs of toxicity could be found. Rats fed large excesses of vitamin A (41 or 34 mg/day) displayed severe vitamin A toxicity and stopped growing. RBP serum levels dropped significantly in both regimens. In contrast to a drop in the RBP level, excess vitamin A administration caused increased vitamin A levels in the serum. The vitamin was mostly found as retinyl esters (Abe *et al.*, 1975). Most of the retinyl esters found in plasma of hypervitaminotic rats were complexed with serum lipoproteins of hydrated densities less than 1.21.

Embryonic skeletal tissue grown in organ culture (Dingle *et al.*, 1972) in the presence of free retinol in this culture system caused destruction of the extracellular matrix of chick limb bone rudiments, as measured by release of chondromucoprotein. Complex formation of retinol with RBP inhibited this lytic effect (Dingle *et al.*, 1972). Mallia *et al.* (1975) suggest that vitamin A toxicity occurs *in vivo* when the amounts of vitamin A in the body are so high that some of the vitamin is not presented to membrane structures as the complex with RBP but possibly as the retinyl esters. In this case, the vitamin presumably causes lysis and release of lysosomal hydrolases (Dingle *et al.*, 1971).

1.13.2. Localization of Retinol and RBP in Rat Liver

Popper (1941, 1944) suggested that vitamin A may be present mostly in Kupffer cells, on the basis of fluorescence microscopy of thin sections of liver tissue. Linder *et al.* (1971) have established that retinol and its esters are actually in parenchymal cells of the liver. This has been accomplished by direct chemical measurements of vitamin A contents of the different isolated cell populations.

Studies have also been conducted on the subcellular concentration of vitamin

A (Nyquist *et al.*, 1971). On homogenization of rat liver tissues and differential centrifugation, the bulk of the vitamin A assayed by spectrofluorometry (Drujan *et al.*, 1968) was recovered from the supernatant and the fraction containing unbroken cells (which usually contains 70–80% of the total protein of homogenate), in agreement with previous work (Krinsky and Ganguly, 1953; Powell and Krause, 1953; Sherman, 1969).

In the particulate membrane fraction, the greatest concentration of vitamin A was found in the Golgi apparatus (1.48 μg/mg Golgi protein.) Plasma membranes of rat liver and kidney contain retinol (liver, 4.4×10^{-3} μg retinol/mg N_2; kidney, approximately 3.25×10^{-3} μg retinol/mg N_2) and retinoic acid, with only a trace of retinyl esters present (Mack *et al.*, 1972), in agreement with findings by Kleiner-Bossaller and DeLuca (1971) that retinol and retinoic acid are the major forms of vitamin A in the kidney of vitamin A deficient rats. Of interest is the fact that hexane extracts only 7% of total vitamin A from membrane preparation, whereas chloroform–methanol (2:1, v/v) extracts more than 90% (Mack *et al.*, 1972).

1.14. Binding Proteins in Tissues Other Than Blood

The search for tissue-specific vitamin A binding proteins has gained momentum in the past few years. Bashor *et al.* (1973) have reported that post microsomal supernatants from testis and liver, incubated with [^{3}H]retinol, yield a [^{3}H]retinol–protein complex which sediments at 2 S. Cold retinol and, to some extent, retinal, but not retinoic acid, can displace [^{3}H]retinol from the tissue-binding sites, but not from the serum-binding site (Bashor *et al.*, 1973). Gel filtration on Sephadex G100 suggests a molecular weight of 18,000 for the tissue-binding protein.

Similar studies conducted in rat testis seminiferous tubules by Gambhir and Ahluwalia (1974) have led to the detection of a binding protein with molecular weight 4,800 as well as one with molecular weight 16,000.

These studies do not exclude the possibility that some of these proteins are degradation products of plasma-binding proteins. Ong and Chytil (1975) have reported that testis, uterus, and lung contain a protein sedimenting at 2 S which binds retinoic acid specifically. The [^{3}H]retinoic acid can be displaced by retinoic acid and not by retinol. Partial separation of the retinol-binding and retinoic acid binding proteins was obtained by chromatography on DEAE-cellulose. A reevaluation of molecular weights gave 14,500 for retinoic acid binding protein and 14,000 for retinol-binding protein from testis (Ong and Chytil, 1975). Tissue concentrations of these proteins are very low. The retinol-binding protein showed endogenous fluorescence with excitation maximum at 350 nm and emission maximum at 480 nm.

The vitamin can be extracted from the protein with organic solvents; thus it is not covalently bound. Ong *et al.* (1975) have also shown the presence of retinoic acid binding protein in human lung carcinoma of the nonkeratinizing epidermoid type, and in breast carcinomas. Suggestive evidence, based mostly on sucrose density gradient analysis, indicates that embryonic tissues and cancer tissues have more retinoic acid binding proteins than the adult human tissue.

1.15. Vitamin A and Transformation

Although vitamin A is necessary for vision (Wald, 1968), less than 1% is engaged in this function. Morphological studies have made clear that the health and functioning of different epithelial tissues depend on proper supply and delivery of the vitamin to the target tissues. A definition of target tissue is, however, quite difficult in the case of vitamin A, because it must be based only on gross visible changes and not on a known biochemical function that is typical of the tissue and is exclusively dependent on the vitamin. Thus it might well be that what we define as primary targets on the basis of morphological changes may turn out to be secondary ones and that the primary effects may occur at sites which do not undergo drastic morphological changes. Nevertheless, it is established that epithelial systems need vitamin A for display of proper morphology and function.

Organ culture studies have been and will continue to be useful in the definition of primary targets. It must, however, be kept in mind that compounds which do not display biological activity in the whole animal may still be active in organ culture systems, where reversal of keratinization is the criterion for biological activity (Sporn *et al.*, 1975). The explanation for this may well be that the biological activity at the target site cannot be displayed *in vivo* because the molecular recognition of transport may differ from that of function, so that the derivative cannot be transported to target issues *in vivo* but displays biological activity when provided directly to the target in the *in vitro* organ culture system. *In vitro* systems for the study of the epithelial function of vitamin A have been developed in several laboratories (Clamon *et al.*, 1974; Lasnitzki, 1962; Sporn *et al.*, 1975; Yuspa and Harris, 1974) and in general have supported the concept that vitamin A is directly involved in maintaining normal phenotypic expression. This concept puts vitamin A in a very special position among nutrients if one considers that most solid tumors arise from epithelial tissues.

Therefore, studies on the protective effects of vitamin A against development of epithelial tumors have been conducted for many years. The earliest recorded experimental production of neoplasms in animals fed a diet deficient in vitamin A was reported by Fibiger in 1913. In 1926, Fujimaki noted that vitamin A deficient rats had a tendency to develop papillomas of the stomach. Cramer (1937) later commented that Fujimaki's figures showed only "an extensive papillomatous hyperplasia" and that there was no clear evidence of any true malignant change. Upon repeat of Fujimaki's work, Cramer (1937) produced very extensive papillomatous hyperplasia in rats on a vitamin A deficient diet, but no malignant changes were found. Moreover, when a refined vitamin A deficient diet was used (Cramer, 1937), the papillomatous hyperplasia could not be reproduced, suggesting that some impurities in the diet were responsible for the lesions and not vitamin A deficiency *per se*. No clear explanation was provided.

Rowe and Gorlin (1959) investigated the effect of vitamin A deficient diet on oral carcinogenesis by 7,12-dimethyl-1,2-benzanthracene in golden hamster. A 5% solution of carcinogen was applied twice weekly for 13 weeks, with a maximum of 21 applications to avoid suppurative reactions in the cheek pouch. The control groups were given the deficient diet plus 400 IU of vitamin A per week. The study involved pair-fed as well as *ad libitum* fed animals to delete possible nutritional

Table VIII. Effect of Retinyl Palmitate on the Incidence of Carcinomas of the Hamster Oral Mucosa[a]

Feeding regimen	Number of animals	Retinyl palmitate	Dead before end of study	Total survivors	Animals with malignancy	Percent with malignancy
Ad libitum	40	0	13	27	16	59
Ad libitum	30	400 IU	13	27	9	33
Pair-feeding	30	400 IU	18	12	3	25

	Animals with tumors	Percent tumor incidence	Normal epithelium	Hyperplastic epithelium	Benign papilloma	Carcinoma in situ	Invasive squamous cell carcinoma
Ad libitum	22	81.5	1	3	7	14	2
Ad libitum	20	74.0	5	4	9	7	2
Pair-feeding	7	58.3	1	2	6	3	0

[a]From Rowe and Gorlin (1959).

differences. It was found that the incidence of neoplastic malignant lesions in hamsters given the vitamin A deficient diet was about 26% higher than in the supplemented *ad libitum* fed hamsters and approximately 34% greater than in hamsters supplemented with the vitamin and pair-fed the deficient diet (Table VIII). These studies involved an equal number of male and female hamsters for each group. The animals were sacrificed 20 weeks after they had been placed on the purified diet. Vitamin A levels were determined on a sample of blood obtained at death. Serum of supplemented animals contained from 26 to 57 μg of vitamin, whereas the serum from unsupplemented animals contained 19 μg/100 ml. It is remarkable in this study that, although signs attributed to vitamin A deficiency appeared within 7–8 weeks after the administration of the deficient diet, the animals were kept on this diet for 20 weeks with an average weight loss of only 5–6%. The possibility that the animals might have somehow received traces of vitamin A or provitamin must be considered.

In a similar animal model, topical vitamin A (retinyl palmitate) applied to cheek pouches with DMBA for 12 weeks caused an increase in the size of squamous cell carcinomas and the degree of anaplasia. The vitamin also caused mucoid metaplasia of the epithelium and of the tumor (Levij and Polliack, 1968).

Chu and Malmgreen (1965) compared the incidence of neoplastic lesions of the alimentary tract of Syrian golden hamsters after oral feeding of 7,12-dimethylbenzanthracene (DMBA) alone or in combination with retinyl palmitate, and benzo [*a*]pyrene alone or in combination with retinyl palmitate. Table IX shows the results of this study. Vitamin A appears to have a marked inhibitory effect on stomach carcinomas. A parallel study on cervical painting of DMBA alone and DMBA plus retinyl palmitate was also conducted. The results in Table X show a marked inhibition of vaginal cancer, but no effect on perineal cancer.

No definite conclusion on the mechanism of prevention by vitamin A can be drawn from these studies, since the vitamin was mixed with the carcinogen before administration and it might have exerted its action directly on the carcinogen or its metabolism.

Saffiotti *et al.* (1967) have investigated the possible preventive action of vitamin A in respiratory carcinogenesis. Their model is the development of respiratory squamous cell carcinoma in Syrian golden hamster with ten intratracheal instillations of benzo [*a*]pyrene. Retinyl palmitate in doses of 5 mg in 0.1 mg corn oil was given by stomach tube twice weekly for life. This treatment began 10 days

Table IX. *Effect of Retinyl Palmitate on Hamster Stomach Carcinomas Induced by 7,12-Dimethylbenzanthracene*[a]

Carcinogen (orally)	Vitamin A retinyl palmitate	Number of animals used	Number of animals with carcinomas	Percent animals with carcinomas
DMBA, 10 mg/wk	0	18	4	22
DMBA, 10 mg/wk	100 mg/wk	15	0	0
BP, 10 mg/wk	0	13	8	62
BP, 10 mg/wk	100 mg/wk	20	0	0

[a]From Chu and Malmgreen (1965).

Table X. Effect of Retinyl Palmitate on the Incidence of Hamster Vaginal Carcinomas Induced by 7,12-Dimethylbenzanthracene[a]

Carcinogen (topically)	Retinyl palmitate (topically)	Number of animals	Number of animals with vaginal carcinoma	Number of animals with perineal skin cancer	Number of animals with cervix cancer
1% DMBA	0	10	9	9	2
1% DMBA	10% olive oil solution	10	2	8	0

[a]From Chu and Malmgreen (1965).

after the last instillation of the carcinogen so as to exclude the possibility that vitamin A might modify early metabolism of the carcinogen. The results shown in Table XI indicate that vitamin A treatment interferes with the mechanism of induction of squamous metaplasia and neoplasia in the tracheobronchial mucosa. Similar results were obtained by Cone and Nettesheim (1973), with marked inhibition of 3-methylcholanthrene-induced squamous metaplasias and early tumors by retinyl acetate. Saffiotti's study also confirmed Chu and Malmgreen's results on the inhibitory effect of retinyl palmitate on the development of forestomach papillomas. Mouse skin papillomas can also be prevented by oral administration of 100 IU of vitamin A per gram of food (Davies, 1967).

Saffiotti's model of inhibition of benzo[*a*]pyrene-induced respiratory carcinogenesis by retinyl acetate has been used with hamsters on a commercial (Smith *et al.*, 1975*a*) and on a semisynthetic diet (Smith *et al.*, 1975*b*). Male weanling Syrian golden hamsters were fed a commercial diet and were administered a mixture of benzo[*a*]pyrene hematite in 12 weekly intratracheal instillations. One week after the last instillation, the hamsters were divided randomly into three groups receiving 100 μg of retinyl acetate (group 1), 1600 μg (group 2), and 2200 μg (group 3) intragastrically, divided into 2 weekly doses for the duration of the experiment. The incidence of respiratory tract tumors was much higher than in previous studies (Saffiotti *et al.*, 1967), probably because of higher doses of benzo[*a*]pyrene: 58% for group 1, 70% for group 2, and 81% for group 3. Saffiotti's studies showed a maximum incidence of respiratory tract tumor of 32%. Another important difference is the early appearance of tumors in the study by Smith and collaborators. Since the vitamin A treatment was started one week after the last instillation of the carcinogen, it is probable that the neoplastic process had already been initiated, before vitamin A could exert its preventive action. The same differences apply to the study conducted on a semipurified diet, with hamsters essentially on the same carcinogen and vitamin A schedule. Animals on 100 μg of retinyl acetate per week had a 60% incidence of respiratory tract tumors compared to 78% in hamsters given 1600 μg or 2400 μg of retinyl acetate per week.

When the hamsters were housed in laminar flow units, the tumor incidence was 71% for group I, 59% for group II, and 57% for group III. The incidence of forestomach papillomas was reduced from 50% in the group with 100 μg of retinyl acetate per week (Group I) to 25%, and 25% in the other two groups. This is in agreement with previous work (Chu and Malmgreen, 1965; Saffiotti *et al.*,

Table XI. Effect of Retinyl Palmitate on the Prevention of Respiratory Tract Lesions Induced by Administration of Benzo[a]pyrene[a]

Treatment	Retinyl palmitate	Number of animals	Animals with					
			Respiratory tumors		Squamous changes		Squamous tumors	
			Number	Percent	Number	Percent	Number	Percent
BP	0	53	17	32	18	34	11	21
BP	2 × 5000 IU/wk p.o.	46	5	11	2	4	1	2

Treatment	Retinyl palmitate	Number of respiratory tumors	Number of squamous tumors	Number of squamous metaplasias	Number of anaplastic tumors
BP	0	20	13	13	1
BP	2 × 5000 IU/wk p.o.	6	1	1	1

Treatment	Retinyl palmitate	Number of adenomatous tumors	Number of polyps	Number of bronchiolar adenomatous lesions
BP	0	4	2	12
BP	2 × 5000 IU/wk p.o.	1	3	9

[a]From Saffiotti *et al.* (1967).

1967) in which a protective effect by both doses of vitamin A in digestive tract carcinogenesis was demonstrated. Development of liver carcinomas in rat upon administration of aflatoxin B_1 was not affected by marginal levels of dietary vitamin A. Colon carcinomas arising from the same treatment were found only in rats fed a marginal vitamin A diet (Newberne and Rogers, 1973).

Rogers *et al.* (1973) studied the effect of high and low dietary retinyl acetate on colon carcinogenesis by 1,2-dimethylhydrazine. High levels of vitamin A (500 IU per gram of diet) did not affect the tumor incidence, but did decrease the number of tumors per rat in the groups given the highest dose of the carcinogen.

Bollag's group has conducted extensive studies on the preventive and therapeutic action of different derivatives of vitamin A on skin cancer. Skin papillomas were induced in mice with 7,12-dimethylbenz[a]anthracene (DMBA) and croton oil as a promotor, and the therapeutic test was conducted when papillomas reached diameters of at least 4 mm. Usually less than 10% of these tumors regressed spontaneously. Carcinomas appeared later (5–8 months after the application of the carcinogen). The vitamin (either as retinyl palmitate or as retinoic acid) was given orally or intraperitoneally (Bollag, 1971). After 2 weeks of treatment with retinoic acid, a 60–86% regression in papilloma volume was estimated (Table XII).

Similar results were obtained with oral administration of retinoic acid or retinyl palmitate. Work by Bollag (1972) on the prophylactic treatment of papillomas and carcinomas has also appeared. Retinoic acid (200 mg/kg every 14 days) was given orally during the promotion phase of carcinogenesis. Retinoic acid delayed the appearance of both papillomas and carcinomas. The volume of papillomas was greatly reduced, as in the previous study (Bollag, 1971). The incidence of carcinomas was also reduced (Table XIII).

No difference in weight of carcinomas between the two groups was noticed. The conclusion of this work is that vitamin A has a prophylactic and therapeutic effect on skin tumor in mice.

Prutkin (1968) has studied the effect of topical applications of retinoic acid on keratoacanthoma induced by DMBA in the rabbit ear. The result of such application was the production of a "viscous mucus type" product. Electron microscopic

Table XII. Results of Treatment of 7,12-Dimethylbenzanthracene-Induced Mouse Skin Papillomas by All-trans-β-Retinoic Acid[a]

Retinoic acid dose	Mean papilloma volume per animal (mm³)		Percent change
	Day 0	Day 4	
0	1272	1841	+44.7
100 mg/kg/wk i.p.	910	356	−60.9
200 mg/kg/wk i.p.	1705	442	−74.1
400 mg/kg/wk i.p.	1132	159	−86.0

[a]From Bollag (1971).

Table XIII. Effect of All-trans-β-Retinoic Acid on the Cumulative Number of
7,12-Dimethylbenzanthracene-Induced Carcinomas in the Mouse[a]

	Days after first DMBA treatment						
	113	127	148	162	176	197	211
Controls, number of carcinomas	0	0	2	3	8	9	11
Retinoic acid mice, number of carcinomas	0	0	0	0	2	4	4
		225	239	253	274	288	
Controls, number of carcinomas		14	15	15	17	18	
Retinoic acid mice, number of carcinomas		5	5	5	6	6	

[a]From Bollag (1972).

studies of the treated tumor revealed the formation of mucigen droplets and more prominent endoplasmic reticulum and Golgi apparatus. The biochemistry of the glycoprotein-producing capacity of this system has been studied by Levinson and Wolf (1972). Glycoprotein synthesis, as measured by incorporation of labeled fucose and glucosamine, into keratoacanthoma glycoproteins is dramatically enhanced by retinoic acid (Levinson *et al.*, 1972).

Yet another experimental approach has been taken by Smith *et al.* (1972). These authors have transplanted fetal lung tissues mixed with 20-methylcholanthrene (MCA) into the thigh muscle of Balb/c mice. The animals were placed on deficient, normal, and excess retinyl palmitate diets. The earliest tumor appeared at the site 64 days after implantation. All carcinomas arose in mice on an adequate diet with or without supplementary retinyl palmitate. No tumors developed in the rats maintained on the vitamin A deficient diet. In these animals, less vigorous growth of the transplanted pulmonary tissue was also observed.

Organ culture systems have also been used in the study of effects of vitamin A in epithelial changes induced by chemical carcinogens.

Cellular pleomorphic or squamous metaplastic epithelia and shrinking of cartilage matrix were induced by benzo[a]pyrene in doses of 10.5 μg/ml in cultures of suckling hamster tracheas (Crocker and Sanders, 1970). A simultaneous increase in DNA synthesis was observed. The occurrence of squamous metaplasia induced by benzo[a]pyrene was prevented by 3–6 μg of 13-*cis*-retinol per milliliter of culture medium. However, vitamin A also had a destructive effect on the cartilage.

Prostate glands of mice were grown for 7–9 days (Lasnitzki and Goodman, 1974). Methylcholanthrene caused hyperplasia of the alveolar epithelium, with subsequent squamous metaplasia or parakeratosis. β-Retinol, β-retinoic acid, α-retinoic acid, and a cyclopentenyl analogue (Fig. 2) were highly active in inhibiting and preventing the effects of methylcholanthrene.

Since two of these compounds have very little or no growth-promoting activity, the authors concluded that the anticarcinogenic activity of vitamin A does not correspond to the growth-promoting activity (Lasnitzki and Goodman, 1974).

1.16. Conclusion

Vitamin A maintains normal differentiation in epithelial tissues. Squamous metaplasia due to vitamin A deficiency resembles morphological events induced by chemical carcinogens *in vivo* and *in vitro*. Administration of different vitamin A compounds *in vitro* and *in vivo* prevents squamous metaplastic and hyperplastic changes due to deficiency or carcinogen exposure.

An analysis of the various systems reveals that not all of them have been chosen to study prevention of neoplastic transformation. Even in the same biological system, using benzo[*a*]pyrene to induce respiratory cancer, a wide variation in the incidence of squamous cell carcinomas was obtained. Thus conditions must be standardized and care must be taken to ensure that the vitamin is used before neoplasia has occurred, if prevention of carcinogenesis is the objective of the study.

Another important point of concern regards the form of vitamin A to be used in preventive studies. The delivery of retinol to target tissues is tightly regulated by the availability of the retinol-binding protein, thus protecting the target tissues from excessive buildup of the vitamin, which is stored in the liver as retinyl palmitate.

On the other hand, retinoic acid and its derivatives may be more effective because they are readily distributed and processed throughout the body without accumulation in the liver, thus possibly allowing a higher concentration at the target site. However, one must also consider the problem of toxicity resulting from excess vitamin A.

Because of these and other considerations, further investigations of the old and new *in vivo* and *in vitro* systems to study inhibition of carcinogenesis by vitamin A are certainly warranted.

ACKNOWLEDGMENTS

I am indebted to Ms. Patricia Hembree for typing the manuscript. My special thanks go to Silvana, Nicholas, and Mara Julia De Luca for their assistance.

1.17. References

Abe, T., Muto Y., and Hosoya, N., 1975, Vitamin A transport in chicken plasma: Isolation and characterization of retinol-binding protein (RBP), prealbumin (PA), and RBP-PA complex, *J. Lipid Res.* **16**:200.

Abrahamson, E. W., and Wiesenfeld, J. R., 1972, The structure, spectra and reactivity of visual pigments, in: *Photochemistry of Vision* (H. G. A. Dartnall, ed.), p. 69, Springer-Verlag, New York.

Adamany, A. M., and Spiro, R. G., 1975, Glycoprotein biosynthesis: Studies on thyroid mannosyltransferases, *J. Biol. Chem.* **250**:2842.

Agranoff, B. W., Eggerer, H., Henning, U., and Lynen, F., 1959, Biosynthesis of terpenes. VII. Isopentenylpyro phosphate isomerase. *J. Am. Chem. Soc.* **81**:1254.

Akhtar, M., Blosse, P. R., and Dewhurst, P. B., 1965, The reduction of a rhodopsin derivative, *Life Sci.* **4**:1221.

Akhtar, M., Blosse, P. R., and Dewhurst, P. B., 1967, The active site of the visual protein, rhodopsin, *Chem. Commun.* **13**:631.

Alworth, W. L., 1972, Biological stereospecificities in the squalene biosynthetic pathways: Resolution of 13 of the 14 points of ambiguity, in *Stereochemistry and Its Application in Biochemistry*, pp. 211–234, Wiley, New York.

Amdur, B. H., Rilling, H., and Block, K., 1957, The enzymatic conversion of mevalonic acid to squalene, *J. Am. Chem. Soc.* **79**:2646.

Arens, J. F., and Van Dorp, D. A., 1946a, Synthesis of some compounds possessing vitamin A activity, *Nature* **157**:190.

Arens, J. K., and Van Dorp, D. A., 1946b, Activity of "vitamin A acid" in the rat, *Nature* **158**:622.

Arigoni, D., 1958, Zur Biogenese pentazyklischer Triterpene in einer hoheren *Pflanze Experientia* **14**:153.

Ball, S., Goodwin, T. W., and Morton, R. A., 1948, Studies on vitamin A. 5. The preparation of retinene—vitamin A aldehyde, *Biochem. J.* **42**:516.

Ball, S., Collins, F. W., Dalvi, P. D., and Morton, R. A., 1949, Studies in vitamin A. II. Reactions of retinene with amino compounds, *Biochem. J.* **45**:304.

Barr, R. M., and De Luca, L. M., 1974, The *in vivo* incorporation of mannose, retinol, and mevalonic acid into phospholipids of hamster liver, *Biochem. Biophys. Res. Commun.* **60**:355.

Bashor, M. M., Toft, D. O., and Chytil, F., 1973, *In vitro* binding of retinol to rat-tissue components, *Proc. Natl. Acad. Sci. USA* **70**:3483.

Becker, R. S., Berger, S., Dalling, D. K., Grant, D. M., and Pugmire, R. J., 1975, Carbon-13 magnetic resonance investigation of retinal isomers and related compounds, *J. Am. Chem. Soc.* **96**:7008.

Behrens, N. H., and Leloir, L. F., 1970, Dolichol monophosphate glucose: An intermediate in glucose transfer in liver, *Proc. Natl. Acad. Sci. USA* **66**:153.

Blake, C. C. F., Swan, I. D., Rerat, C., Berthou, J., Laurent, A., and Rerat, B., 1971, An X-ray study of the subunit structure of prealbumin, *J. Mol. Biol.* **61**:217.

Blaurock, A. E., and Stoeckenius, W., 1971, Structure of the purple membrane, *Nature (London) New Biol.* **233**:152.

Bloch, K., Chaykin, S., Phillips, A. H., and DeWaard, A., 1959, Mevalonic acid pyrophosphate and isopentenylpyrophosphate, *J. Biol. Chem.* **234**:2595.

Bollag, W., 1971, Therapy of chemically induced skin tumors of mice with vitamin A palmitate and vitamin A acid, *Experientia* **37**:90.

Bollag, W., 1972, Prophylaxis of chemically induced benign and malignant epithelial tumors by vitamin A acid (retinoic acid), *Eur. J. Cancer* **8**:689.

Bonanni, F., and De Luca, L. M., 1974, Vitamin A-dependent fucose-glycopeptide from rat tracheal epithelium. *Biochim. Biophys. Acta* **343**:632.

Bonanni, F., Levinson, S., Wolf, G., and De Luca, L. M., 1973, Glycoproteins from the hamster respiratory tract and their response to vitamin A, *Biochim. Biophys. Acta* **297**:441.

Bownds, D., 1967, Site of attachment of retinal in rhodopsin, *Nature (London)* **216**:1178.

Bownds, D., and Wald, G., 1965, Reaction of the rhodopsin chromophore with sodium borohydrate, *Nature (London)* **205**:254.

Bownds, D., Daves, J., Miller, J., and Stahlman, M., 1972, Phosphorylation of frog photoreceptor membranes induced by light, *Nature (London) New Biol.* **237**:125.

Bownds, D., Brodie, A., Robinson, W. E., Palmer, D., Miller, J., and Sheldovsky, A., 1974, Physiology and enzymology of frog photoreceptor membranes, *Exp. Eye Res.* **18**:253.

Branch, W. T., Jr., Robbins, J., and Edelhoch, H., 1971, Thyroxine-binding prealbumin, *J. Biol. Chem.* **246**:6011.

Broda, E. E., and Goodeve, C. F., 1941, The behaviour of visual purple at low temperature, *Proc. R. Soc. A London Ser.* **179**:151.

Budowski, P., Ascarelli, I., Gross, J., and Nir, I., 1963, Provitamin A from lutein, *Science* **142**:969.

Buggy, M. J., Britton, G., and Goodwin, T. W., 1969, Stereochemistry of phytoene biosynthesis by isolated chloroplasts, *Biochem. J.* **114**:641.

Burgos-Gonzales, J., and Morton, R. A., 1962, The intracellular distribution of dolichol in pig liver, *Biochem J.* **82**:454.

Burgos-Gonzales, J., Hemming, F. W., Pennock, J. F., and Morton, R. A., 1963, Dolichol; a naturally occurring C_{100} isoprenoid alcohol, *Biochem. J.* **88**:470.

Caccam, J. F., Jackson, J. J., and Eylar, E. H., 1969, Biosynthesis of mannose-containing glycoproteins: Possible lipid intermediate, *Biochem. Biophys. Res. Commun.* **35**:505.

Cahn, R. S., Ingold, C. K., and Prelog, V., 1956, The specification of asymmetric configuration in organic chemistry, *Experientia* **12**:81.

Cama, H. R., Dalvi, P. D., Morton, R. A., and Salah, M. K., 1952a, Studies in vitamin A. 20 and 21. Properties of retinene and vitamin, *Biochem. J.* **52**:540.

Cama, H. R., Steinberg, G. R., and Stubbs, A. L., 1952b, Studies on vitamin A. 19. Preparation and properties of retinene. *Biochem. J.* **52**:535.

Chen, T. T., 1957, On the formation of a phosphorylated derivative of mevalonic acid, *J. Am. Chem. Soc.* **79**:6344.

Chichester, C. O., Yokoyama, H., Makayama, T., Lukton, A., and Mackinney, G., 1959, Leucine metabolism and carotene biosynthesis. *J. Biol. Chem.* **234**:598.

Chu, E. W., and Malmgreen R. A., 1965, An inhibitory effect of vitamin A on the induction of tumors of forestomach and cervix in the Syrian golden hamster by carcinogenic polycyclic hydrocarbons, *Cancer Res.* **25**:884.

Clamon, G. H., Sporn, M. B., Smith, J. M., and Saffiotti, U., 1974, Alpha- and beta-retinyl acetate reverse metaplasias of vitamin A deficiency in hamster trachea in organ culture, *Nature (London)* **250**:64.

Collins, F. D., 1953, Rhodopsin and indicator yellow, *Nature (London)* **171**:469.

Collins, F. D., 1954, The chemistry of vision, *Biol. Rev.* **29**:453.

Collins, F. D., and Morton, R. A., 1950a, Studies on rhodopsin. 1. Methods of extraction and the absorption spectrum, *Biochem. J.* **47**:3.

Collins, F. D., and Morton, R. A., 1950b, Studies on rhodopsin. 2. Indicator yellow, *Biochem. J.* **47**:10.

Cone, M. V., and Nettesheim, P., 1973, Effects of vitamin A on 3-methylcholanthrene-induced squamous metaplasias and early tumors in the respiratory tract of rats, *J. Natl. Cancer Inst.* **50**:1599.

Cornforth, J. W., and Popják, G., 1954, Studies on the biosynthesis of cholesterol, *Biochem. J.* **58**:403.

Cornforth, J. W., and Popják, G., 1959, Mechanism of biosynthesis of squalene from sesquiterpenoids, *Tetrahedron Lett.* **19**:29.

Cornforth, R. H., and Popják, G., 1969, Chemical synthesis of substrates of sterol biosynthesis, *Methods Enzymol.* **4**:359.

Cornforth, J. W., Cornforth, R. H., Donninger, C., and Popják, G., 1966, Studies on the biosynthesis of cholesterol. XIX. Steric course of hydrogen eliminations and of C—C bond formations in squalene biosynthesis, *Proc. R. Soc. London Ser. B* **163**:492.

Coward, W. A., Howell, J. M., Thompson, J. N., and Pitt, G. A. J., 1969, The retinol requirements of rats for spermatogenesis and vision, *Br. J. Nutr.* **23**:619.

Crain, F. D., Lotspeich, F. J., and Krause, R. F., 1967, Biosynthesis of retinoic acid by intestinal enzymes of the rat, *J. Lipid Res.* **8**:249.

Cramer, W., 1937, Papillomatosis in the forestomach of the rat and its bearing on the work of Fibiger, *Am. J. Cancer* **31**:537.

Crescitelli, F., Mommaerts, W. F. H. M., and Shaw, T. I., 1966, Circular dichroism of visual pigments in the visible and ultraviolet spectral regions, *Proc. Natl. Acad. Sci. USA* **56**:1729.

Crocker, T. T., and Sanders, L. L., 1970, Influence of vitamin A and 3,7-dimethyl-2,6-octadienal (Citral) on the effect of benzo[*a*]pyrene on hamster trachea in organ culture, *Cancer Res.* **30**:1312.

Davies, R. E., 1967, Effect of vitamin A on 7,12-dimethylbenz[*a*]anthracene-induced papilloma in rhino mouse skin, *Cancer Res.* **27**:237.

De Grip, W. J., Bonting, S. L., and Daemen, F. J. M., 1975, Biochemical aspects of the visual process, *Biochim. Biophys. Acta* **396**:104.

Delmelle, M., and Pontus, M., 1974, Magnetic resonance study of spin-labeled rhodopsin, *Biochim. Biophys. Acta* **365**:47.

De Luca, L. M., and Wolf, G., 1972, Mechanism of action of vitamin A in differentiation of mucus-secreting epithelia, *Agr. Food Chem.* **20**:474.

De Luca, L. M., and Yuspa, S. H., 1974, Altered glycoprotein synthesis in mouse epidermal cells treated with retinyl acetate *in vitro*, *Exp. Cell Res.* **86**:106.

De Luca, L. M., Little, E. P., and Wolf, G., 1969, Vitamin A and protein synthesis by rat intestinal mucosa, *J. Biol. Chem.* **244**:701.

De Luca, L. M., Schumacher, M., Wolf, G., and Newberne, P. M., 1970*a*, Biosynthesis of a fucose-containing glycopeptide from rat small intestine in normal and vitamin A deficient conditions, *J. Biol. Chem.* **245**:4551.

De Luca, L. M., Rosso, G. C., and Wolf, G., 1970*b*, The biosynthesis of a mannolipid that contains a polar metabolite of [15-^{14}C]retinol, *Biochem. Biophys. Res. Commun.* **41**:615.

De Luca, L. M., Kleinman, H. K., Little, E. P., and Wolf, G., 1971*a*, RNA metabolism in rat intestinal mucosa of normal and vitamin A deficient rats, *Arch. Biochem. Biophys.* **145**:332.

De Luca, L. M., Schumacher, M., and Nelson, D. P., 1971*b*, Localization of the retinol-dependent fucose-glycopeptide in the goblet cell of the rat small intestine, *J. Biol. Chem.* **246**:5762.

De Luca, L. M., Maestri, N., Bonanni, F., and Nelson, D. P., 1972, Maintenance of epithelial cell differentiation: The mode of action of vitamin A, *Cancer* **30**:1326.

De Luca, L. M., Maestri, N., Rosso, G. C., and Wolf, G., 1973, Retinol glycolipids, *J. Biol. Chem.* **248**:641.

De Luca, L. M., Silverman-Jones, C. S., and Barr, R. M., 1975, Biosynthetic studies on mannolipids and mannoproteins of normal and vitamin A depleted hamster livers, *Biochim. Biophys. Acta* **409**:342.

De Luca, L. M., Frot-Coutaz, J. P., Silverman-Jones, C. S., and Roller, P. R., 1977, Chemical synthesis of phosphorylated retinoids: Their mannosyl acceptor activity in rat liver membranes, *J. Biol. Chem.* **252**:2574.

Deshmukh, D. S., and Ganguly, J., 1967, Oxidation and reduction of retinal in rat intestine, *Indian J. Biochem.* **4**:18.

Deshmukh, D. S., Murthy, S. K., Mahdevan, S., and Ganguly, J., 1965, Metabolism of vitamin A: Absorption of retinal (vitamin A aldehyde) in rats, *Biochem. J.* **96**:377.

Dingle, J. T., and Lucy, J. A., 1965, Vitamin A, carotenoids and cell function, *Biol. Rev.* **40**:422.

Dingle, J. T., Barrett, A. J., and Weston, P. D., 1971, Cathepsin D: Characteristics of immunoinhibition and the confirmation of a role in cartilage breakdown, *Biochem. J.* **123**:1.

Dingle, J. T., Fell, H. B., and Goodman, D. S., 1972, The effect of retinol and of retinol binding protein on embryonic skeletal tissue in organ culture, *J. Cell Sci.* **11**:393.

Donner, K. P., and Reuter, T., 1969, The photoproducts of rhodopsin in the isolated retina of the frog, *Vision Res.* **9**:815.

Dowling, J. E., and Wald, G., 1960, The biological function of vitamin A acid, *Proc. Natl. Acad. Sci. USA* **46**:587.

Drujan, B. D., Castillon, R., and Guerrero, E., 1968, Application of fluorometry in the determination of vitamin A, *Anal. Biochem.* **23**:44.

Drummond, J. C., 1920, The nomenclature of the so called accessory food factors (vitamins), *Biochem. J.* **14**:660.

Dunagin, P. E., Jr., Zachman, R. D. and Olson, J. A., 1964, Identification of free and conjugated retinoic acid as a product of retinal (vitamin A aldehyde) metabolism in the rat *in vivo*, *Biochim. Biophys. Acta* **90**:432.

Dunagin, P. E., Jr., Meadows, E. H., Jr., and Olson, J. A., 1965, Retinoyl-beta glucuronic acid: A major metabolite of vitamin A in rat bile, *Science* **148**:86.

Dunagin, P. E., Jr., Zachman, R. D., and Olson, J. A., 1966, The identification of metabolites of retinal and retinoic acid in rat bile, *Biochim. Biophys. Acta.* **124**:71.

Eberle, M., and Arigoni, D., 1960, Absolute Konfiguration des Mevalonlactons, *Helv. Chim. Acta* **43**:1508.

Ebrey, T., 1967, The thermal decay of the intermediates of rhodopsin *in situ*, thesis, University of Michigan.

Eisenberg, F., Jr., 1972, Gas chromatography of carbohydrates as butaneboronic acid esters, *Methods Enzymol.* **28B**:168.

Emerick, R. J., Zile, M., and DeLuca, H. F., 1967, Formation of retinoic acid from retinol in the rat, *Biochem. J.* **102**:606.

Erhardt, F., Ostry, S. E., and Abrahamson, E. W., 1966, Protein configuration changes in the photolysis of rhodopsin. I. The thermal decay of cattle lumirhodopsin *in vitro*, *Biochim. Biophys. Acta* **112**:256.

Ernster, L., and Orrenius, S., 1965, Substrate induced synthesis of the hydroxylating enzyme system of liver microsomes, *Fed. Proc.* **24**:1190.

Evans, H. M., 1928, The effects of inadequate vitamin A on the sexual physiology of the female, *J. Biol. Chem.* **77**:651.

Evans, H. M., 1932, Testicular degeneration due to inadequate vitamin A in cases where E is adequate, *Am. J. Physiol.* **99**:477.

Evans, H. M., and Bishop, K. S., 1922, On an invariable and characteristic disturbance of reproductive function in animals reared on a diet poor in fat soluble vitamin A, *Anat. Rec.* **23**:17 (abstr. 23).

Fell, H. B., 1957, Effect of excess vitamin A on cultures of embryonic chicken skin explanted at different stages of differentiation, *Proc. R. Soc. London Ser. B.* **146**:242.

Fell, H. B., 1962, Influence of hydrocortisone on the metaplastic action of vitamin A on the epidermis of embryonic chicken skin in organ culture, *J. Embryol. Exp. Morphol.* **10**:389.

Fibiger, J., 1913, Experimental cancer: Parasites as a cause of cancer, in: *The Riddle of Cancer* (C. Oberling, ed.), p. 66, Yale University Press, New Haven, Conn., 1952.

Fidge, N. H., and Goodman, D. S., 1968, The enzymatic reduction of retinal to retinol in rat intestine, *J. Biol. Chem.* **243**:4372.

Fidge, N. H., Shiratori, T., Ganguly, J., and Goodman, D. S., 1968, Pathways of absorption of retinol and retinoic acid in the rat, *J. Lipid Res.* **9**:103.

Frank, R. M., and Besinger, R. E., 1974, Rhodopsin and light-sensitive kinase activity of retinal outer segments, *Exp. Eye Res.* **18**:271.

Fridericia, L. S., and Holm, E., 1925, Relation between night blindness and malnutrition-influence of deficiency of fat-soluble A vitamin in the diet on the visual purple in the eyes of rats, *Am. J. Physiol.* **73**:63.

Frot-Coutaz, J. P., Silverman-Jones, C. S., and De Luca, L. M., 1976, Isolation, characterization and biological activity of retinyl phosphate from hamster intestinal epithelium. *J. Lipid Res.* **17**:220.

Fujimaki, Y., 1926, Formation of gastric carcinoma in albino rats fed on deficient diets, *J. Cancer Res.* **10**:69.

Gaede, K., and Rodriguez, P., 1973, Formation of retinol $(\alpha,\beta-^{32}P)$pyrophosphate with $(\gamma-^{32}P)$ATP catalyzed by whole homogenates of rat thyroid, *Biochem. Biophys. Res. Commun.* **54**:76.

Gambhir, K. K., and Ahluwalia, B. S., 1974, A smaller molecular weight retinol binding protein in rat testis seminiferous tubules, *Biochem. Biophys. Res. Commun.* **61**:501.

Ganguly, J., and Murthy, S. K., 1967. Vitamin A. VI. Biogenesis of vitamin A and carotenes, in: *The Vitamins* (W. H. Sebrell, Jr., and R. S. Harris, eds.), pp. 125–153, Academic Press, New York.

Geison, R. L., and Johnson, B. C., 1970, Studies on the *in vivo* metabolism of retinoic acid in the rat, *Lipids* **5**:371.

Glover, J., Jay, C., and White, G. H., 1974, Distribution of retinol-binding protein in tissues, *Vitamins Hormones* **32**:215.

Gonzales, G., and Offord, R. E., 1971, The subunit structure of prealbumin, *Biochem. J.* **125**:309.

Goodman, D. S., and Huang, H. S., 1965, Biosynthesis of vitamin A with rat intestinal enzymes, *Science* **149**:879.

Goodman, D. S., and Olson, J. A., 1969, The conversion of all-*trans*-β-carotene into retinal, *Methods Enzymol.* **15**:463.

Goodman, D. S., and Raz, A., 1972, Extraction and recombination studies of the interaction of retinol with human plasma retinol-binding protein, *J. Lipid Res.* **13**:338.

Goodman, D. S., Huang, H. S., and Shiratori, T., 1965, Tissue distribution and metabolism of newly absorbed vitamin A in the rat, *J. Lipid Res.* **6**:390.

Goodman, D. S., Blomstrand, R., Werner, B., Huang, H. S. and Shiratori, T., 1966, The intestinal absorption and metabolism of vitamin A and β-carotene in man, *J. Clin. Invest.* **45**:1615.

Goodwin, T. W., 1959, The biosynthesis and function of the carotenoid pigments, in: *Advances in Enzymology*, Vol. 21 (F. F. Nord, ed.), p. 295, Interscience, New York

Goodwin, T. W., 1971, Biosynthesis of carotenoids and plant triterpenes: The Fifth Ciba Medal Lecture, *Biochem. J.* **123**:293.

Goodwin, T. W., and Williams, R. J. H., 1966, The stereochemistry of phytoene biosynthesis, *Proc. R. Soc. London Ser. B* **163**:515.

Gough, D. P., and Hemming, F. W., 1970, The characterization and stereochemistry of biosynthesis of dolichols in rat liver, *Biochem. J.* **118**:163.

Grellman, K. H., Livingstone, R., and Pratt, D., 1962, A flash photolytic investigation of rhodopsin at low temperatures, *Nature (London)* **193**:1258.

Grobb, E., and Buttler, R., 1954, Über die Biosyntheses des β-Carotins bei Mucor hiemalis: Die Beteiligung der Essigsäure am Aufbau des Carotinmoleküls, untersucht mit Hilfe C^{14}-markierter Essignsäure, *Experientia* **10**:250.

Hagins, W. A., 1957, Rhodopsin in the mammalian retina, thesis, University of Cambridge.

Hara, T., and Hara, R., 1965, New photosensitive pigment found in the retina of the squid *Ommastrephes*, *Nature (London)* **206**:1331.

Harris, C. C., Sporn, M. B., Kaufman, D. G., Smith, J. M., Jackson, F. E., and Saffiotti, U., 1972, Histogenesis of squamous metaplasia in the hamster tracheal epithelium caused by vitamin A deficiency or benzo (a)pyrene-ferric oxide, *J. Natl. Cancer Inst.* **48**:743.

Heller, J., and Horwitz, J., 1973, Conformation changes following interaction between retinol isomers and human retinol-binding protein and between the retinol-binding protein and prealbumin, *J. Biol. Chem.* **248**:6308.

Helting, T., and Peterson, P. A., 1972, Galactosyl transfer in mouse mastocytomas: Synthesis of galactose containing polar metabolite of retinol, *Biochem. Biophys. Res. Commun.* **46**:429.

Hemming, F. W., 1970, Polyprenols, *Biochem. Soc. Symp.* **29**:105.

Henbest, H. B., Jones, E. R. H., and Owen, T. C., 1955, Conversion of vitamin A_1 into vitamin A_2, *J. Chem. Soc.* **3**:2765.

Henning, U., Moslein, E. M., and Lynen, F., 1959, Biosynthesis of terpenes. V. Formation of 5-pyrophosphomevalonic acid by phosphomevalonic acid kinase, *Arch. Biochem. Biophys.* **83**:259.

Higashi, Y., Strominger, J. L., and Sweeley C. C., 1967, Structure of a lipid intermediate in cell-wall peptidoglycan synthesis: A derivative of a C_{55} isoprenoid alcohol, *Proc. Natl. Acad. Sci. USA* **57**:1878.

Higashi, Y., Strominger, J. L., and Sweeley, C. C., 1970, Biosynthesis of the peptidoglycan of bacterial cell wall, *J. Biol. Chem.* **245**:3697.

Holm, E., 1929, Demonstration of vitamin A in retinal tissue and a comparison with the vitamin content of brain tissue, *Acta Ophthalmol.* **7**:146.

Holmes, H. N., and Corbett, R. E., 1937, The isolation of crystalline vitamin A, *J. Am. Chem. Soc.* **59**:2042.

Horwitz, J., and Heller, J., 1973, Interaction of all-*trans*-9-, 11-, and 13-*cis*-retinal, all-*trans*-retinyl acetate, and retinoic acid with human retinol-binding protein and prealbumin, *J. Biol. Chem.* **248**:6317.

Howell, J. M., and Thompson, J. N., 1967a, Lesions associated with the development of ataxia in vitamin A deficient chicks, *Br. J Nutr.* **21**:741.

Howell, J. M., and Thompson, J. N., 1967b, Observations on the lesions in vitamin A deficient adult fowls with particular reference to changes in bone and central nervous system, *Br. J. Exp. Pathol.* **48**:450.

Howell, J. M., Thompson, J. N., and Pitt, G. A. J., 1963, Histology of the lesions produced in the reproductive tract of animals fed a diet deficient in vitamin A alcohol but containing vitamin A acid. I. The male rat, *J. Reprod. Fertil.* **5**:159.

Howell, J. M., Thompson, J. N., and Pitt, G. A. J., 1964, Histology of the lesions produced in the reproductive tract of animals fed a diet deficient in vitamin A alcohol but containing vitamin A acid. II. The female rat, *J. Reprod. Fertil.* **7**:251.

Howell, J. M., Thompson, J. N., and Pitt, G. A. J., 1967, Changes in the tissues of guinea-pigs fed on a diet free from vitamin A but containing methyl retinoate, *Br. J. Nutr.* **21**:37.

Huang, H. S., and Goodman, D. S., 1965, I. Intestinal absorption and metabolism of ^{14}C-labeled vitamin A alcohol and β-carotene in the rat, *J. Biol. Chem.* **240**:2839.

Hubbard, R., and Wald, G., 1952, Cis-trans isomers of vitamin A and retinene in the rhodopsin system, *J. Gen. Physiol.* **36**:269.

Hubbard, R., Bownds, D., and Yoshizawa, T., 1965, The chemistry of visual photoreception, *Cold Spring Harbor Symp. Quant. Biol.* **30**:301.

Ingbar, S. H., 1963, Observations concerning the binding of thyroid hormones by human serum prealbumin, *J. Clin. Invest.* **42**:143.

Inhoffen, H. H., Pommer, H., and Bohlman, F., 1950a, Syntheses of β-carotene, *Ann. Chem.* **569**:237.

Inhoffen, H. H., Bohlman, F., Bartram, K., Rummert G., and Rommer, H., 1950b, Syntheses in the carotenoid series XV, *Ann. Chem.* **570**:54.

Inhoffen, H. H., Pommer, H., and Westphal, F., 1950c, An additional synthesis of β-carotene, *Ann. Chem.* **570**:69.

Isler, O., Huber, W., Ronco, A., and Kofler, M., 1947, Synthese des Vitamin A, *Helv. Chim. Acta* **30**:1911.

Isler, O., Klaui, H., and Solms, U., 1967, Vitamin A. III. Industrial preparation and production, in: *The Vitamins* (W. H. Sebrell, Jr., and R. S. Harris, eds.), p. 101, Academic Press, New York.

Isler, O., Solms, U., and Wursch, J., 1970, in: *Fat Soluble Vitamins* (R. A. Morton, ed.), p. 99, Pergamon Press, Oxford.

Ito, T., Zile, M., DeLuca, H. F., and Ahrens, H. M., 1974, Metabolism of retinoic acid in vitamin A-deficient rats, *Biochim. Biophys. Acta* **369**:338.

IUPAC, 1960, Commission on the nomenclature of biological chemistry, *J. Am. Chem. Soc.* **82**:5581.

IUPAC-IUB, 1966, Commission on biochemical nomenclature, tentative rules, *J. Biol. Chem.* **241**:2987.

Johnson, B. C., Kennedy, M., and Chiba, N., 1969, Vitamin A and nuclear RNA synthesis, *Am. J. Clin. Nutr.* **22**:1048.

Johnston, E. M., and Zand, R., 1973, Extrinsic Cotton effects in retinaldehyde Schiff's bases, *Biochemistry* **12**:4631.

Juneja, H. S., Murthy, S. K., and Ganguly, J., 1964, Effect of retinoic acid on the reproductive performances of male and female rats, *Indian J. Exp. Biol.* **2**:153.

Jungalwala, F. B., and Porter, J. W., 1967, Biosynthesis of phytoene from isopentenyl and farnesyl-pyrophosphates by a partially purified tomato enzyme system, *Arch. Biochem. Biophys.* **119**:209.

Kanai, M., Raz, A., and Goodman, D. S., 1968, Retinol-binding protein: The transport protein for vitamin A in human plasma, *J. Clin. Invest.* **47**:2025.

Karrer, P., and Eugster, C. H., 1950, Synthese von Carotinoiden. II. Totalsynthese des β-Carotins I, *Helv. Chim. Acta* **33**:1172.

Karrer, P., and Morf, R., 1933, Synthese des Perhydrovitamins-A, *Helv. Chim. Acta* **16**:557.

Karrer, P., Helfenstein, A., Wehrli, H., and Wettstein, A., 1930, Pflanzenfarbstoffe. XXV. Ueber die Konstitution des Lycopins und Carotins, *Helv. Chim. Acta* **13**:1084.

Karrer, P., Morf, R,. and Schopp, K., 1931, Zur Kenntnis des Vitamins-A aus Fischtranen, *Helv. Chim. Acta* **14**:1036.

Kaufman, D. G., Baker, M. S., Smith, J. M., Henderson, W. R., Harris, C. C., Sporn, M. B., and Saffiotti, U., 1972, RNA metabolism in tracheal epithelium: Alteration in hamster deficient in vitamin A, *Science* **177**:1105.

Kayushin, L. P., and Skulachev, V. P., 1974, Bacteriorhodpsin as an electrogenic proton pump: Reconstitution of bacteriorhodopsin proteoliposomes generating $\Delta\psi$ and ΔpH, *FEBS Lett.* **39**:39.

Kito, Y., and Takezaki, M., 1966, Optical rotation of irradiated rhodopsin solution, *Nature (London)* **211**:197.

Kleiner-Bossaller, A., and DeLuca, H. F., 1971, Formation of retinoic acid from retinol in the kidney, *Arch. Biochem. Biophys.* **142**:371.

Kleinman, H., and Wolf, G., 1974, Extraction and characterization of a native vitamin A-sensitive glycoprotein from rat intestine, *Biochim. Biophys. Acta* **359**:90.

Kochhar, D. M., and Aydelotte, M. B., 1974, Susceptible stages and abnormal morphogenesis in the developing mouse limb, analyzed in organ culture after transplacental exposure to vitamin A, *J. Embryol. Exp. Morphol.* **31**:721.

Krinsky, N. I., and Ganguly, J., 1953, Intracellular distribution of vitamin A ester and vitamin A alcohol in rat liver, *J. Biol. Chem.* **202**:227.

Krishnamurthy, S., Bieri, J. G., and Andrews, E. L., 1963, Metabolism and biological activity of vitamin A acid in the chick, *J. Nutr.* **79**:503.

Kropf, A., and Hubbard, R., 1970, The photoisomerization of retinal, *Photochem. Photobiol.* **12**:249.

Kuhn, H., 1974, Light-dependent phosphorylation of rhodopsin in living frogs, *Nature (London)* **250**:588.

Kuhn, R., and Lederer, E., 1933, Uber die Farbstoffe des Hummers (*Astacus gammarus* L.) und ihre Stammsubstanz, das Astacin, *Berichte* **66**:488.

Kushwaha, S. C., Suzue, G., Subbarayan, C., and Porter, J. W., 1970, The conversion of phytoene-^{14}C to acyclic, monocyclic and dicyclic carotenes and the conversion of lycopene-15-15'-^{3}H to mono and dicyclic carotenes by soluble enzyme systems obtained from plastids of tomato fruits, *J. Biol. Chem.* **245**:4708.

Lahav, M., Chiu, T. H., and Lennarz, W. J., 1969, Biosynthesis of mannan in *Micrococcus lysodeikticus*. II. Enzymic synthesis of mannosyl-1-phosphoryl undecaprenol, *J. Biol. Chem.* **244**:5890.

Langdon, R. G., and Bloch, K., 1953*a*, The biosynthesis of squalene, *J. Biol. Chem.* **200**:129.

Langdon, R. G., and Bloch, K., 1953*b*, The utilization of squalene in the biosynthesis of cholesterol, *J. Biol. Chem.* **200**:135.

Lasnitzki, I., 1962, Hypovitaminosis A in the mouse prostate gland cultured in chemically defined medium, *Exp. Cell Res.* **28**:40.

Lasnitzki, I., 1963, The effect of excess vitamin A on the embryonic rat oesophagus in culture, *J. Exp. Med.* **118**:1.

Lasnitzki, I., and Goodman, D. S., 1974, Inhibition of the effects of methylcholanthrene on mouse prostate in organ culture by vitamin A and its analogs, *Cancer Res.* **34**:1564.

Lee, T. C., and Chichester, C. O., 1969, Geranylgeranyl pyrophosphate as the condensing unit for enzymatic synthesis of carotenes, *Phytochemistry* **8**:603.

Levij, I. S., and Polliack, A., 1968, Potentiating effect of vitamin A on 9-10 dimethyl 1-2 benzanthracene carcinogenesis in the hamster cheek pouch, *Cancer* **22**:300.

Levinson, S. S., and Wolf, G., 1972, The effect of vitamin A acid on glycoprotein synthesis in skin tumors (keratoacanthomas), *Cancer Res.* **32**:2248.

Levy, J. A., Carminatti, H., Cantarella, A. I., Behrens, N. H., Leloir, L. F., and Tabora, E., 1974, Mannose transfer to lipid linked di-*N*-acetylchitobiose, *Biochem. Biophys. Res. Commun.* **60**:118.

Lin, R. L., 1969, Metabolism of retinoic acid, Ph.D. dissertation, Oklahoma State University.

Linder, M. C., Anderson, G. H., and Ascarelli, I., 1971, Quantitative distribution of vitamin A in Kupffer cell and hepatocyte populations of rat liver, *J. Biol. Chem.* **246**:5538.

Lippel, K., and Olson, J. A., 1968, Origin of some derivatives of retinoic acid found in rat bile, *J. Lipid Res.* **9**:580.

Lippel, K., Manyan, D. R., and Llewellyn, A., 1970, Retinoic acid decarboxylation and activation of retinoate and other branched long-chain acids *in vitro*, *Arch. Biochem. Biophys.* **139**:421.

Lynen, F., Eggerer, H., Henning, U., and Kessel, I., 1958, Farnesyl Pyrophosphat und 3-methyl Δ^3-Butenyl-1-Pyrophosphat, die biologischen Vorstufen des Squalens, *Angew. Chem.* **70**:739.

Lythgoe, R. J., 1937, The absorption spectra of visual purple and of indicator yellow, *J. Physiol.* (*London*) **89**:331.

Lythgoe, R. J., and Quilliam, J. P., 1938, The relation of transient orange to visual purple and indicator yellow, *J. Physiol.* (*London*) **94**:390.

Mack, J. P., Lui, N. S. T., Roels, O. A., and Anderson, O. R., 1972, The occurrence of vitamin A in biological membranes, *Biochim. Biophys. Acta* **288**:203.

Malathi, P., Subba Rao, K., Seshadri Sastri, P., and Ganguly, J., 1963, Studies on metabolism of vitamin A, *Biochem. J.* **87**:305.

Mallia, A. K., Smith, J. E., and Goodman, D. S., 1975, Metabolism of retinol-binding protein and vitamin A during hypervitaminosis A in the rat, *J. Lipid Res.* **16**:180.

Martin, H. G., and Thorne, K. J. K., 1974, The involvement of endogenous dolichol in the formation of lipid-linked precursors of glycoproteins in rat liver, *Biochem. J.* **138**:281.

Mason, K. E., 1932, Differences in testes injury and repair after vitamin A deficiency, vitamin E deficiency and inanition, *Am. J. Anat.* **51**:153.

Mason, K. E., 1933, Differences in testes injury and repair after vitamin E deficiency and inanition, *Am. J. Anat.* **52**:153.

Matthews, R. G., Hubbard, R., Brown, P. K., and Wald, G., 1963, Tautomeric forms of metarhodopsin, *J. Gen. Physiol.* **47**:215.

Mayer, H., and Isler, O., 1971, Total syntheses of carotenoids in: *Carotenoids* (O. Isler, ed.), p. 328, Birkhauser Verlag, Basel.

McCollum, E. V., and Davies, M., 1913, The necessity of certain lipins in the diet during growth, *J. Biol. Chem.* **15**:167.

McCollum, E. V., and Davies, M., 1915, The nature of the dietary deficiency of rice, *J. Biol. Chem.* **23**:181.

McCollum, E. V., and Simmonds, N., 1917, A biological analysis of pellagra producing diets. II. The minimum requirements of the two unidentified dietary factors for maintenance as contrasted with growth, *J. Biol. Chem.* **32**:181.

Mellanby, E., 1944, Nutrition in relation to bone growth and the nervous system, *Proc. R. Soc. London Ser. B* **132**:28.

Milas, N. A., Davis, P., Belic, I., and Fles, D., 1950, Synthesis of β-carotene, *J. Am. Chem. Soc.* **72**:4844.

Miller, J. A., and Paulsen, R., 1975, Phosphorylation and dephosphorylation of frog rod outer segment membranes as part of the visual process, *J. Biol. Chem.* **250**:4427.

Mitchell, P., 1961, Coupling of phosphorylation to electron and hydrogen transfer by a chemi-osmotic type of mechanism, *Nature (London)* **191**:144.

Moore, L. A., 1939, Relationship between carotene, blindness due to constriction of the optic nerve, papillary edema and nyctalopia in calves, *J. Nutr.* **17**:443.

Moore, L. A., Hoffman, C. F., and Duncan, C. W., 1935, Blindness in cattle associated with a constriction of the optic nerve and probably of nutritional origin, *J. Nutr.* **9**:533.

Moore, T., 1930, Vitamin A and carotene. V. The absence of the liver oil vitamin A from carotene. VI. The conversion of carotene to vitamin A *in vivo*, *Biochem. J.* **24**:692.

Moore, T., 1957, *Vitamin A*, Elsevier, Amsterdam.

Morgan, F. J., Canfield, R. E., and Goodman, D. S., 1971, The partial structure of human plasma prealbumin and retinol-binding protein, *Biochim. Biophys. Acta* **236**:798.

Morriss, G. M., 1973, The ultrastructural effects of excess maternal vitamin A on the primitive streak stage rat embryo, *J. Embryol. Exp. Morphol.* **30**:219.

Morton, R. A., 1944, Chemical aspects of the visual process, *Nature (London)* **153**:69.

Morton, R. A., 1972, The chemistry of the visual pigments, in: *Photochemistry of Vision* (H. J. A. Dartnall, ed.), p. 33, Springer-Verlag, New York.

Morton, R. A., and Goodwin, T. W., 1944, Preparation of retinene *in vitro*, *Nature (London)* **153**:405.

Morton, R. A., and Pitt, G. A. J., 1955, pH and the hydrolysis of indicator yellow, *Biochem. J.* **59**:128.

Morton, R. A., and Pitt, G. A. J., 1957, Visual pigments, *Fortschr. Chem. Org. Naturst.* **14**:244.

Muto, Y., and Goodman, D. S., 1972, Vitamin A transport in rat plasma, *J. Biol. Chem.* **247**:2533.

Muto, Y., Smith, J. E., Milch, P. O., and Goodman, D. S., 1972, Regulation of retinol-binding protein metabolism by vitamin A status in the rat, *J. Biol. Chem.* **247**:2542.

Nelson, E. C., Mayberry, M., Reid, R., and John, K. V., 1971, The decarboxylation of retinoic acid by horseradish peroxidase and an acetone–butanol–ether–dried liver powder, *Biochem. J.* **121**:731.

Newberne, P. M., and Rogers, A. E., 1973, Rat colon carcinomas associated with aflatoxin and marginal vitamin A, *J. Natl. Cancer Inst.* **50**:439.

Nilsson, S. F., and Peterson, P. A., 1971, Evidence for multiple thyroxine-binding sites in human prealbumin, *J. Biol. Chem.* **246**:6098.

Nyquist, S. E., Crane, F. L., and Morré, D. J., 1971, Vitamin A: Concentration in the rat liver Golgi apparatus, *Science* **173**:939.

Oesterhelt, D., 1971a, Effect of organic solvents on the retinal-opsin complex from *Halobacterium halobium*, *Fed. Proc.* **30**:1188.

Oesterhelt, D., 1971b, The binding site of retinal in the purple membrane of *Halobacterium halobium*, *Abstr. Comm. 7th Meet. Eur. Biochem. Soc.*, p. 205.

Oesterhelt, D., 1974, Bacteriorhodopsin as a light driven proton pump, in: *Membrane Proteins in Transport and Phosphorylation* (G. S. Azzone, M. Klingernberg, and N. Siliprandi, eds.), pp. 79–84, Elsevier, Amsterdam.

Oesterhelt, D., 1975, The purple membrane of *Halobacterium halobium*: A new system for light energy conversion, *Ciba Found. Symp.* **31**:147.

Oesterhelt, D., and Hess, B., 1973, Reversible photolysis of the purple complex in the purple membrane of *Halobacterium halobium*, *Eur. J. Biochem.* **37**:316.

Oesterhelt, D., and Stoeckenius, W., 1971, Rhodopsin-like protein from the purple membrane of *Halobacterium halobium*, *Nature (London) New Biol.* **233**:149.

Oesterhelt, D., and Stoeckenius, W., 1973, Functions of a new photoreceptor membrane, *Proc. Natl. Acad. Sci. USA* **70**:2853.

Oesterhelt, D., and Stoeckenius, W., 1974, Isolation of the cell membrane of *Halobacterium halobium* and its fractionation into red and purple membrane, *Methods Enzymol.* **31**:667.

Oesterhelt, D., Meentzen, M., and Schuhmann, L., 1973, Reversible dissociation of the purple complex in bacteriorhodopsin and identification of 13-*cis* and all-*trans* retinal as its chromophores, *Eur. J. Biochem.* **40**:453.

Olson, J. A., 1961, The conversion of radioactive β-carotene into vitamin A by the rat intestine *in vivo*, *J. Biol. Chem.* **236**:349.

Olson, J. A., 1964, The effect of bile and bile salts on the uptake and cleavage of β-carotene into retinol ester (vitamin ester) by intestinal slices, *J. Lipid Res.* **5**:402.

Olson, J. A., 1967, The metabolism of vitamin A, *Pharmacol. Rev.* **19**:559.

Olson, J. A., and Hayashi, O., 1965, Enzymic cleavage of β-carotene into vitamin A by soluble enzymes of rat liver and intestine, *Proc. Natl. Acad. Sci. USA* **54**:1364.

Ong, D. E., and Chytil, F., 1975, Retinoic acid-binding protein in rat tissue, *J. Biol. Chem.* **250**:6113.

Ong, D. E., Page, D. L., and Chytil, F., 1975, Retinoic acid binding protein in human tumors, *Science* **190**:60.

Oppenheimer, J. H., 1968, Role of plasma proteins in the binding, distribution and metabolism of the thyroid hormones, *N. Engl. J. Med.* **278**:1153.

Oppenheimer, J. H., and Surks, M. I., 1964, Determination of free thyroxine in human serum: A theoretical and experimental analysis, *J. Clin. Endocrinol. Metab.* **24**:785.

Oroshnick, W., 1956, The synthesis and configuration of neo-b-vitamin A and neoretinene-b, *J. Am. Chem. Soc.* **78**:2651.

Oroshnick, W., and Mebane, A. D., 1954, Isoprenoid polyenes containing sterically hindered *cis*-configurations, *J. Am. Chem. Soc.* **76**:5719.

Oroshnick, W., Brown, P. K., Hubbard, R., and Wald, G., 1956, Hindered *cis*-isomers of vitamin A and retinene: The structure of the neo-b-isomer, *Proc. Natl. Acad. Sci. USA* **42**:578.

Ostroy, S. E., Erhardt, F., and Abrahamson, E. W., 1966, Protein configuration changes in the photolysis of rhodopsin. II. The sequence of intermediates in the thermal decay of cattle rhodopsin *in vitro*, *Biochim. Biophys. Acta* **112**:265.

Pages, R. A., Robbins, J., and Edeloch, H., 1973, Binding of thyroxine and thyroxine analogs to human serum prealbumin, *Biochemistry* **12**:2773.

Papastephanou, C., Barnes, F. J., Briedis, A. V., *et al.*, 1973, Enzymatic synthesis of carotenes by cell-free preparations of fruit of several genetic selections of tomatoes, *Arch. Biochem. Biophys.* **157**:415.

Pennock, J. G., Hemming, F. W., and Morton, R. A., 1960, Dolichol: A naturally occurring isoprenoid alcohol, *Nature (London)* **186**:470.

Peterson, P. A., 1971*a*, Characteristics of a vitamin A transporting protein complex occurring in human serum, *J. Biol. Chem.* **246**:34.

Peterson, P. A., 1971*b*, Studies on the interaction between prealbumin, retinol-binding protein, and vitamin A, *J. Biol. Chem.* **246**:44.

Peterson, P. A., and Berggard, I., 1971, Isolation and properties of a human retinal-transporting protein, *J. Biol. Chem.* **246**:25.

Peterson, P. A., Rask, L., Ostberg, L., Andersson, L., Kamwends, F., and Pertoft, H., 1974*a*, Studies on the transport and cellular distribution of vitamin A in normal and vitamin A-deficient rats with special reference to the vitamin A-binding plasma protein, *J. Biol. Chem.* **248**:4009.

Peterson, P. A., Nilsson, S. F., Ostberg, L., Rask, L., and Vahlquist, A., 1974*b*, Aspects of the metabolism of retinol-binding protein and retinol, *Vitamins Hormones* **32**:181.

Poincelot, R. P., and Abrahamson, E. W., 1970, Phospholipid composition and extractability of bovine rod outer segments and rhodopsin micelles, *Biochemistry* **9**:1820.

Popják, G. and Cornforth, J. W., 1966, Substrate stereochemistry in squalene biosynthesis, *Biochem. J.* **101**:553.

Popják, G., Goodman, D. S., Cornforth, J. A., Cornforth, R. H., and Ryhage, R., 1961. Studies on the biosynthesis of cholesterol. XV. Mechanism of squalene biosynthesis from farnesyl pyrophosphate and from mevalonate, *J. Biol. Chem.* **236**:1934.

Popják, G., Cornforth, J. W., Cornforth, R. H., Ryhage, R., and Goodman, D. S., 1962, Studies on the biosynthesis of cholesterol XVI. Chemical synthesis of 1-^{3}H$_2$-C^{14}- and 1-D-2-^{14}C-*trans-trans*-farnesyl pyrophosphate and their utilization in squalene biosynthesis, *J. Biol. Chem.* **237**:56.

Popper, H., 1941, Histological distribution of vitamin A in human organs under normal and under pathologic conditions, *Arch. Pathol.* **31**:766.

Popper, H., 1944, Distribution of vitamin A in tissue as visualized by fluorescence microscopy, *Physiol. Rev.* **24**:205.

Powell, L. T., and Krause, R. F., 1953, Vitamin A distribution in the rat liver cell, *Arch. Biochem. Biophys.* **44**:102.

Prutkin, L., 1968, The effect of vitamin A acid on tumorigenesis and protein production, *Cancer Res.* **28**:1021.

Qureshi, A. A., Kim, M., Qureshi, N., and Porter, J. W., 1974, The enzymatic conversion of *cis*-[^{14}C]-phytofluene, *trans*-[^{14}C]phytofluene, and *trans*-β-[^{14}C]carotene to poly-*cis*-acyclic carotenes by a cell-free preparation of tangerine tomato fruit plastids, *Arch. Biochem. Biophys.* **162**:108.

Racker, E., and Stoeckenius, W., 1974, Reconstitution of purple membrane vesicles catalyzing light-driven proton uptake and adenosine triphosphate formation, *J. Biol. Chem.* **249**:662.

Rapp, J., Wiesenfeld, J. R., and Abrahamson, E. W., 1970, The kinetics of intermediate processes in the photolysis of rhodopsin. I. A reexamination of the decay of bovine lumirhodopsin, *Biochim. Biophys. Acta* **201**:119.

Rask, L., Peterson, P. A., and Nilsson, S. F., 1971, The subunit structure of human thyroxine-binding prealbumin, *J. Biol. Chem.* **246**:6087.

Raz, A., and Goodman, D. S., 1969, The interaction of thyroxine with human plasma prealbumin and with the prealbumin-retinol-binding protein complex, *J. Biol. Chem.* **244**:3230.

Raz, A., Shiratori, T., and Goodman, D. S., 1970, Protein–protein and protein–ligand interactions involved in retinol transport in plasma, *J. Biol. Chem.* **245**:1093.

Richards, J. B., and Hemming, F. W., 1972, The transfer of mannose from guanosinediphosphate mannose to dolicholphosphate and protein by pig liver endoplasmic reticulum, *Biochem. J.* **130**:77.

Rietz, P., Wiss, O., and Weber, F., 1974, Metabolism of vitamin A and the determination of vitamin A status, *Vitamins Hormones* **32**:237.

Rilling, H. C., and Bloch, K., 1959, On the mechanism of squalene biogenesis from mevalonic acid, *J. Biol. Chem.* **234**:1424.

Roberts, A. B., and DeLuca, H. F., 1967, Pathways of retinol and retinoic acid metabolism in the rat, *Biochem. J.* **102**:600.

Roberts, A. B., and DeLuca, H. F., 1968a, Decarboxylation of retinoic acid in tissue slices from rat kidney and liver, *Arch. Biochem. Biophys.* **123**:279.

Roberts, A. B., and DeLuca, H. F., 1968b, Oxidative decarboxylation of retinoic acid in microsomes of rat liver and kidney, *J. Lipid Res.* **9**:501.

Roberts, A. B., and DeLuca, H. F., 1969a, Metabolism of retinol and retinoic acid, in: *The Fat Soluble Vitamins*, Proceedings of a Symposium in Honor of Harry Steenbock (H. F. DeLuca and J. W. Suttie, eds.), University of Wisconsin Press, Madison, Wis.

Roberts, A. B., and DeLuca, H. F., 1969b, Effect of DPPD on the decarboxylation of retinoic acid *in vitro* and *in vivo*, *Arch. Biochem. Biophys.* **129**:290.

Robeson, C. D., Blum, W. P., Dieterle, J. M., Cawley, J. D., and Baxter, J. G., 1955, Geometrical isomers of vitamin A aldehyde and an isomer of its α-ionone analog, *J. Am. Chem. Soc.* **77**:4120.

Rodriguez, P., Bello, O., and Gaede, K., 1972, Formation of retinol glycosides with nucleotide sugars in the presence of whole homogenates of rat thyroid, *FEBS Lett.* **28**:133.

Rogers, A. E., Herndon, B. J., and Newberne, P. M., 1973, Induction by dimethylhydrazine of intestinal carcinoma in normal rats and rats fed high or low levels of vitamin A, *Cancer Res.* **33**:1003.

Rosso, G. C., De Luca, L. M., Warren, R., and Wolf, G., 1975, Enzymatic synthesis of mannosyl retinyl phosphate from retinyl phosphate and guanosine diphosphate mannose, *J. Lipid Res.* **16**:235.

Rowe, N. H., and Gorlin, R. J., 1959, The effect of vitamin A deficiency upon experimental oral carcinogenesis, *J. Dent. Res.* **38**:72.

Saffiotti, U., Montesano, R., Sellakumar, A. R., and Borg, S. A., 1967, Experimental cancer of the lung: Inhibition by vitamin A of the induction of tracheobronchial squamous metaplasia and squamous cell tumors, *Cancer* **20**:857.

Schwenk, E., Todd, D., and Fish, C. A., 1954, The biosynthesis of cholesterol. VI. Companions of cholesterol-^{14}C in liver perfusions, including squalene-^{14}C, as possible precursors in its biosynthesis, *Arch. Biochem. Biophys.* **49**:187.

Scher, M., Lennarz, W. J., and Sweeley, C. C., 1968, The biosynthesis of mannosyl-1-phosphoryl polyisoprenol in *Micrococcus lysodeikticus* and its role in mannan synthesis, *Proc. Natl. Acad. Sci. USA* **59**:1313.

Schwieter, U., and Isler, O., 1967, Vitamin A. II. Chemistry, in: *The Vitamins*, Vol. 1 (W. H. Sebrell, Jr., and R. S. Harris, eds.), Academic Press, New York.

Sherman, S., 1969, Autoradiographic localization of ^{3}H-vitamin A in rat liver, *Int. J. Vitamin Res.* **39**:111.

Shichi, H., 1971, Biochemistry of visual pigments. II. Phospholipid requirement and opsin conformation for regeneration of bovine rhodopsin, *J. Biol. Chem.* **246**:6178.

Shichi, H., and Somers, R. L., 1974, Possible involvement of retinylidene phospholipid in photoisomerization of all-*trans*-retinal to 11-*cis*-retinal, *J. Biol. Chem.* **249**:6570.

Shichi, H., Somers, R. L., and O'Brien, P. J., 1974, Phosphorylation of rhodopsin: Most rhodopsin molecules are not phosphorylated, *Biochem. Biophys. Res. Commun.* **61:**217.

Shneour, E. A., and Zabin, I., 1957, Biosynthetic relation between lycopene and colorless polyenes in tomatoes, *J. Biol. Chem.* **226:**861.

Silverman-Jones, C. S., Frot-Coutaz, J. P., and De Luca, L. M., 1976, Separation of mannosylretinyl phosphate from dolichylmannosylphosphate by solvent extraction, *Anal. Biochem.* **75:**664.

Smith, Yudkin, Kriss, and Zimmerman, 1931, Vitamin A content of retinal and choroidal tissue, *J. Biol. Chem.* **92:** Proc. XCII.

Smith, D. M., Rogers, A. E., Herndon, B. J., and Newberne, P. M., 1975a, Vitamin A (retinyl acetate) and benzo[a]pyrene-induced respiratory tract carcinogenesis in hamsters fed a commercial diet, *Cancer Res.* **35:**11.

Smith, D. M., Rogers, A. E., and Newberne, P. M., 1975b, Vitamin A and benzo[a]pyrene carcinogenesis in the respiratory tract of hamsters fed a semi-synthetic diet, *Cancer Res.* **35:**1485.

Smith, F. R., and Goodman, D. S., 1971, The effects of diseases of the liver, thyroid, and kidneys on the transport of vitamin A in human plasma, *J. Clin. Invest.* **50:**246.

Smith, F. R., Raz, A., and Goodman, D. S., 1970, Radioimmunoassay of human plasma retinol-binding protein, *J. Clin. Invest.* **49:**1754.

Smith, J. E., Milch, P. O., Muto, Y., and Goodman, D. S., 1973, The plasma transport and metabolism of retinoic acid in the rat, *Biochem. J.* **132:**821.

Smith, W. E., Yazdi, E., and Miller, L., 1972, Carcinogenesis in pulmonary epithelia in mice on different levels of vitamin A, *Environ. Res.* **5:**152.

Sporn, M. B., Clamon, G. H., Dunlop, N. M., Newton, D. L., Smith, J. M., and Saffiotti, U., 1975, Activity of vitamin A analogs in cell cultures of mouse epidermis and organ cultures of hamster tracheas, *Nature (London)* **253:**47.

Stam, C. H., and MacGillavry, C. H., 1963, The crystal structure of the triclinic modification of vitamin A acid, *Acta Crystallogr.* **16:**62.

Steele, W. J., and Gurin, S., 1970, Biosynthesis of β-carotene in *Euglena gracilis*, *J. Biol. Chem.* **235:** 2778.

Steenbock, H., Sell, M. T., Nelson, E. M., and Buell, M. V., 1921, The fat-soluble vitamin, *Proc. Am. Soc. Biol. Chem. J. Biol. Chem.* **46:** Proc. XXXII.

Stepp, W., 1909, Versuche über Futterung mit Lipoid freier Ernährung, *Biochem. Ztschr.* **22:**452.

Sundaresan, P. R., and Sundaresan, G. M., 1973, Studies on the urinary metabolites of retinoic acid in the rat, *Int. J. Vitamin Nutr. Res.* **43:**61.

Suzue, G., and Porter, J. W., 1969, Enzymic synthesis of lycopene from isopentenyl-4-^{14}C-pyrophosphate, *Biochim. Biophys. Acta* **176:**653.

Tavormina, P. A., Gibbs, M. H., and Huff, J. W., 1956, The utilization of β-hydroxy-β-methyl-8-valerolactone in cholesterol biosynthesis, *J. Am. Chem. Soc.* **78:**4498.

Thompson, J. M., Howell, J. M., and Pitt, G. A. J., 1964, Vitamin A and reproduction in rats, *Proc. R. Soc. London Ser. B* **159:**510.

Thompson, J. N., 1969, The role of vitamin A in reproduction, in: *The Fat Soluble Vitamins* (H. F. DeLuca and J. W. Suttie, eds.), University of Wisconsin Press, Madison, Wis.

Thompson, S. Y., Braude, R., Coates, M. E., Cowie, A. T., Ganguli, J., and Kon, S. K., 1950, Further studies of the conversion of β-carotene to vitamin A in the intestine, *Br. J. Nutr.* **4:**398.

Thompson, W. A., Howell, J. M., Thompson, J. N., and Pitt, G. A. J., 1969, The retinol rquirements of rats for spermatogenesis and vision, *Br. J. Nutr.* **23:**619.

Tkacz, J. S., Herscovics, A., Warren, C. D., and Jeanloz, R. W., 1974, Mannosyltransferase activity in calf microsomes; formation from GDP-[^{14}C]mannose of a [^{14}C]mannolipid with properties of a dolichyl mannosylphosphate, *J. Biol. Chem.* **249:**6372.

Vahlquist, A., and Peterson, P. A., 1972, Comparative studies on the vitamin A transporting protein complex in human and cynomolgus plasma, *Biochemistry* **11:**4526.

Van Dorp, D. A., and Arens, J. F., 1946, Synthesis of "vitamin A acid," a biologically active substance, *Rec. Trav. Chim.* **65:**338.

van Jaarsveld, P. P., Branch, W. T., Edelhoch, H., and Robbins, J., 1973, Polymorphism of Rhesus monkey serum prealbumin: Molecular properties and binding of thyroxine and retinol binding protein, *J. Biol. Chem.* **248:**4706.

Virmaux, N., Weller, M., Mandel, P., and Trayhurn, P., 1975, Localization of the major site of light stimulated phosphorylation in a region of rhodopsin distinct from the chromophore binding site, *FEBS Lett.* **53**:320.

von Euler, H., von Euler, B., and H. Hellstrom, 1928, A-Vitamin Wirkungen der Lipochrome, *Biochem. Ztschr.* **203**:370.

Von Planta, C., Schwieter, U., Chopard-Dit-Jean, L., Ruegg, R., Kofler, M., and Isler, O., 1962, Synthesen in der Vitamin A-Reihe. 4. Physikalische Eigenschaften von isomeren Vitamin-A und Vitamin-A-Verbindungen, *Helv. Chim. Acta* **45**:548.

Waggoner, A. S., and Stryer, L., 1971, Induced optical activity of the metarhodopsins, *Biochemistry* **10**:3250.

Wagner, H., Wyler, F., Rinde, G., and Bernhard, K., 1960, Resorption und Umwandlung von ^{14}C-β-Carotin bei der Ratte, *Helv. Physiol. Pharmacol. Acta* **18**:438.

Wald, G., 1933, Vitamin A in the retina, *Nature (London)* **132**:316.

Wald, G., 1934, Carotenoids and the vitamin A cycle in vision, *Nature (London)* **134**:65.

Wald, G., 1935*a*, Vitamin A in eye tissue, *J. Gen. Physiol.* **18**:905

Wald, G., 1935*b*, Carotenoids and the visual cycle, *J. Gen. Physiol.* **19**:351.

Wald, G., 1935*c*, Pigments of the bullfrog retina, *Nature (London)* **136**:832.

Wald, G., 1936*a*, Pigments in the retina. I, *J. Gen. Physiol.* **19**:781.

Wald, G., 1936*b*, Pigments of the retina. II, *J. Gen. Physiol.* **20**:45.

Wald, G., 1937, Visual purple system in fresh water fishes, *Nature (London)* **139**:1017.

Wald, G., 1939, The porphyropsin visual system, *J. Gen. Physiol.* **22**:775.

Wald, G., 1953, Vision, *Fed. Proc.* **12**:606.

Wald, G., 1968, The molecular basis of visual excitation, *Nature (London)* **219**:800.

Wald, G., Brown, P. K., Hubbard, R., and Oroshnik, W., 1955, Hindered *cis* isomers of vitamin A and retinene: The structure of the neo-b-isomer, *Proc. Natl. Acad. Sci. USA* **41**:438.

Weissman, G., 1961, Alterations in connective tissue and intestine produced by hypervitaminosis A in *Xenopus laevis*, *Nature (London)* **192**:235.

Weller, M., Virmaux, N., and Mandel, P., 1975, Light-stimulated phosphorylation of rhodopsin in the retina: The presence of a protein kinase that is specific for photobleached rhodopsin, *Proc. Natl. Acad. Sci. USA* **72**:381.

West, I. C., and Mitchell, P., 1974, The proton-translocating ATPase of *Escherichia coli*, *FEBS Lett.* **40**:1.

Williams, R. J. H., Britton, G., Charlton, J. M., and Goodwin, T. W., 1967, The stereospecific biosynthesis of phytoene and polyunsaturated carotenes, *Biochem. J.* **104**:767.

Woeber, K. A., and Ingbar, S. H., 1968, The contribution of thyroxine-binding prealbumin to the binding of thyroxine in human serum, as assessed by immunoadsorption, *J. Clin. Invest.* **47**:1710.

Wolbach, S. B., and Howe, P. R., 1925, Tissue changes following deprivation of fat-soluble A vitamin, *J. Exp. Med.* **42**:753.

Wolf, D. E., Hoffman, C. H., Aldrich, P. E., Skeggs, H. R., Wright, L. D., and Folkers, K., 1956, β-Hydroxy-β-methyl-8-valerolactone (mevalonic acid), a new biological factor, *J. Am. Chem. Soc.* **78**:4499.

Wolf, D. E., Hoffman, C. H., Aldrich, P. E., Skeggs, H. R., Wright, L. D., and Folkers, K., 1957, Determination of structure of β,8-dihydroxy-β-methylvaleric acid, *J. Am. Chem. Soc.* **79**:1486.

Wright, A., Dankert, M., Fennessey, P., and Robbins, P. W., 1967, Characterization of polyisoprenoid compound functional in O-antigen biosynthesis, *Proc. Natl. Acad. Sci. USA* **57**:1798.

Wright, L. D., Cresson, E. L., Skeggs, H. R., McRae, G. D. E., Hoffman, C. H., Wolf, D. E., and Folkers, K., 1956, Isolation of a new acetate-replacing factor, *J. Am. Chem. Soc.* **78**:5273.

Wu, C. W., and Stryer, L., 1972, Proximity relationships in rhodopsin, *Proc. Natl. Acad. Sci. USA* **69**:1104.

Yogeeswaran, G., Laine, R. A., and Hakomori, S., 1975, Participation of retinyl phosphate sugar glycolipid in glycosylation of glycosphingolipid glass complex on cell contact, *Fed. Proc.* **35**:abstr. 2421.

Yokoyama, H., Nakayama, T. O. M., and Chichester, C. O., 1962, Biosynthesis of β-carotene by cell-free extracts of *Phycomyces blakesleeanus*, *J. Biol. Chem.* **237**:681.

Yoshizawa, T., and Horiuchi, S., 1972, Studies on interaction of visual pigments by absorption spectra at liquid helium temperature, and circular dichroism at low temperatures, in: *Biochemistry and Physiology of the Visual Pigments* (H. Langer, ed.), pp. 69–81, Springer-Verlag, New York.

Yoshizawa, T., and Wald, G., 1963, Prelumirhodopsin and the bleaching of visual pigments, *Nature (London)* **197**:1279.

Yuan, C., and Bloch, K., 1959, Synthesis of 3-methyl-3-butenyl-1-pyrophosphate, *J. Biol. Chem.* **234**:2605.

Yuspa, S. H., and Harris, C. C., 1974, Altered differentiation of mouse epidermal cells treated with retinyl acetate *in vitro*, *Exp. Cell Res.* **86**:95.

Zachman, R. D., 1967, The stimulation of RNA synthesis *in vivo* and *in vitro* by retinol (vitamin A) in the intestine of vitamin A deficient rats, *Life Sci.* **6**:2207.

Zachman, R. D., and Olson, J. A., 1963, The uptake of ^{14}C-β-carotene and its conversion to retinol ester (vitamin A ester) by the isolated perfused rat liver, *J. Biol. Chem.* **238**:541.

Zachman, R. D., Dunagin, P. E., Jr., and Olson, J. A., 1966, Formation and enterohepatic circulation of metabolites of retinol and retinoic acid in bile duct-cannulated rats, *J. Lipid Res.* **7**:3.

Zile, M., and DeLuca, H. F., 1968, Retinoic acid: Some aspects of growth-promoting activity in the albino rat, *J. Nutr.* **94**:302.

Zile, M., and DeLuca, H. F., 1970, Vitamin A and ribonucleic acid synthesis in rat intestine, *Arch. Biochem. Biophys.* **140**:210.

Zile, M., Emerick, R. J., and DeLuca, H. F., 1967, Identification of 13-*cis* retinoic acid in tissue extracts and its biological activity in rats, *Biochim. Biophys. Acta* **141**:639.

Chapter 2

Vitamin D

H. F. DeLuca

2.1. Introduction

Whether or not vitamin D can be considered a true vitamin is a matter of considerable debate. Certainly vitamin D is unique among the vitamins inasmuch as it is the only vitamin known to be a precursor of a hormone and its action is truly hormonal in nature. Furthermore, it can be readily surmised that vitamin D is not required in the diet if the appropriate organism such as man is exposed to sufficient quantities of ultraviolet light. It is believed that under the influence of approximately 300 nm ultraviolet light, vitamin D is produced from the 7-dehydrocholesterol which exists in the epidermis of skin. Because man wears clothing and lives inside buildings much of his lifetime and because the atmosphere in and around major cities is contaminated heavily with ultraviolet-absorbing material, insufficient amounts of vitamin D are produced in skin. This is compounded in some areas by a low incidence of sunny days because of climatic conditions. As a result, man in the northern and most southern reaches of the globe has become dependent on dietary sources of the vitamin for survival and reproduction. It is on this basis, therefore, that we now consider vitamin D a vitamin; but it must also be considered as a special case inasmuch as it is the precursor of at least one hormone whose function it is to regulate calcium and phosphorus metabolism. This chapter will attempt to bring together historical features of vitamin D, some pertinent present chemistry of vitamin D, what is known concerning its hormonal action, and finally what is known concerning its mechanism of action in the target tissues. In no way will this chapter attempt to cover the numerous scholarly contributions to the vitamin D literature; rather, it will attempt to bring together those salient features drawing upon literature when necessary to make the points believed to be true by the author.

H. F. DeLuca • Department of Biochemistry, College of Agricultural and Life Sciences, University of Wisconsin—Madison, Madison, Wisconsin 53706.

2.2. Historical

2.2.1. Discovery of Vitamin D

Although it has been reported that the symptoms of vitamin D deficiency appear in the ancient literature, this contention is difficult to document. However, a clear description of the disease rickets appeared in 1645 by the English physician Whistler (Smerdon, 1950). This disease, which afflicts children, results from a failure of hydroxyapatite mineral to deposit in the skeleton, giving rise to soft and pliable bones which then undergo gross deformation because of the weight-bearing and functional processes of the skeleton. This disease was described several times during the ensuing centuries, but the basis for the disease and methods of treatment were unclear until the experiments in 1919 by Sir Edward Mellanby (Mellanby, 1919*a,b*). In 1913 E. V. McCollum described a fat-soluble substance which promoted growth in animals maintained on a semichemically defined diet (McCollum and Davis, 1913). He attributed this growth-promoting activity to a fat-soluble substance found in cod liver oil, which he described as "fat-soluble vitamin A" (McCollum *et al.*, 1916). Similar investigations by McCollum and elsewhere by Osborne and Mendel also established the essentiality of a water-soluble dietary component for growth of animals which was described as "water-soluble vitamin B" (McCollum *et al.*, 1916; Osborne and Mendel, 1917). Undoubtedly this work on accessory food substances essential for life and reproduction must have inspired Sir Edward Mellanby to attempt the experimental production of rickets by dietary means. Perhaps another inspiration was the marked increase in incidence in the disease rickets associated with the industrial revolution. Likely, the appearance of extensive amounts of smoke and environmental pollution because of industrialization contributed to a failure of ultraviolet activation of 7-dehydrocholesterol in the skin (DeLuca *et al.*, 1971). These factors then led Sir Edward Mellanby to feed dogs a diet of oatmeal, which, when the animals were kept away from sunlight, caused marked skeletal deformities that appeared identical to rickets. Mellanby quickly chose to prevent and cure this disease by the administration of cod liver oil, and, because of E. V. McCollum's experience with cod liver oil, Mellanby concluded that this activity was due to the fat-soluble vitamin A discovered by McCollum. McCollum, who had since moved from the University of Wisconsin to Johns Hopkins, being familiar with the chemistry of the vitamin A compounds, quickly realized that the material which prevented rickets in dogs could not be the fat-soluble vitamin A. He therefore carried out experiments in which he destroyed the vitamin A material by aeration and heating of the cod liver oil (McCollum *et al.*, 1922*a*). The cod liver oil under these conditions still retained the antirachitic or vitamin D-like activity; thus McCollum had shown the existence of a new vitamin necessary for calcification of bone, which he called "fat-soluble vitamin D."

In 1919, a group of physicians in Vienna made the important observation that not only cod liver oil but also sunlight and artificially produced ultraviolet light could cure rickets in children (Huldshinsky, 1919). This, therefore, led to the apparent discrepancy that cod liver oil and sunlight were equal in their ability to cure rickets. Goldblatt and Soames carried out an important investigation in which they demonstrated that the livers taken from rachitic rats irradiated with ultraviolet light could cure rickets in other rats which were not irradiated, suggesting the production of the antirachitic substance by ultraviolet light (Goldblatt and Soames, 1923). It was Steenbock and Black, however, who provided the important breakthrough which considerably clarified the antirachitic vitamin (Steenbock, 1924; Steenbock and Black, 1924). They were able to produce antirachitic activity not only by irradiating the animals but also by irradiating their diet. Steenbock and Black (1925) were quick to demonstrate further that the activatable material which was converted to the antirachitic vitamin was found in the fat-soluble portion of the diet. This discovery, which was also made somewhat later by Hess *et al.* (1925), provided the basis for the elimination of rickets as a major medical problem and the theoretical basis for the isolation and subsequent identification of the D vitamins. By irradiating such foods as milk, bread, meats, and butter, it was possible to induce vitamin D activity. This practice led to the essential eradication of the disease rickets from the human population in western civilization (Sebrell and Harris, 1954). The discovery that a fat-soluble substance present in diets and other biological materials could be activated to vitamin D led to the preparation of large quantities of vitamin D from the plant sterols, leading ultimately to the isolation of vitamin D_2 by Askew *et al.* (1931) and somewhat later by Windaus *et al.* (1932). Windaus and his colleagues had also isolated what was believed to be vitamin D_1; however, it later was demonstrated that vitamin D_1 was an adduct of vitamin D_2 and the photoisomer lumisterol (Windaus *et al.*, 1932). The structure of the isolated vitamin D_2 compound was determined by both groups at the turn of the 1930 decade. Steenbock *et al.* (1932) noted, however, that rachitic chickens did not react appropriately to irradiated ergosterol but reacted appropriately to irradiated cholesterol preparations and cod liver oil, demonstrating their lack of equivalence. Waddell (1934) was the first to conclude on the basis of the chick insensitivity to the irradiated ergosterol that the cholesterol material must contain an additional antirachitic precursor. In 1936 Windaus and collaborators chemically synthesized 7-dehydrocholesterol (Windaus *et al.*, 1935) and shortly thereafter isolated vitamin D_3 from an irradiation mixture of the synthetic 7-dehydrocholesterol (Windaus *et al.*, 1936). Later identification of the completely synthetic vitamin D_4 compound or 22,23-dihydroergocalciferol by Windaus and Trautman (1937) essentially ended the story of isolation and identification of the D vitamins until the discovery of the metabolites of vitamin D in the mid to late 1960s. The production of crystalline vitamin D_2 led to its use in the fortification of foods and diets of experimental animals, replacing the direct irradiation method of fortification, which produced off flavors in the irradiated food (Sebrell and Harris, 1954).

2.2.2. Physiology of Vitamin D Action

Following the discovery of rickets in dogs, considerable work was also carried out in producing rickets in rats. It was clear from the work of McCollum *et al.* (1925) and that of Steenbock and Black (1925) that rickets in rats cannot be produced unless the diet is high in calcium and low in available phosphorus.

Shipley *et al.* (1925) and Howland and Kramer (1921) studied the rachitogenic process and concluded that rickets is a disease of the blood rather than a disease of bone. Their experimental results were very simply that slices of rachitic cartilage incubated in rachitic rat serum failed to calcify whereas the same cartilage slices calcified quite well when placed in normal serum or in inorganic salt solutions resembling serum but containing adequate amounts of calcium and phosphorus. This led them to believe that the calcium × phosphorus product of the blood was the determining feature in production of rickets. This oversimplified view was, nevertheless, correct in principle and was largely ignored for many decades. Orr *et al.* (1923) were the first investigators to demonstrate that vitamin D plays an essential role in the absorption of calcium. They noted the loss of large amounts of calcium in the feces of animals deficient in vitamin D. This could be reversed or corrected by the administration of the vitamin, and therefore they concluded that vitamin D plays an important role in promoting absorption of calcium. This position was criticized and was later firmly established by the conclusive work of Nicolaysen (1937*a,b*). He demonstrated that vitamin D increased the absorption of calcium and had no effect on the secretion of calcium into the intestinal tract. Nicolaysen also demonstrated that vitamin D improved the rate of absorption of calcium from isolated loops. With the use of radioisotopes, the work of Nicolaysen was reconfirmed many times, demonstrating a primary role of vitamin D in the intestinal absorption of calcium (Wasserman, 1963).

Because rickets could be observed in rats only when the diet was low in phosphorus, Steenbock and colleagues concerned themselves with the availability of dietary phosphorus under the influence of vitamin D. Because vitamin D improved the availability of phytate, the search for improved phytate hydrolysis took place (Lowe and Steenbock, 1936; Krieger and Steenbock, 1940). In the 1950s, however, it was demonstrated clearly that vitamin D increased the utilization of phosphorus even in the absence of phytate and the role of vitamin D in phosphorus metabolism was one of a systemic nature rather than dietary (Pileggi *et al.*, 1955).

In 1952 Carlsson provided an important breakthrough in our understanding of the function of vitamin D, demonstrating that in addition to improving intestinal absorption of calcium, vitamin D plays an important role in the mobilization of calcium from bone (Carlsson, 1952). Later in the 1950s, Schachter and his colleagues obtained *in vitro* evidence to demonstrate that vitamin D improved the active transport of calcium across the intestinal wall (Schachter and Rosen, 1959), whereas Rasmussen *et al.* (1963) and Harrison *et al.* (1958) were able to demonstrate that the bone calcium mobilization effect of vitamin D was both vitamin D and parathyroid dependent. Harrison and Harrison (1961) in following the transport of calcium across intestinal segments also noted that vitamin D improved the transport of phosphate across distal small intestine. Therefore, it could be readily shown by about 1960 that vitamin D plays an essential role in stimulating the active transport of calcium, the transport of phosphate across small intestine, and the mobilization of calcium from bone.

2.2.3. Vitamin D Metabolism

With radioactive vitamin D of high specific activity (Neville and DeLuca, 1966) it could be demonstrated that vitamin D is converted to biologically active metabolites before the responses in the intestine and bone can be observed (Lund and DeLuca, 1966; Norman *et al.*, 1964). This led to the isolation and identification of 25-hydroxyvitamin D_3 (25-OH-D_3) in 1968 (Blunt *et al.*, 1968) together with its chemical synthesis in 1969 (Blunt and DeLuca, 1969), ushering in the era of vitamin D metabolism and the development of the concept that vitamin D is a precursor of at least one hormone functioning in calcium and phosphorus metabolism (DeLuca, 1974). The isolation and chemical identification of the true hormone which is derived from 25-OH-D_3 took place in late 1970 and early 1971 (Holick *et al.*, 1971*a,b*). Its chemical structure was unequivocally demonstrated at that time by mass spectrometry and specific chemical reactions, but the true assignment of the confirguation of the 1-hydroxyl position of 1,25-dihydroxyvitamin D_3 [1,25-$(OH)_2D_3$] could not be accomplished until 1972, when chemical synthesis of the metabolite was accomplished (Semmler *et al.*, 1972). There will undoubtedly be additional new chapters in our understanding of the metabolism and mechanism of action of vitamin D; this contribution is meant as a milestone poised between the recent and the expected new developments.

2.3. Absorption of Vitamin D

Being lipid soluble, vitamin D might be expected to be absorbed in the small intestine via the lacteal system in the chylomicrons with a wide diversity of fats and fat-soluble material. Unfortunately, little is known concerning the absorption of vitamin D. Studying the distribution of intestinal vitamin D following the oral administration of radioactive vitamin D led Kodicek (1956) and Norman and DeLuca (1963*a*) to believe that vitamin D is likely absorbed primarily in the distal small intestine. Schachter and his colleagues, incubating segments of small intestine with solutions of vitamin D and bile salts, concluded that the proximal small intestine is the primary site of absorption (Schachter *et al.*, 1964). This question has not yet been resolved, but it is possible that vitamin D dissolved in lipids and oils might be absorbed primarily in the distal small intestine whereas vitamin D and other more available solutions might be absorbed in the proximal small intestine. Presumably vitamin D is absorbed via the lacteal system into the chylomicrons, which then make their appearance in the thoracic duct and into the bloodstream. Certainly the absorption of vitamin D has long been known to be dependent on bile salts to solubilize the lipid material (Greaves and Schmidt, 1933). However, beyond that much remains to be studied in this important area of vitamin D utilization. It is well known that steatorrhea and other conditions such as sprue and celiac disease in which there is an interference with absorption of fats also precipitate vitamin D deficiency.

2.4. Production of Vitamin D in the Skin

The skin is in possession of a potent sterol-biosynthesizing system (DeLuca *et al.*, 1971). This system also produces large amounts of 7-dehydrocholesterol, which are readily found in the epidermis (Idler and Baumann, 1952). The biosynthesis of the sterols in skin has already been reviewed and need not be repeated here

(DeLuca *et al.*, 1971). Ultraviolet light of approximately 300 nm wavelength readily penetrates skin to the level of the epidermis (Daniels, 1964) where 7-dehydrocholesterol is known to exist. Although the actual reaction in the skin has never been adequately studied, ultraviolet light is believed to induce a photolysis of the B ring of 7-dehydrocholesterol, giving rise to vitamin D_3 and other photoisomers. This is based entirely on the *in vitro* demonstration of ultraviolet irradiation of 7-dehydrocholesterol in organic solvents giving rise to a variety of photolysis products. The sequence of events during photolysis in organic solvents has been studied but not yet entirely worked out. However, the work of Vida (1971) and Havinga (1973) has established that previtamin D is the product rather than vitamin D_3. The previtamin D_3 then undergoes conversion to vitamin D_3 in an equilibrium which is temperature dependent. At very low temperatures, the equilibrium yields somewhere in the neighborhood of 3% previtamin D_3, whereas at higher temperatures the equilibrium provides 13% of the previtamin D. Intermediates in the photolysis reaction are not entirely known, but lumisterol and tachysterol are believed to be involved, as shown in Fig. 1 (Vida, 1971; Havinga, 1973).

Ultraviolet irradiation of skin from rachitic animals is known to cause an appearance of antirachitic material (Knudson and Benford, 1938). This antirachitic material has recently been identified positively as vitamin D_3 (Esvelt *et al.*, 1977). It is not known whether there is any enzymatic machinery or template which might render specificity to the photoisomerization of 7-dehydrocholesterol to vitamin D_3 without other photolysis by-products known to be produced in organic solvent irradiation systems. Again, additional investigation is required in this important area before the biochemistry of the conversion in the skin can be written with confidence. Also, exactly how much vitamin D can be produced in skin is un-

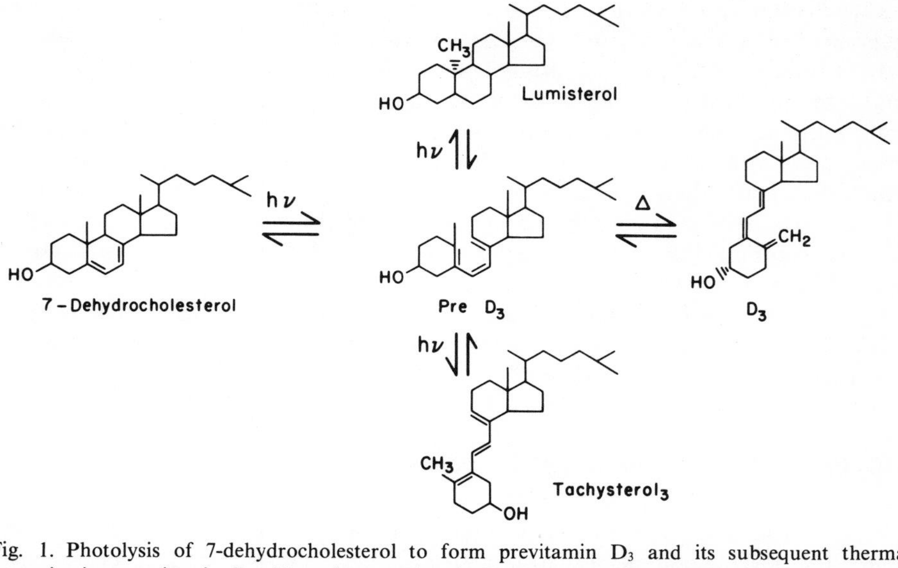

Fig. 1. Photolysis of 7-dehydrocholesterol to form previtamin D_3 and its subsequent thermal isomerization to vitamin D_3. Note that previtamin D_3 can, upon further irradiation, proceed to lumisterol or to tachysterol.

known, and, furthermore, there is no information available as to whether or not the conversion in skin is a regulated process. No report of toxicity due to hypervitaminosis D induced by excessive ultraviolet light has made its appearance. This might suggest that some control mechanism exists; on the other hand, there may not be sufficient amounts of 7-dehydrocholesterol available to permit the production of toxic amounts of vitamin D_3. The best estimates of vitamin D production in skin have been by Bekemeier (1958), and these values appear to be quite high. Clearly, much more quantitative information is necessary in this area.

Of interest is the demonstration with the use of the binding assays for 25-OH-D_3 in recent years that excessive sunbathing does cause an increase in the circulating level of 25-OH-D_3, giving an indication that vitamin D_3 must be the product of ultraviolet irradiation and also giving the impression that large amounts of vitamin D_3 can be produced by ultraviolet light exposure (Haddad and Stamp, 1974).

2.5. Occurrence of Vitamin D Naturally

Vitamin D is not abundantly distributed in nature. Table I is a compilation of the availability of vitamin D from a variety of food sources. The best sources of vitamin D_3 from naturally occurring sources are fish liver and fish liver oils (Sebrell and Harris, 1954). Extremely high levels have been found in tuna liver oils, shark liver oils, etc. The reason for the existence of the large amounts of vitamin D in the fish livers is an unsolved biological question. Bills (1954) carried out experiments using catfish in which he concluded that they must possess a non-

Table I. Occurrence of Vitamin D in Foods

Source	IU/100 g	or IU/quart
Nonfortified milk	4.4	40
Fortified milk (USA)	44	400
Fish	100–500	
Salmon	500	
Sardines	300	
Mackerel	275	
Liver oils		
Tuna	400,000	
Mackerel	600,000	
Halibut	120,000	
Cod	10,000	
Barracuda	470,000	
Egg yolk	265	
Egg white	0	
Butter	92	
Cheese	30	
Beef	10	
Liver	10–40	
Vegetables	<5	
Plants, seeds, etc.	<1	

photochemical system for preparing vitamin D inasmuch as catfish maintained for 6 months without a source of vitamin D and in the absence of ultraviolet light accumulated vitamin D in their oils. Fortunately, this experiment could not exclude small amounts of dietary vitamin D accumulating in an animal which may not rapidly metabolize and excrete it. Another possibility is that vitamin D accumulates in the lipid stores of such animals as sharks and tuna, which are at the end of the food chain. There has been one report of a nonphotochemical process of conversion of 7-dehydrocholesterol to vitamin D_3 in Atlantic striped bass liver homogenates (Blondin *et al.*, 1967). However, the investigators themselves were not convinced, and conclusive evidence for such a nonphotochemical process is lacking.

Very little vitamin D had been found in the plant world until the recent demonstration that the South American plant *Solanum glaucophyllum* has in its leaves a conjugated form of 1,25-$(OH)_2D_3$ and 1α-hydroxyvitamin D_3 (1α-OH-D_3) (Napoli *et al.*, 1977; Wasserman *et al.*, 1976). Exactly how this plant accumulates this active form of vitamin D and the basic reason for its existence in the plant world remain unsolved. However, the ingestion of this plant in South America and ingestion of *Cestrum diurnum* in the southern United States have resulted in a calcemic disease called "enteque seco" and "espichamento" (Wasserman, 1975).

The only reasonable and consistent sources of vitamin D from the diet are, therefore, fortified foods. Milk is fortified in the United States to a level of 400 IU/ quart or 10 μg/quart. Other countries use a lower level of fortification, and, although some cereals are fortified with vitamin D, many other foods which were previously fortified are no longer supplemented. Fortification of foods in the United States likely is responsible for the lack of a large number of rachitic children in our population, and despite claims to the contrary, fortification of the food of culturally and economically underprivileged children is essential to prevent the widespread occurrence of vitamin D deficiency in that segment of our population. In addition, borderline deficiency of vitamin D may be an important contributing factor to degenerative bone disease in our senior population. Although textbooks sometimes state that vitamin D is not required by adults, this notion is dangerously false. Vitamin D is an essential building block for an essential hormone which functions to regulate calcium and phosphorus metabolism and is therefore required throughout life. The recommended daily allowance of vitamin D is of the order of 400 IU or 10 μg of vitamin D_3 per day (*Recommended Dietary Allowances*, 1974). Although less is likely sufficient, up to 1000–2000 units/day is well tolerated and can be ingested with safety except under disease conditions.

2.6. Structure and Physical Constants of the D Vitamins and Their Precursors

There are only two D vitamins of significance nutritionally and biologically; these are vitamins D_2 and D_3, which are also known as ergocalciferol and cholecalciferol, respectively. Vitamin D_2 is prepared by the photolysis of the plant sterol ergosterol, while vitamin D_3 is derived from irradiation of the animal sterol 7-dehydrocholesterol. The structures of the known vitamin D's are given in Fig. 2 and those of the three principal provitamins in Fig. 3. Figure 4 illustrates the numbering system for vitamin D_3. Table II gives the physical constants of vitamin D_2 and vitamin D_3, including their mass fragments on mass spectrometry and the

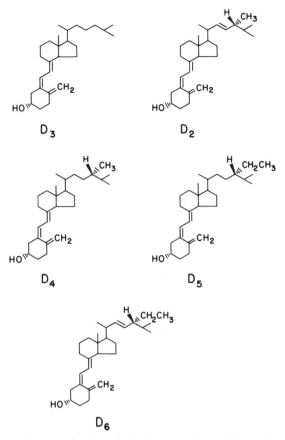

Fig. 2. The D vitamins. Note that vitamin D_3 is also known as cholecalciferol and vitamin D_2 is known as ergocalciferol. These are the only two commonly known D vitamins. The others are known from the literature as biochemical curiosities.

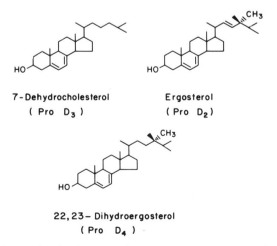

Fig. 3. Precursors of the D vitamins. Note especially that 7-dehydrocholesterol is a precursor of vitamin D_3 and ergosterol is a precursor of vitamin D_2.

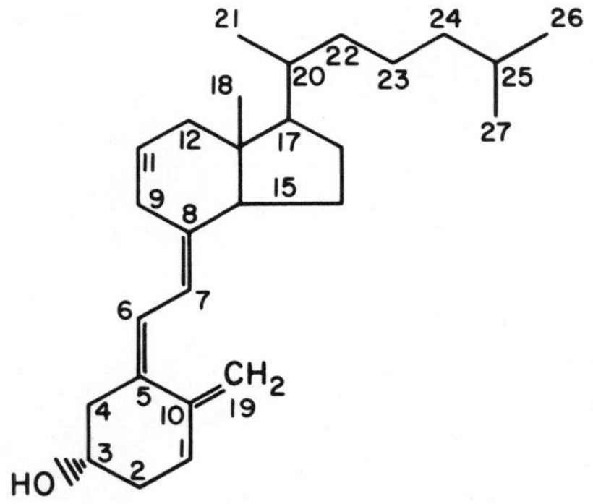

Fig. 4. Numbering system used for the vitamin D carbons.

principal constants derived from nuclear magnetic resonance measurements. Table III gives the current nomenclature.

The important aspects of vitamin D chemistry center predominantly about the *cis*-triene structure. In all other respects the chemistry of these compounds is similar to that of the steroid nuclei characteristic of all other steroid hormones (Sebrell and Harris, 1954; Fieser and Fieser, 1959; Holick and DeLuca, 1974; DeLuca and Schnoes, 1976). The *cis*-triene structure imparts to the D vitamins a characteristic ultraviolet absorption at 265 nm with a minimum at 228 nm. An index of purity of the vitamin D compounds is when the optical density at 264 nm is 1.8 times that at its minimum 228 nm. Deviation from this maximum/minimum ratio indicates the presence of ultraviolet-absorbing impurities. The 265 nm absorption band is somewhat surprising for a *cis*-triene structure. However, the 10,19-methylene group is out of plane, which may account for its unexpectedly low absorption maximum.

The *cis*-triene portion of the molecule imparts to the vitamin D structure an instability to oxidation. It is readily oxidized, although the products have never been completely identified. Vitamin D suspensions in aqueous solutions in the presence of dissolved oxygen oxidize so rapidly that one can readily observe the oxidation by the disappearance of absorbance at 265 nm. Another aspect of the vitamin D molecule dependent on the *cis*-triene is its instability to acid. Under conditions of even mild acidity, the vitamin D molecule isomerizes to form the 5,6-*trans* isomer and the isotachysterol isomer (Fieser and Fieser, 1959; Holick and DeLuca, 1974; DeLuca and Schnoes, 1976). Furthermore, upon heating, the 5,6-*trans* isomer is readily isomerized to the isovitamin D structure, all of which are shown in Fig. 5. Inasmuch as the 5,6-*trans*-vitamin D, the isovitamin D, and the isotachysterol all possess much less biological activity than vitamin D_3 itself, disappearance of biological activity is readily observed under acid conditions.

Concern is always expressed regarding vitamin D stability, but the vitamin D structure is quite stable provided that it is dissolved in an organic solvent or an oil which is nonacidic and which contains an antioxidant such as α-tocopherol in an atmosphere of nitrogen and in the absence of light, preferably in a dark bottle.

Vitamin D is always in equilibrium with its isomer, previtamin D_3, as shown in Fig. 1. Velluz has determined the proportion of vitamin D_3 vs. previtamin D_3 at

Table II. Chemical Properties of the Vitamin D-Active Compounds

Compound	Melting point (°C) Vitamin	Melting point (°C) Dinitrobenzoate	Optical rotation	Ultraviolet γ_{max}	Ultraviolet ϵ	Mass spectra (m/e)	NMR peaks (ppm)
Cholecalciferol (vitamin D_3)	84–85	142–150	$[\alpha]^{20} = +84.6°$ (acetone)	265	18,200	384, 271, 136, 118	0.87 doublet ($J = 6.5$); 0.54 singlet; 0.93 doublet ($J = 5.0$); 4.81 singlet; 5.03 singlet; 6.02 doublet ($J = 11.5$); 6.24 doublet ($J = 10.5$)
Ergocalciferol (vitamin D_2)	121	104–105	$[\alpha]^{20} = +106°$ (alcohol)	265	19,400	396, 271, 136, 118	0.54 singlet; 0.87 doublet ($J = 7.0$); 0.98 doublet ($J = 6.0$); 0.79 doublet ($J = 6.5$); 0.81 doublet ($J = 7.0$); 4.81 singlet; 5.01 singlet; 5.20 multiplet; 6.02 doublet ($J = 11.5$); 6.24 doublet ($J = 10.5$)
22,23-Dihydro-ergocalciferol	96–98	127–128	$[\alpha]_D^{20} = +89.3°$	265	19,400		1.20 singlet; 0.54 singlet; 0.93 doublet ($J = 5.7$)
25-OH-D_3	82–83	—		265	18,200	400, 271, 136, 118, 59	4.81 singlet; 5.03 singlet; 6.02 doublet ($J = 11.5$); 6.24 doublet ($J = 10.5$); 1.20 singlet; 0.93 doublet ($J = 5$)
25-OH-D_2	—	—		265	19,400	136, 412, 271, 118, 59	0.54 singlet; 0.79 doublet ($J = 6.5$); 0.81 doublet ($J = 7.0$); 1.20 singlet; 4.81 singlet; 5.01 singlet; 5.20 multiplet; 6.02 doublet ($J = 11.5$); 6.24 doublet ($J = 10.5$)

Table III. Current Vitamin D Nomenclature

Current trivial names	Systematic name (current)
Vitamin D₃, cholecalciferol	(5Z,7E)-9,10-*seco*-5,7,10(19)-Cholestatrien-3β-ol
25-Hydroxyvitamin D₃, 25-hydroxycholecalciferol	(5Z,7E)-9,10-*seco*-5,7,10(19)-Cholestatrien-3β,25-diol
1α,25-Dihydroxyvitamin D₃, 1,25-dihydroxycholecalciferol	(5Z,7E)-9,10-*seco*-5,7,10(19)-Cholestatrien-1α,3β,25-triol
Vitamin D₂, ergocalciferol	(5Z,7E,22E)-9,10-*seco*-5,7,10(19),22-Ergostatetraen-3β,ol
1α,25-Dihydroxyvitamin D₂, 1α,25-dihydroxyergocalciferol	(5Z,7E,22F)-9,10-*seco*-5,7,10(19),22-Ergostatetraen-1α,3β,25-triol

various temperatures. At low temperatures the equilibrium favors vitamin D₃, whereas at higher temperatures a greater proportion of previtamin D₃ is found. One can rapidly isomerize previtamin D to vitamin D by merely heating the solution, although the equilibrium is not so favorable for vitamin D₃ under these conditions (Velluz and Amiard, 1949a,b,c; Velluz *et al.*, 1949). Previtamin D₃ is approximately one-half as active as vitamin D₃ when given to experimental animals (J. Lund and H. F. DeLuca, unpublished results). It is not clear whether the biological activity of the previtamin D₃ is the result of isomerization to the vitamin D₃ molecule as it enters the warm body temperatures. In any case, previtamin D can readily be separated from vitamin D by means of silicic acid column chromatography, thin-layer chromatography, and hydroxyalkoxypropyl Sephadex chromatography (Norman and DeLuca, 1963b; DeLuca *et al.*, 1969; Jones *et al.*, 1975). Also useful is high-pressure liquid chromatography, utilizing silica gel columns (Jones and DeLuca, 1975).

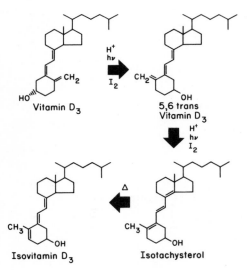

Fig. 5. Conversion of vitamin D to 5,6-*trans*-vitamin D₃, isotachysterol, and isovitamin D₃ under the catalysis of either iodine, acidic conditions, or light.

2.7. Vitamin D Deficiency

Classically, the discovery of vitamin D was on the basis of the deficiency disease rickets which afflicts the young, resulting in inadequate calcification of newly forming bone and epiphyseal cartilage (Sebrell and Harris, 1954; Reed *et al.*, 1939; DeLuca, 1967; Dent, 1977). Subsequent to this was the realization that an inadequate calcification of bone occurs in the adult in vitamin D deficiency, giving rise to osteomalacia. This disease is also an inadequate mineralization of newly laid down collagen matrix of bone related to the remodeling process. Failure to mineralize the organic matrix of newly forming bone in the adult gives rise to wide osteoid seams, again characterized by a failure of mineralization. In the adult, endochondral growth of long bones no longer takes place; therefore, failure to mineralize endochondral cartilage is no longer a factor. However, the basic underlying mechanisms of these two diseases are a failure to mineralize the organic matrix, which is made in approximately normal fashion. Because of these basic observations of vitamin D deficiency, it has been assumed that vitamin D has one basic role, namely to induce the mineralization of bone at a normal rate. However, it is now quite clear that there are several deficiency diseases in which an absence of vitamin D or its metabolites participates.

The adjustment of serum calcium concentration to the normal range preventing hypocalcemic tetany is clearly a major role of the vitamin. Furthermore, the presence of adequate amounts of vitamin D or its active metabolites is necessary to aid in the prevention of bone degeneration such as osteoporosis and secondary hyperparathyroidism. Finally, there is clear evidence that vitamin D plays a role in muscle metabolism inasmuch as muscle weakness is an obvious vitamin D deficiency symptom long recorded by clinical scientists interested in vitamin D deficiency disease (Dent, 1977; Dent and Smith, 1969). It is possible that following the unraveling of the vitamin D story we may understand that several pathologies will result from a failure to supply adequate amounts of vitamin D or a failure of its metabolism to its active forms under appropriate signals. This section will describe the deficiency diseases and what is known concerning their pathophysiology.

2.7.1. Rickets and Osteomalacia

There are two types of mechanisms for bone growth; one is the intramembranous bone growth phenomenon in which there is increased bone synthesis by osteoblasts. Such bone growth is found in the cranium and periosteal bone or in general found in bones which do not appreciably lengthen during the course of development. Long bones, on the other hand, grow primarily because of the endochondral growth mechanism. The epiphyseal plate, composed of chondroblasts, begins as resting chondroblasts immediately under the epiphyseal tip of the bone (Fig. 6). The resting chondroblasts then proliferate and become hypertrophic. The proliferating and hypertrophic cartilage cells then secrete the organic matrix composed of chondrocyte collagen and mucopolysaccharide. In the organic matrix immediately adjacent to the hypertrophic cartilage cells there is, in response to adequate amounts of calcium and phosphorus and to vitamin D, a deposit of mineral.

Fig. 6. Histological section of an epiphyseal end of bone from a normal growing rat. Note resting cartilage at the top, hypertrophic cartilage in the center, and spongeosa or trabecular bone at the bottom of the section. Courtesy of Dr. Jenifer Jowsey, Mayo Clinic, Rochester, Minnesota.

The mechanism whereby the mineralization process occurs is still a matter of debate (Anderson, 1969; Anderson and Reynolds, 1973; Glimcher, 1959). Initially Robison believed that glycogen accumulated in the hypertrophic cartilage cells. At the appropriate time, it was glycolyzed, giving rise to organic phosphate esters which were elaborated into the organic matrix and were hydrolyzed by alkaline phosphatase, giving local high concentrations of phosphate resulting in mineralization (Robison and Soames, 1924). This simplified hypothesis did not, however, stand the test of time, and it seems certain that this mechanism is not operative. Nevertheless, the presence of alkaline phosphatase in large amounts in the mineralizing matrix area is well known, and it is known that large amounts are produced by the chondroblasts in preparation for mineralization.

Another suggested mechanism is that the fluid bathing the collagen fibrils is supersaturated inasmuch as it represents or reflects the extracellular fluid compartment calcium and phosphorus (Neuman and Neuman, 1958). Mineralization is brought about by the depolymerization of the mucoprotein coating the collagen matrix, thus exposing specific groups which catalyze the deposit of calcium and phosphorus in the form of an amorphous calcium hydrogen phosphate precipitate (Glimcher, 1959). More recently, because of the work of Anderson and others, there has been the suggestion that chondrocytes take up calcium and phosphorus from the extracellular fluid and package the calcium and phosphorus in the form of mineral deposits in vesicles. These vesicles are then elaborated into the matrix area, finding their particular sites and initiating mineralization (Anderson, 1969; Anderson and Reynolds, 1973). There is adequate evidence for the existence of the packaged calcium and phosphorus vesicles, but the actual mechanism of mineralization remains largely unknown.

The question to be considered here is whether vitamin D plays a role in the mineralization process. Because the administration of phosphorus to rachitic rats will bring about the calcification of the cartilage matrix in approximately normal amounts even in the absence of vitamin D, it is suggested that the calcification process in the chondrocyte system is not directly vitamin D dependent (McCollum *et al.*, 1922*b*; Steenbock and Herting, 1955; Fraser *et al.*, 1967; Pileggi *et al.*, 1956). In support of this, it has been adequately demonstrated that serum from rachitic rats will not support *in vitro* chondrocyte-mediated calcification, whereas serum from normal animals or serum from rachitic animals adjusted to have calcium and phosphorus concentrations approximating those of normal serum will support this mineralization. Thus evidence that vitamin D is directly involved in this process is lacking, and the current weight of evidence is contrary to this belief. In any case, the final result of a vitamin D deficiency is that the collagen which is elaborated by the hypertrophic cartilage cells fails to mineralize and thus the mature chondroblasts last considerably longer but eventually die, leaving behind large areas of uncalcified chondrocyte-elaborated matrix. Furthermore, in the normal sequence of bone growth, osteoblasts and blood vessels invade the hypertrophic zone of mineralization, and the chondrocyte matrix and mineral are resorbed and redeposited as true bone by the osteoblasts, giving rise to trabecular bone characteristic of the spongeosa immediately below the epiphyseal plate (Fig. 6). In the absence of mineralization of the hypertrophic zone, osteoblasts and blood vessels invade but there is no mineral to be resorbed and instead at least some of the matrix which has been laid down by the chondrocytes is dissolved and re-formed by the osteoblasts. The mineralization of this osteoblast-elaborated matrix does not take place, giving rise to wide areas of uncalcified organic matrix of osteoblast origin called osteoid.

Because of the rapid growth of the long bones in certain phases of life and because of the high demands for calcium and phosphorus by these growing regions of bone, they are the most obviously uncalcified in vitamin D deficiency. It should be borne in mind that failure of general calcification of osteoblast-elaborated collagen fibrils occurs everywhere. Thus intramembranous bone also fails to mineralize in a normal fashion, as do periosteal bone and other sites of newly forming bone tissue. Thus there is a generalized deficiency of mineralization of bone in vitamin D deficiency whether it is chondrocyte mediated or osteoblast mediated and regardless of the site at which the mineralization is taking place. It should be pointed out that the actual phase change which takes place during the mineralization process, even that occurring on osteoblast-elaborated collagen, remains largely unknown, although it has been speculated that it occurs in an amorphous state. The presence of vesicles which mediate mineralization of bone has also been demonstrated and represents a suggested mechanism. Another mechanism is that phosphate groups presumably present on the serine hydroxyls of collagen catalyze the mineralization or nucleation of the supersaturated solution presumed to be bathing the collagen fibrils (Glimcher, 1959). Undoubtedly, the depolymerization of mucoprotein must occur prior to mineralization, and so far a clear mechanism has not been worked out. Perhaps the mechanism involves (1) the accumulation of mineral by the osteoblasts and (2) the depolymerization or preparation of the collagen fibrils for mineralization. It is well known that collagen must mature before it is ready for mineralization. Cross-linking of collagen occurs, and it has been suggested that vitamin D plays a role in the cross-linking and maturation of collagen, a hypothesis which has yet to receive adequate support (Gonnerman *et al.*, 1976; Barnes *et al.*, 1973). There is no doubt that collagen does not mature adequately in the vitamin D deficient state, but it is not clear whether the maturation of collagen is the direct result of a function of vitamin D or results from the adjustment of the mineral components of the extracellular fluid and hence fluid in the osteoblasts to the normal range, which is in turn necessary to provide the proper enzymatic environment for the maturation of collagen in preparation for mineralization. Once the collagen has been adequately matured, it is unknown whether there is a pumping of calcium and phosphorus into the regions undergoing mineralization by osteoblasts so that a supersaturated solution is provided for mineralization. It is also uncertain if the mineral is crystallized into hydroxyapatite by specific nucleation centers as described above or whether the calcium and phosphorus are first packaged in vesicles and elaborated into the matrix area in preparation for mineralization. Again there is inadequate evidence to support the idea that vitamin D might play a role in any of these processes. Certainly in vitamin D deficiency the major, if not only, problem in the failure of mineralization is the supply of calcium and phosphorus to the mineralization sites.

Thus the mineralization process fails in vitamin D deficiency at all centers. However, because the long bones are growing most rapidly and are under the most weight-bearing stress, it is these bones that present the obvious deformities of rickets, which is the bending and twisting of bone as will be described below. Osteomalacia gives rise to weak bones with a high degree of bone pain and a high incidence of fractures and microfractures.

From a pathological point of view, there have been three suggested phases of rachitogenesis in children, as described by Fraser *et al.* (1967) (Fig. 7). During the initial phases of vitamin D depletion there is inadequate absorption of calcium, giving rise to hypocalcemia, which stimulates the parathyroids to hypertrophy and to elaborate large amounts of parathyroid hormone. During this initial rachitogenic phase, no overt bone disease can be detected by X-ray or by histological means. Once the parathyroid hypertrophy has taken place to a sufficient degree and large amounts of parathyroid hormone are elaborated, there is a marked mobilization of calcium from bone, correcting the hypocalcemia to give rise to normal calcemia. The parathyroid hormone on the other hand causes a block in renal tubular reabsorption of phosphorus, giving a phosphaturia and diminishing plasma phosphorus to below normal values. It is at this stage that the mineralization process begins to fail and the rachitic bone lesions and osteomalacia begin to appear. The third stage of rickets is a complete depletion of vitamin D, giving a skeletal resistance to the parathyroid hormone induced mobilization of calcium from bone. The patient then becomes hypocalcemic. The presence of large amounts of parathyroid hormone also causes a phosphaturia, giving rise to hypophosphatemia. Furthermore, the absence of vitamin D prevents adequate absorption of

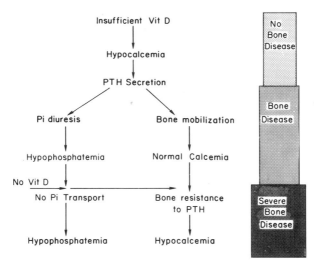

Fig. 7. The three phases of rachitogenesis as described by Fraser *et al.* (1967). Note that the earliest stage of rachitogenesis is hypocalcemia, which brings about enlarged parathyroid glands and excessive parathyroid hormone secretion. With incomplete vitamin D depletion, the excessive parathyroid hormone will mobilize calcium from bone, correcting the hypocalcemia. However, the hypersecretion of parathyroid hormone brings about excessive loss of phosphate in the urine, giving rise to hypophosphatemia. At this stage histological and radiological rickets appears. The third stage of rachitogenesis is a resistance of the skeleton to parathyroid hormone because of the absence of vitamin D and its metabolites, giving rise to hypocalcemia as well as to hypophosphatemia, the latter caused by parathyroid hormone induced phosphaturia. At this stage there is drastic and florid bone disease.

phosphorus from the small intestine. Thus hypocalcemia and hypophosphatemia together are associated with a florid rachitic condition in which severe bone lesions can be observed. In the rat, rickets can be produced only when both vitamin D and phosphorus are lacking (Steenbock and Herting, 1955; Tanaka and DeLuca, 1974*a*). Rachitic lesions in both man and rats can be healed by phosphate infussions. Thus it appears that phosphorus is the main lacking mineral element in the failure of bone mineralization under conditions of rachitogenesis.

The disease rickets can best be illustrated by the photograph of a rachitic child shown in Fig. 8. X-rays of the heads of long bones under rachitic conditions are shown in Fig. 9, while Fig. 10 illustrates a histological section of the end of a long bone from a rachitic animal vs. that from an animal which has received normal amounts of vitamin D. The wide osteoid seams characteristic of osteomalacia are shown in Fig. 11A as compared to normal (Fig. 11B). A thorough pathological description of rickets is available elsewhere (Sebrell and Harris, 1954; Harris, 1956; Maximow and Bloom, 1957), but in brief one can observe the bending and twisting of the long bones, displacement of the cartilage of the ends of long bones, and the softening of the cranium. Also characteristic is the rachitic rosary rib, which was used as a diagnostic indication of the disease rickets in the era of limited radiology. The long bones are the most affected, although the rib cage becomes very severely distorted and in severe rickets this can impair the functioning of the internal organs. Often the patient or rachitic animal will succumb to secondary

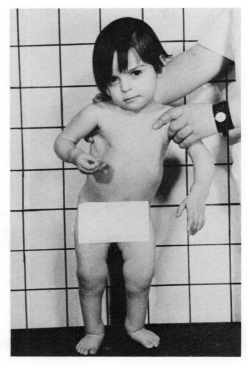

Fig. 8. A child suffering from a mild degree of rickets. Courtesy of Dr. Sonia Balsan, Hopital des Enfants Malades, Paris, France.

pulmonary infection which is the direct result of impaired pulmonary function because of the skeletal changes. Osteomalacia does not result in the bending and twisting of bones as illustrated for rickets simply because there are no long and wide areas of uncalcified organic matrix as in the epiphyseal plate region, but instead there are patches of newly formed osteoid tissue which remain uncalcified because of the lack of mineral. These patients suffer severe bone pain and can suffer fractures and immobilization.

Rickets in rats can be produced only when the diet is high in calcium and low in phosphorus in addition to being vitamin D deficient. Under conditions of phosphate depletion, therefore, the rachitic lesions illustrated above result. On the other hand, rats made vitamin D deficient on a low-calcium diet develop hypocalcemia and tetany but their bones do not evidence rachitic lesions and instead evidence an osteoporotic condition. Rats fed a normal-calcium and normal-phosphorus diet deficient in vitamin D also do not develop rickets, but instead develop a hypocalcemic tetany but not osteoporosis. It is therefore incorrect to refer to a vitamin D deficient animal as having rickets unless the correct mineral imbalance is also associated. In the case of the chicken, dog, and other species, rickets develops in the absence of vitamin D even if the diet is not low in phosphorus. These details must be kept in mind when one experimentally approaches the various conditions to be studied.

2.7.2. Hypocalcemic Tetany as a Disease of Vitamin D Deficiency

The early history of vitamin D investigation carries with it many discussions of "rachitis" and "tetany" which were often described in children suffering from vitamin D deficiency. In fact, that is a common experience among clinical investigators who now observe the rare case of rickets. One observation which has been made is that upon treatment of rickets with large doses of vitamin D some care must be exercised inasmuch as there may be an initial hypocalcemic dip resulting in tetany prior to correction of the hypocalcemia by the vitamin D. It is not clear why this is experienced, although a dip in serum calcium concentration has also been reported in rats by Carlsson and Hollunger (1954). In any case, it has been suggested that vitamin D is not required by adults and is not required under conditions of normal calcium and phosphorus intake in animals. These statements are both clearly incorrect. If rats are maintained on a normal-calcium and normal-phosphorus diet deficient in vitamin D, they will develop hypocalcemia at about the third to fourth week which becomes more and more severe with each ensuing week.

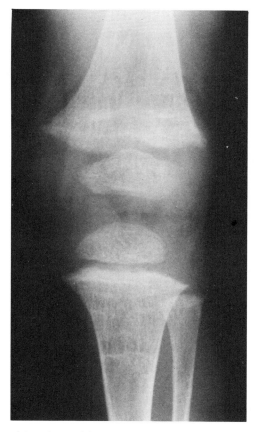

Fig. 9. X-ray of the knee joint of a rachitic child. Note the wide epiphyseal plate with large uncalcified areas. Courtesy of Dr. Donald Fraser, Hospital for Sick Children, Toronto, Ontario, Canada.

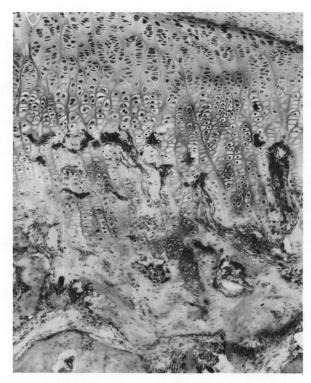

Fig. 10. Section of the epiphyseal growth plate taken from a rachitic rat. Notice the wide area of uncalcified bone or osteoid. Note the disorganization and large numbers of hypertrophic cartilage cells and the wide epiphyseal proliferating cartilage zone. Courtesy of Dr. Jenifer Jowsey, Mayo Clinic, Rochester, Minnesota.

At approximately 8–12 weeks on this diet, the animals will go into spontaneous tetany, especially if some predisposing situation such as a loud noise occurs. In the author's experience, large groups of extremely vitamin D deficient, severely hypocalcemic rats have been known to succumb to the slamming of a door. An analysis of their plasma levels of vitamin D metabolites revealed that their 25-OH-D$_3$ and 1,25-(OH)$_2$D$_3$ disappear in measurable quantities from their plasma at about 4 weeks of feeding a vitamin D free diet when they are obtained at weaning from a breeder maintaining a colony of rats on marginal vitamin D intakes. In any case, it is quite clear that hypocalcemia is one of the most critical conditions which results from vitamin D deficiency. The prevention or correction of hypocalcemia is the responsibility of vitamin D and the parathyroid gland. In the absence of vitamin D, the parathyroid glands become hypertrophied and secrete large amounts of parathyroid hormone, but the hypocalcemia cannot be corrected. Vitamin D is essential for the parathyroid hormone to bring about an elevation of serum calcium concentration at the expense of the bone fluid compartment (Rasmussen *et al.*, 1963; Harrison *et al.*, 1958). Thus it is clear from the early work of Harrison and Harrison and from the author's laboratory that the function of the parathyroid hormone in the elevation of serum calcium and prevention of hypocalcemic tetany

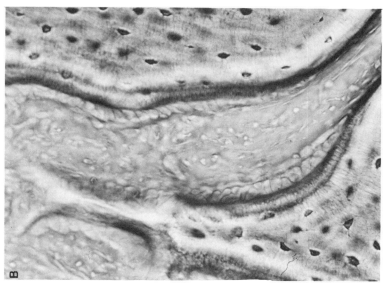

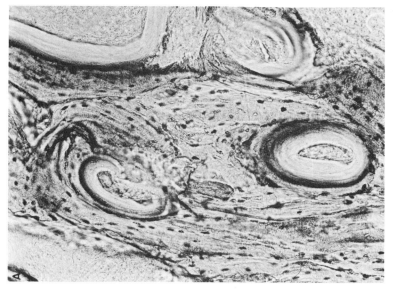

Fig. 11. A: Undemineralized section of bone taken from a patient suffering from severe osteomalacia. This section was unstained; it reveals no mineralization front and very wide inactive osteoid. This is to be compared with B, which represents a section taken from normal human bone undemineralized and unstained, which shows a small osteoid border, osteoblasts, and an active mineralization front. Courtesy of Dr. Jenifer Jowsey, Mayo Clinic, Rochester, Minnesota.

is dependent on the presence of vitamin D. Using this argument, it is obvious that normal calcium and phosphorus levels in the diet cannot prevent the need for vitamin D, and it is also quite evident that the functioning of the parathyroid system requiring the presence of vitamin D dictates that vitamin D is a requirement throughout life.

The basis for hypocalcemic tetany remains unknown. It is clear that the neuromuscular conjunction fails to operate normally under conditions of low ionized calcium of the blood. There is a continual excitation of the skeletal muscles, bringing about a continuous state of tetany, which is an extremely serious condition and must be corrected immediately. Low-calcium tetany can obviously result from a deficiency of parathyroid hormone or of vitamin D, or it can result from a condition of milk formation, as in the case of parturient paresis in dairy cattle, or any other serious and rapid loss of calcium from the plasma. Under normal conditions, because of the acuteness of this disease, the physiology of mammals is so constructed that the skeleton will be sacrificed to prevent this disease. The sacrifice of the skeleton or the mobilization of calcium from the skeleton is usually under the control of vitamin D and parathyroid hormone and perhaps to some degree the sex hormones. Thus inadequate intake of calcium as well as insufficient amounts of vitamin D preventing adequate calcium absorption will predispose the organism to mobilization of calcium from the skeleton, giving rise to bone deterioration or osteoporosis. Thus another problem with marginal or partial vitamin D deficiency may be the development of osteoporosis because of the inadequate absorption of calcium to meet the physiological needs.

Finally, another function of vitamin D which is poorly understood is that which occurs at the muscle. Primarily from clinical investigators has come the report of extreme muscle weakness under conditions of vitamin D deficiency or disrupted vitamin D metabolism (Dent, 1977; Dent and Smith, 1969). Muscle weakness in rachitic children is very rapidly corrected by the administration of vitamin D, and thus it is obvious that some change must occur in muscle function. It is not clear whether the change in muscle function results from an improvement in calcium and phosphorus homeostasis or calcium and phosphorus fluxes into the muscle cell, but nevertheless the end effect is indeed obvious. Biochemists have not yet approached the problem as to the molecular mechanism whereby vitamin D improves muscle strength and tone, and it is to be expected that much new work will appear in this area in the next several years.

2.7.3. *Physiological Functions of Vitamin D in the Prevention of the Deficiency Diseases*

From the above discussion, it is quite clear that vitamin D prevents hypocalcemic tetany, provides for the mineralization of bone and cartilage, increases intestinal absorption of calcium contributing to the prevention of osteoporosis, and functions in some manner to improve muscle strength and tone. It is well known that blood plasma is supersaturated with regard to the hydroxyapatite mineral component of bone (DeLuca, 1967; Neuman and Neuman, 1958). To support normal mineralization of bone it is essential that calcium and phosphorus be maintained at the supersaturating levels. Furthermore, to prevent hypocalcemic tetany, it is evident that the serum calcium level must be maintained in the normal range of about 10 mg/100 ml, and finally it is clear that to prevent osteoporosis the intestine must be able to absorb calcium according to the needs of the body for skeletal and other functions. In regard to the latter, it has been known since the early work of Nicolaysen that the intestine is able to adjust its efficiency of absorption according to the skeletal needs for calcium (Nicolaysen *et al.*, 1953). Thus animals and man placed on a low-calcium diet develop highly efficient intestinal absorption of calcium whereas those on a high-calcium diet develop low rates of intestinal calcium absorption. Furthermore, in human subjects depleted of calcium for long periods of time the high efficiencies of calcium absorption remain until the skeleton is recalcified (Nicolaysen *et al.*, 1953). Nicolaysen postulated the existence of the "endogenous factor," which would provide the intestine with this information, and he also demonstrated clearly that this endogenous factor requires vitamin D to be present before it can either function or exist. Thus the basic function of the vitamin D molecule is to elevate the plasma calcium and phosphorus to normal concentrations and to provide sufficient amounts of calcium absorption in the intestine to satisfy the skeletal and body needs.

Obviously, a major contributor to the function of vitamin D is its importance in stimulating intestinal calcium absorption (Nicolaysen, 1937a,b; Nicolaysen and Eeg-Larsen, 1953; Omdahl and DeLuca, 1973). Phosphate is the normal accompanying anion to calcium being transported against an electrochemical potential gradient (Omdahl and DeLuca, 1973; Martin and DeLuca, 1969), but, in addition, vitamin D stimulates a phosphate transport mechanism in small intestine by a process quite independent of calcium (Harrison and Harrison, 1961; Wasserman and Taylor, 1973; Chen *et al.*, 1974).

For a number of years it had been believed that vitamin D functions solely in the small intestine, but it was the work of Carlsson (1952) which demonstrated clearly that vitamin D plays an essential role in the utilization of calcium from the skeleton. Additional work by Harrison *et al.* (1958) and Rasmussen *et al.* (1963) illustrated that the elevation of serum calcium concentration by parathyroid hormone requires the presence of vitamin D. This was further clarified when it was demonstrated that only small amounts of vitamin D could serve a permissive role in allowing the parathyroid hormone to mobilize skeletal calcium for the support of serum calcium concentration (Harrison *et al.*, 1958). It is therefore evident that the mobilization of calcium and hence phosphorus from the fluid compartment requires both vitamin D and the parathyroid hormone (Omdahl and DeLuca, 1973). Thus another source of calcium and phosphorus made available by vitamin D for saturation of plasma with calcium and phosphorus is provided from the skeleton.

In addition to this mechanism, it had been suggested by Harrison and Harrison (1941) that vitamin D may play a role in improving renal reabsorption of phosphorus. There has been much recent work, and so far it remains unsettled whether or not vitamin D does improve renal reabsorption of phosphorus under some conditions (Puschett *et al.*, 1975; Bonjour and Fleisch, 1977). Current evidence is that the control of renal reabsorption of phosphorus is primarily by a vitamin D independent mechanism involving the parathyroid hormone on one hand and some unknown factor on the other (Steele *et al.*, 1975). There is some evidence that vitamin D does improve renal reabsorption of calcium, but so far detailed renal physiological experiments have not been carried out (Steele *et al.*, 1975). It is known that the parathyroid hormone does improve renal reabsorption of calcium as well (Kleeman *et al.*, 1961), but so far the experiment which would demonstrate that vitamin D is necessary for parathyroid improvement of renal reabsorption of calcium has not been carried out. Bonjour and Fleisch (1977) have demonstrated that a metabolite of vitamin D might be involved in parathyroid hormone induced phosphate diuresis, but these experiments are still in the preliminary stage and will require additional investigation before they can be accepted at the textbook level.

Thus the overall function of vitamin D is to improve the active transport of calcium and phosphorus in the small intestine by two independent mechanisms, to bring about the mobilization of calcium and hence phosphorus from the bone fluid compartment and to improve renal reabsorption of calcium and perhaps phosphorus. This results in an elevation of serum calcium and phosphorus concentration to a point that will prevent hypocalcemic tetany and provide for normal mineralization of bone. The ability of the intestine to adjust its rate of calcium absorption to meet the skeletal needs is also under vitamin D control, as will be discussed in detail later. Finally, it must be kept in mind that so far no direct evidence has been presented to support the idea that vitamin D might be involved in the mineralization process of the endochondral cartilage on one hand and osteoblastic-mediated mineralization on the other. Thus despite the appealing intellectual idea that vitamin D may be involved directly in the mineralization process, this concept must be held in reserve until clear experimental evidence is provided.

2.8. Vitamin D Metabolism

During the course of the past 15 years it has become clear that vitamin D does not act directly on its target tissues of intestine, bone, and kidney but must be metabolically altered before it can function (DeLuca, 1974; DeLuca and Schnoes, 1976; Omdahl and DeLuca, 1973). When radioactive vitamin D is injected into vitamin D deficient animals, it very rapidly accumulates in the liver, with as much as 60–80% of the dose appearing in that organ within the first 60 min (Olson *et al.*, 1976; Ponchon and DeLuca, 1969). This is unique among the vitamin D compounds because injection of 25-OH-D_3 does not result in rapid accumulation in the hepatic tissues. The reason for the sequestering of vitamin D by the liver is unknown, but it has led to the suggestion that the liver is the prime storage site for vitamin D (Kodicek, 1956). So far that idea has not received a great deal of support, and it has been shown more recently by Rosenstreich *et al.* (1971) that the primary storage site of chronically dosed radioactive vitamin D is in the adipose tissues. Exactly what controls the mobilization of vitamin D from the adipose tissues, however, has not been investigated.

Vitamin D is converted to esters of long-chain fatty acids with the 3-hydroxyl group of the vitamin molecule (Lund *et al.*, 1967; Fraser and Kodicek, 1965). The degree of conversion to the esters is indeed small but nevertheless significant and has been studied extensively by Fraser and Kodicek (1965). Because the amount of esterified vitamin D found in the tissues at any one time is relatively small, it is unlikely that this is a storage mechanism but rather is probably a biochemical curiosity. The storage form appears to be primarily the free vitamin D sterol.

Once vitamin D_3 is in the liver it is converted in its first obligatory reaction to 25-OH-D_3, which was isolated, identified, and chemically synthesized in 1968–69 (Blunt *et al.*, 1968; Blunt and DeLuca, 1969) (Fig. 12). The enzymatic machinery which carries out this hydroxylation has received only slight attention. It is clear, however, that the reaction is primarily microsomal or occurs in the endoplasmic reticulum, but exactly which cells in the liver carry out this reaction remain unknown (Horsting and DeLuca, 1969; Bhattacharyya and DeLuca, 1974). The reaction requires NADPH, molecular oxygen, and magnesium ions. It is not blocked by cytochrome P_{450} inhibitors, it is not induced by phenobarbital, and it is not blocked by lipid peroxidation inhibitors such as diphenylparaphenylene-diamine (Horsting, 1970). The activity of this enzyme is suppressed in animals given vitamin D, and it is believed that the hepatic product, 25-OH-D_3, provides a suppression of the hydroxylase and that as 25-OH-D_3 is removed from the liver the hydroxylase is free to make additional amounts of product (Bhattacharyya and De-Luca, 1973). The degree of regulation by the 25-OH-D_3 is quite limited, as is obvious from the fact that giving large amounts of vitamin D brings about high circulating levels of 25-OH-D_3 in the plasma (Blunt *et al.*, 1968; Haddad and Stamp, 1974). It is unknown how the overcoming of the feedback mechanism on the 25-hydroxylase occurs, and two possibilities are now visualized. One is that at physiological concentrations the suppression mechanism operates, but when large amounts of vitamin D are given, sufficient amounts then are hydroxylated by the nonspecific cholesterol 25-hydroxylase also located in the liver. This would bypass the feedback suppression by 25-OH-D_3. An alternative is that vitamin D_3 competes with 25-OH-D_3, releasing the hydroxylase to make additional supplies of 25-OH-D_3. Whatever the mechanism, it is clear that giving large amounts of vitamin D will

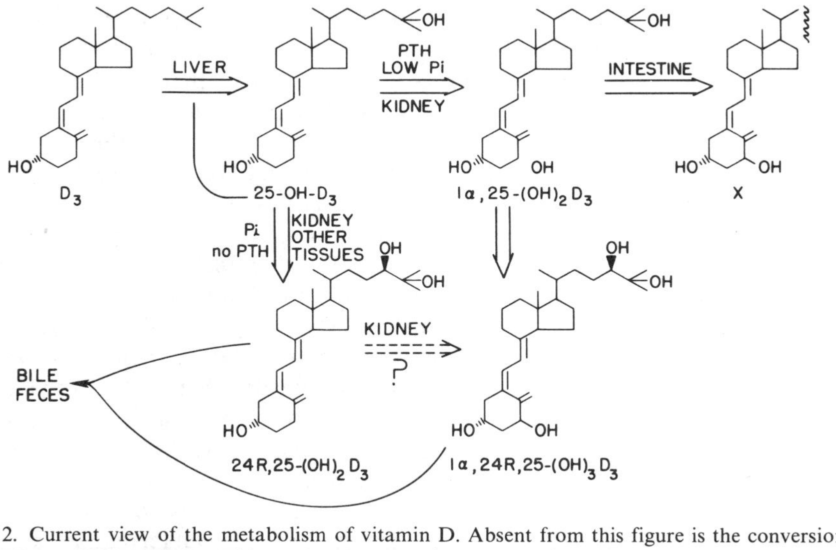

Fig. 12. Current view of the metabolism of vitamin D. Absent from this figure is the conversion of 25-OH-D_3 to 25,26-(OH)$_2$$D_3$. This product has been isolated and identified, but so far its site of bio-genesis has not been determined, nor has its function. The major arrows represent major pathways of conversion. Some of the controlling factors are written over the arrows, as are the known or expected sites of conversion.

bring about elevated plasma levels of 25-OH-D$_3$. There have been suggestions that induction of hepatic microsomes with such agents as phenobarbital and dilantin brings about increased metabolism of vitamin D to inactive products (Hahn *et al.*, 1972*a,b*). So far the experiments are preliminary and additional investigation is required before this conclusion can be made.

The 25-OH-D$_3$ is about 2–5 times more active than vitamin D$_3$ in the mineralization of bone, elevation of serum calcium, and improvement of calcium transport (Tanaka *et al.*, 1973*a*). However, it apparently does not act directly at physiological concentrations on any target tissue. Thus nephrectomy prevents 25-OH-D$_3$ given at physiological doses from elevating serum calcium concentration (Holick *et al.*, 1972*a*) or from initiating intestinal calcium absorption in rachitic rats (Boyle *et al.*, 1972*a*). Instead, the 25-OH-D$_3$ is transported to the kidney, where a specific reaction in the kidney converts 25-OH-D$_3$ to 1,25-(OH)$_2$D$_3$. The 1,25-(OH)$_2$D$_3$ is currently believed to be the active hormone derived from vitamin D which acts on calcium and phosphorus metabolism. The 1,25-(OH)$_2$D$_3$ was isolated in pure form in 1970–1971 and its structure was unequivocally identified by mass spectrometry and specific chemical reactions (Holick *et al.*, 1971*a,b*). It was subsequently chemically synthesized to establish its structure as 1α,25-(OH)$_2$D$_3$ (Semmler *et al.*, 1972). Fraser and Kodicek (1970) in Great Britain first discovered that this metabolite is synthesized exclusively in the kidney. They demonstrated that nephrectomy prevented the production of what was then called the "peak P metabolite," characterized by a loss of 1α-^{3}H. They were further able to demonstrate that homogenates of chick kidneys could carry out the conversion to "peak P" *in vitro*. This important and basic observation was quickly confirmed by Gray *et al.* (1971), who in addition demonstrated that the nephrectomy did not prevent the conversion because of resultant uremia inasmuch as ureteric-ligated animals, which have a high degree of uremia, clearly convert 25-OH-D$_3$ to the 1,25-(OH)$_2$D$_3$. Of great significance is the fact that 1-hydroxylation does not occur in any other organ but the kidney, at least as has been presently determined. Continued work has illustrated that the 25-OH-D$_3$-1-hydroxylase system is found exclusively in the mitochondria of kidney and can be supported in the intact mitochondria by a Krebs cycle substrate, molecular oxygen, and magnesium ions (Gray *et al.*, 1972; Ghazarian and DeLuca, 1974). However, in intact mitochondria oxidative phosphorylation is essential to the hydroxylation. If the mitochondria are subject to swelling conditions which would permit penetration by the reduced pyridine nucleotides, it can be demonstrated that NADPH is specifically required and that the oxidative phosphorylation requirement in intact mitochondria is probably because of the need for energy-linked transhydrogenation which would transfer electrons from NADH to NADP, which can be used for the hydroxylation reaction. Additional work has revealed that the hydroxylation is carbon monoxide, glutethimide, and metyrapone sensitive (Ghazarian *et al.*, 1974). Furthermore, white light or light of about 450 nm can reverse the carbon monoxide inhibition (Ghazarian *et al.*, 1974). Final proof that this system is a cytochrome P$_{450}$ dependent reaction was provided when it was successfully solubilized and the cytochrome P$_{450}$ was isolated and combined with beef adrenal ferredoxin and beef adrenal ferredoxin reductase together with NADPH, magnesium ions, and radioactive 25-OH-D$_3$ (Ghazarian *et al.*, 1974; Pedersen *et al.*, 1976). This system

convincingly gave 1,25-$(OH)_2D_3$, and more recently the iron sulfur protein called
"renal ferredoxin" has been isolated in pure form and has a molecular weight of
12,500. It functions to accept electrons from the renal ferredoxin reductase and
transfers them to cytochrome P_{450}, which inserts molecular oxygen into the 1
position of 25-OH-D_3 to yield 1,25-$(OH)_2D_3$, as shown in Fig. 13. Furthermore,
oxygen-18 experiments demonstrated that the 1α-hydroxyl oxygen of 1,25-
$(OH)_2D_3$ is derived from molecular oxygen, providing conclusive proof that the 1-
hydroxylase system is a mixed-function monooxygenase (Ghazarian *et al.*, 1973).

As pointed out previously, nephrectomy prevents the conversion of
25-OH-D_3 to 1,25-$(OH)_2D_3$. Using the tool of nephrectomy, it has been possible to
show that this maneuver prevents the response of intestinal calcium transport
(Boyle *et al.*, 1972*a*), intestinal phosphate transport (Chen *et al.*, 1974), and
mobilization of calcium from bone (Holick *et al.*, 1972*a*) to 25-OH-D_3, while 1,25-
$(OH)_2D_3$ produces these responses equally well in intact and nephrectomized rats.
Thus we can conclude on the basis of these experiments and those with actinomycin
D (Tanaka *et al.*, 1971) that 1,25-$(OH)_2D_3$ or a further metabolite must be the
metabolically active form of vitamin D in stimulating intestine, bone, and perhaps
kidney to carry out the functions described earlier. Since 1,25-$(OH)_2D_3$ is made
exclusively in the kidney and has its target in intestine and bone, it can certainly be
regarded as a hormone (DeLuca, 1974; Omdahl and DeLuca, 1973).

The question of whether 1,25-$(OH)_2D_3$ must be metabolized further before it
can function has been examined. Using 1,25-$(OH)_2D_3$ labeled in the 26 and 27 positions, it has been possible to show that at the time the intestine and bone respond to
the 1,25-$(OH)_2D_3$, only 1,25-$(OH)_2D_3$ and no other metabolite can be detected in
the lipid-soluble extracts taken from these target tissues (Frolik and DeLuca, 1971,
1972). Thus the tentative conclusion is that, for intestinal calcium transport at
least, 1,25-$(OH)_2D_3$ is the metabolically active form. Unfortunately this is not an
airtight conclusion since some aqueous soluble radioactivity was observed and
more recent results have suggested that 1,25-$(OH)_2D_3$ rapidly undergoes an
additional reaction in which a portion of the side chain is lost including the 26 and
27 label. In short, it has been demonstrated with 1,25-$(OH)_2$-[26,27-^{14}C]D_3 that as
much as 30% of an injected dose of 1,25-$(OH)_2D_3$ is oxidatively converted to an
unknown metabolite and the 26 and 27 positions appear as $^{14}CO_2$ (Kumar *et al.*,
1976; Harnden *et al.*, 1976). 1-25-$(OH)_2D_3$ is the required intermediate in this reaction in that the carbon dioxide derived from these positions from 25-OH-D_3 does
not appear in the expired carbon dioxide in a nephrectomized animal. The nature

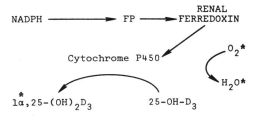

Fig. 13. Mechanism of 1α-hydroxylation of 25-OH-D_3 in the mitochondria of renal cells. The asterisks
indicate isotopically labeled oxygen.

and meaning of the side-chain oxidation reaction remain unknown, and this reaction is the subject of current investigation to evaluate its possible importance in function. The initiation of the side-chain oxidation reaction occurs 3 or 4 hr after injection, which is early enough to be of physiological significance, at least in the initiation of intestinal phosphate transport. Thus the matter of further metabolism of 1,25-(OH)$_2$D$_3$ prior to function is not yet totally resolved.

When radioactive vitamin D is given to normal animals, another dihydroxy metabolite of vitamin D appears in the chromatographic profiles. This compound has been isolated in pure form and identified as 24(R),25-dihydroxyvitamin D$_3$ [24(R),25-(OH)$_2$D$_3$] (Holick *et al.*, 1972*b*; Tanaka *et al.*, 1975*a*). This compound has not only been isolated and chemically characterized but also synthesized in quantity, including both the *S* and *R* configurations about the 24-carbon (Tanaka *et al.*, 1975*a*; Partridge *et al.*, 1977). Testing of biological activity of these compounds has revealed that the 24(R),25-(OH)$_2$D$_3$ is almost as active as the 25-OH-D$_3$ in rats in support of calcification, mobilization of calcium from bone, and initiation of intestinal calcium absorption (Tanaka *et al.*, 1975*a*; Boyle *et al.*, 1973). However, in the chicken this compound is much less active than 25-OH-D$_3$, which strongly suggests that it may well be the initial event in the inactivation of the vitamin D molecule (Holick *et al.*, 1976*a*). At least one site of biogenesis of the 24(R),25-(OH)$_2$D$_3$ is in the kidney and the reaction is mitochondrial; it requires molecular oxygen and internally generated NADPH (Knutson and DeLuca, 1974). So far it has not been demonstrated to be a cytochrome P$_{450}$ reaction, although clear evidence has now been provided that it is a mixed-function monooxygenase inasmuch as molecular oxygen is inserted in the 24(R) position rather than oxygen derived from water (Madhok *et al.*, 1977). The 24-hydroxylase system is fully capable of hydroxylating not only 25-OH-D$_3$ but also 1,25-(OH)$_2$D$_3$ (Tanaka *et al.*, 1977). It is converted to 1,24(R),25-trihydroxyvitamin D$_3$ [1,24(R),25-(OH)$_3$D$_3$], which has been isolated and its structure chemically identified (Holick *et al.*, 1973*a*; Tanaka *et al.*, 1977). Recently there has been chemical synthesis of both the *S* and *R* isomers, and, unlike 24(R),25-(OH)$_2$D$_3$, the *S* and *R* isomers are approximately equal in activity, suggesting that the discrimination against the 24S,25-(OH)$_2$D$_3$ is at the level of the 1-hydroxylase, which converts it to the 1,24,25-(OH)$_3$D$_3$. This has recently been confirmed with ^{3}H-labeled *S* and *R* isomers of 24-hydroxyvitamin D$_3$ (24-OH-D$_3$) (Tanaka *et al.*, 1976*a*). The 1,24,25-(OH)$_3$D$_3$ is about 60% as active as 1,25-(OH)$_2$D$_3$ in stimulating intestinal calcium absorption and bone calcium mobilization in the rat (Holick *et al.*, 1973*a*), but is much less active than 1,25-(OH)$_2$D$_3$ in the chick (Holick *et al.*, 1976*a*). The exact function of the 1,24,25-(OH)$_3$D$_3$ in the physiology of vitamin D function remains unknown. It is certainly much less active than the 1,25-(OH)$_2$D$_3$, which strongly suggests that 24-hydroxylation may be a mechanism of inactivation of this potent hormone. Nephrectomy does not prevent 24-hydroxylation of either 24-OH-D$_3$ or 1,25-(OH)$_2$D$_3$, which strongly suggests that there is an extrarenal site of 24-hydroxylation (Tanaka *et al.*, 1977). So far this site has not yet been detected.

Both the 24(R)-hydroxylase and the 1α-hydroxylase will function on any vitamin D molecule provided that the *cis*-triene structure is present and provided that there is a hydroxyl in the 25-position. Neither enzyme system will act on a 1α-OH-D$_3$ or on 24-OH-D$_3$ (Gray *et al.*, 1972; Tanaka *et al.*, 1977).

The 25-OH-D$_3$-24-hydroxylase is virtually absent in the vitamin D deficient

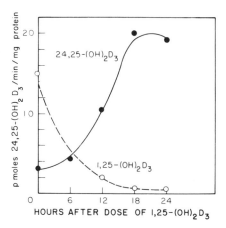

Fig. 14. Induction of the 24-hydroxylase in vitamin D deficient chickens by 1,25-(OH)₂D₃. Rachitic chickens were given a single injection of 1,25-(OH)₂D₃ at 0 time. At various times thereafter animals were sacrificed and the hydroxylases in their kidney homogenates were determined. Note that 24-hydroxylation appears at about 6 hr after injection of 1,25-(OH)₂D₃.

animal (Boyle *et al.*, 1971; Tanaka and DeLuca, 1974*b*). Administration of vitamin D brings about its appearance, and specifically the 1,25-(OH)₂D₃ appears to be the most active in this induction (Tanaka *et al.*, 1975*b*). As illustrated in Fig. 14, it is clear that the injection of nonradioactive 1,25-(OH)₂D₃ to rachitic chickens brings about the appearance of the 25-OH-D₃-24-hydroxylase within 6 hr after injection. Likely this is a nuclear response to 1,25-(OH)₂D₃ inasmuch as 1,25-(OH)₂D₃ has been shown to stimulate nuclear RNA synthesis in the vitamin D deficient animal (Chen and DeLuca, 1973*a*), and this response has been reported to be sensitive to α-aminitin (Larkins *et al.*, 1974). Because 1,25-(OH)₂D₃ specifically induces the 24-hydroxylase and because it is absent in the vitamin D deficient state, it would appear that the physiological mechanism by which 1,24,25-(OH)₃D₃ arises is by 1,25-(OH)₂D₃ being converted by the 24-hydroxylase rather than the 24,25-(OH)₂D₃ being 1-hydroxylated. Nevertheless, both pathways are possible but require the shifting of enzymatic activity in an appropriate manner to bring about biogenesis.

Besides the 1-hydroxylation, the 25-hydroxylation, and the 24-hydroxylation, another hydroxylation is known to occur *in vivo*: 25-OH-D₃ is known to be converted to 25,26-dihydroxyvitamin D₃ [25,26-(OH)₂D₃] (Suda *et al.*, 1970*a*). So far the configuration of the asymmetrical carbon on that molecule has not been determined in the natural product. Furthermore, the function of the 25,26-(OH)₂D₃ remains unknown, although biologically it has slight intestinal calcium transport activity and little activity on bone mobilization and mineralization (Lam *et al.*, 1975). Mawer (1977) has suggested that this compound may be a major metabolite in man, but this suggestion requires more direct definition that the metabolite detected by Mawer and colleagues is in fact 25,26-(OH)₂D₃.

2.9. *Regulation of Vitamin D Metabolism: Definition of the Vitamin D Endocrine System*

In 1971 Boyle *et al.* (1971, 1972*b*) demonstrated for the first time that the production or accumulation of 1,25-(OH)₂D₃ is under the control of dietary and serum calcium concentration. They clearly showed that by feeding a low-calcium diet, production of 1,25-(OH)$_2$D$_3$ *in vivo* from 25-OH-D$_3$ occurred at high rates. On the other hand, high-calcium diets with attendant high serum calcium levels brought about a suppression of production of the 1,25-(OH)$_2$D$_3$ and instead stimulated production of 24,25-(OH)$_2$D$_3$. Thus the need for calcium stimulated production of the calcium-mobilizing hormone, 1,25-(OH)$_2$D$_3$. Boyle and colleagues were further able to demonstrate that the *in vivo* accumulation of 1,25-(OH)$_2$D$_3$ in blood, intestine, and bone is directly related to serum calcium concentration, as is illustrated in Fig. 15. Thus under conditions of normal calcemia the animal converts 25-OH-D$_3$ to both 1,25-(OH)$_2$D$_3$ and 24,25-(OH)$_2$D$_3$. Calcium deprivation, as illustrated by even slight hypocalcemia, markedly stimulates the conversion to 1,25-(OH)$_2$D$_3$ and suppresses 24-hydroxylation. Thus the need for calcium, as illustrated by hypocalcemia, stimulates production of 1,25-(OH)$_2$D$_3$, a calcium-mobilizing hormone. This fundamental observation was shown subsequently to be the result of a suppression or an elevation of the 25-OH-D$_3$-1-hydroxylase in the mitochondria of chickens (Omdahl *et al.*, 1972; Henry *et al.*, 1974). This basic understanding led to an understanding of the basis for strontium-induced rickets in animals given vitamin D (Omdahl and DeLuca, 1971, 1972). The feeding of strontium in the diet completely blocked 25-OH-D$_3$-1-hydroxylase and stimulated the 24-hydroxylase. The block in intestinal calcium absorption brought about by strontium could be completely overcome by the administration of exogenous 1,25-(OH)$_2$D$_3$, illustrating that the basis of strontium-induced rickets is at least in part the suppression of the 25-OH-D$_3$-1-hydroxylase system.

Because the parathyroid glands serve as the hypocalcemic detection organ, it became clear very early that the parathyroid gland actually monitors hypocalcemia and in response secretes parathyroid hormone, which in turn stimulates production of 1,25-(OH)$_2$D$_3$. This important observation was first discovered by Garabedian *et al.* (1972) and was confirmed by Fraser and Kodicek (1973). Since that time, this important relationship has been verified in a variety of species, including man (Haussler *et al.*, 1976; DeLuca, 1977). Thus the need for calcium

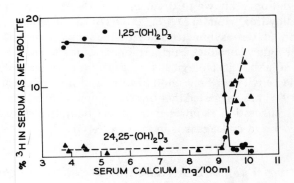

Fig. 15. Relationship of serum calcium concentration to accumulation of two vitamin D metabolites from their precursor, 25-OH-D$_3$, in rats. Rats were fed various diets for 2 weeks to bring about the indicated level of serum calcium concentration. They were then injected with radioactive 25-OH-D$_3$ and 12 hr later their serum was taken, extracted, and chromatographed to reveal the indicated metabolites. Normal serum calcium concentration in this population of rats is 9.5 mg/100 ml.

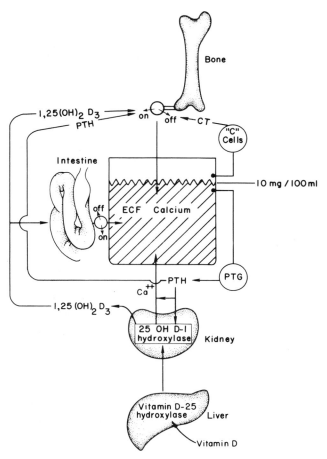

Fig. 16. Mechanisms which regulate serum (ECF or extracellular fluid) calcium concentration. Parathyroid glands monitor hypocalcemia while the parafollicular cells or "C" cells of the thyroid monitor hypercalcemia. In response to hypocalcemia, parathyroid hormone is secreted, which stimulates the kidney to produce 1,25-(OH)₂D₃. 1,25-(OH)₂D₃ initiates (turns on) intestinal calcium absorption by itself and together with the parathyroid hormone stimulates (turns on) the mobilization of calcium from bone. In addition, both parathyroid hormone and 1,25-(OH)₂D₃ improve renal reabsorption of calcium. These mechanisms therefore elevate serum calcium concentration, shutting off secretion of parathyroid hormone. The parafollicular cells secrete calcitonin (CT) in response to hypercalcemia, which then inhibits the mobilization of calcium from bone, thereby bringing about a suppression of serum calcium concentration.

brings about parathyroid hormone secretion, which in turn stimulates production of 1,25-(OH)₂D₃. It is 1,25-(OH)₂D₃ which then stimulates the intestine to absorb calcium. Together with the parathyroid hormone, the 1,25-(OH)₂D₃ brings about mobilization of calcium from bone, and both 1,25-(OH)₂D₃ and the parathyroid hormone bring about increased renal reabsorption of calcium. These three sources of calcium then restore serum calcium to normal, which suppresses parathyroid hormone secretion and shuts down the production of 1,25-(OH)₂D₃. This important calcium homeostatic mechanism is illustrated in Fig. 16.

It is immediately obvious that the parathyroid hormone's role in increasing intestinal calcium absorption is entirely mediated by its stimulating the production of 1,25-$(OH)_2D_3$ (Garabedian *et al.*, 1974). It is also apparent that both 1,25-$(OH)_2D_3$ and the parathyroid hormone are required for the mobilization of calcium from bone (Garabedian *et al.*, 1974).

This important series of relationships has permitted an explanation of Nicolaysen's "endogenous factor" and has provided a clear insight into how the intestine can adjust its efficiency of calcium absorption to meet the needs for calcium in bone and other organs (Boyle *et al.*, 1972*b*; Ribovich and DeLuca, 1975). It is now clear that hypocalcemia brought about by a lack of calcium in the diet or by deposit of calcium in the skeleton brings about an increased secretion of parathyroid hormone. The parathyroid hormone then stimulates production of 1,25-$(OH)_2D_3$. Increased levels of 1,25-$(OH)_2D_3$ then stimulate the intestine to become more efficient at absorbing calcium (Ribovich and DeLuca, 1976). Thus animals on a low-calcium diet have high rates of calcium absorption whereas those on a high-calcium diet have low rates of absorption. To prove this point, it has been clearly demonstrated that the administration of constant and exogenous sources of 1,25-$(OH)_2D_3$ eliminates the ability of the intestine to adjust the efficiency of absorption of dietary calcium (Ribovich and DeLuca, 1975; Omdahl and DeLuca, 1973). Similarly, parathyroidectomized animals given a constant and exogenous source of parathyroid hormone lose their ability to adjust their intestinal absorption of dietary calcium (Ribovich and DeLuca, 1976). Thus the endogenous factor is a combination of the parathyroid gland and the 1,25-$(OH)_2D_3$ endocrine systems operating to provide the intestine with the information regarding the need for calcium. Although reports that the intestine can adjust its absorption of calcium in the absence of parathyroid glands have appeared, these experiments have not been convincing simply because it is difficult to be sure that there is a complete absence of parathyroid glands in parathyroidectomized animals (Kimberg *et al.*, 1961; Favus *et al.*, 1974). In fact, it would be difficult to understand how a completely parathyroidectomized animal could survive on a low-calcium diet unless some source of parathyroid hormone were available.

2.10. Regulation of the Vitamin D System by the Need for Phosphorus

It has been adequately demonstrated that phosphate depletion brings about an elevated intestinal calcium absorption (Morrissey and Wasserman, 1971; Tanaka *et al.*, 1973*b*) and an increased mobilization of calcium from previously formed bone (Baylink *et al.*, 1971; Castillo *et al.*, 1975). The mechanism of this has not been completely elucidated, but at least in part it now appears that phosphate depletion brings about an elevation of 1,25-$(OH)_2D_3$ accumulation in blood, intestine, and bone (Tanaka *et al.*, 1973*b*; Castillo *et al.*, 1975; Tanaka and DeLuca, 1973). However, the administration of exogenous supplies of 1,25-$(OH)_2D_3$ does not completely eliminate the ability of the animal to increase its intestinal calcium absorption under conditions of phosphate depletion (Ribovich and DeLuca, 1975). Thus some additional effect of phosphate depletion must be responsible for the elevation of intestinal calcium absorption. Nevertheless, phosphate depletion does stimulate

1,25-(OH)₂D₃ production and accumulation (Baxter and DeLuca, 1976). Young growing rats have serum inorganic phosphorus levels of 9 mg %. In the thyroparathyroidectomized state they have little ability to make 1,25-(OH)₂D₃ and instead these animals produce 24,25-(OH)₂D₃ (Garabedian *et al.*, 1972) (Fig. 17). If serum inorganic phosphorus is lowered to values below 8 mg/100 ml, then 1,25-(OH)₂D₃ biogenesis occurs together with a suppression of the 25-OH-D₃-24-hydroxylase. Thus 1-hydroxylation can be stimulated by phosphate depletion even in the absence of the parathyroid gland (Tanaka and DeLuca, 1973; Baxter and DeLuca, 1976). Because 1,25-(OH)₂D₃ specifically stimulates the elevation of serum inorganic phosphorus (Tanaka and DeLuca, 1974*a*), the mobilization of inorganic phosphorus from bone (Castillo *et al.*, 1975), and the intestinal phosphate transport mechanism (Chen *et al.*, 1974), it seems reasonable to expect that its biogenesis would be regulated by serum inorganic phosphorus. These results make it likely that 1,25-(OH)₂D₃ is a phosphate-regulating hormone as well as a calcium-regulating hormone (DeLuca, 1974).

It might appear difficult to understand how 1,25-(OH)₂D₃ can serve a dual capacity as a phosphate-mobilizing hormone and a calcium-mobilizing hormone (DeLuca, 1974). However, the answer becomes immediately clear when one realizes that the mobilization of calcium occurs only under conditions where parathyroid hormone is secreted. As illustrated in Fig. 18, it can be shown that hypocalcemia brings about parathyroid hormone secretion, which in turn stimulates 1,25-(OH)₂D₃ production. The 1,25-(OH)₂D₃ will stimulate the bone to mobilize calcium and phosphorus, and it will stimulate the intestine to transport both calcium and phosphorus, but of great importance is the fact that the parathyroid hormone in a vitamin D independent reaction (Forte *et al.*, 1976) causes a phosphate diuresis. This negates the effect of 1,25-(OH)₂D₃ in stimulating intestinal

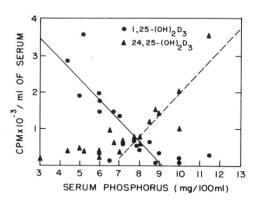

Fig. 17. Relationship between serum phosphorus concentration and accumulation of 24,25-(OH)₂D₃ and 1,25-(OH)₂D₃ in thyroparathyroidectomized rats. Thyroparathyroidectomized rats were fed a variety of diets and were given a source of vitamin D. After they had adapted to their diet, they were given a single injection of radioactive 25-OH-D₃ and the accumulation of the two metabolites in the plasma was determined following extraction and chromatography. Normal serum phosphorus concentration in the young growing rat is about 9 mg/100 ml. In the thyroparathyroidectomized state, serum phosphorus concentration at this level suppresses 1,25-(OH)₂D₃ accumulation and stimulates 24-hydroxylation. Hypophosphatemia will stimulate 1-hydroxylation and suppress 24-hydroxylation even in the absence of the parathyroid hormone.

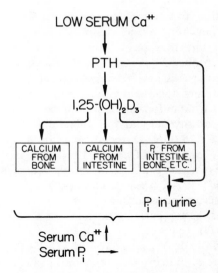

Fig. 18. Sequence of events following low calcium stimulation of 1,25-(OH)₂D₃ production. Note that the total net effect is to elevate serum calcium concentration without affecting serum phosphorus concentration.

phosphate absorption and the mobilization of phosphate from bone. The net effect under hypocalcemic conditions, therefore, is elevation of serum calcium without elevation of serum phosphorus concentration (DeLuca, 1974).

If one considers that hypophosphatemia brings about the stimulation of 1,25-(OH)₂D₃ production without the parathyroid hormone being secreted, it is clear that the mobilization of calcium from bone is blunted by the absence of parathyroid hormone. However, intestinal calcium transport as well as intestinal phosphate transport will be increased by 1,25-(OH)₂D₃, and because there is no parathyroid hormone to cause a phosphate diuresis and in fact the kidneys reabsorb all

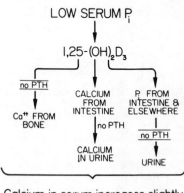

Fig. 19. Sequence of events following hypophosphatemic stimulation of 1,25-(OH)₂D₃ production. This stimulation results in only slight change in serum calcium concentration with a marked elevation of phosphorus in the plasma.

the filtered phosphorus (Steele and DeLuca, 1976), this sequence of events results in an elevation of serum phosphorus without appreciable elevation of serum calcium concentration (Fig. 19). Thus 1,25-$(OH)_2D_3$ can act as a phosphate-mobilizing hormone as well as a calcium-mobilizing hormone by virtue of its inter-action with the parathyroid system.

2.11. Regulation of Vitamin D Metabolism by the Sex Hormones and by Other Endocrine Systems

One of the most demanding calcium metabolic systems is that which exists in birds at the time of eggshell formation. It is known that prior to eggshell formation, calcium is deposited in a form of bone known as medullary bone (Bell and Freeman, 1971). At the time of eggshell and egg production, medullary bone is utilized to form the eggshells. Thus large calcium fluxes are experienced in and out of bone, and furthermore, large calcium fluxes across the shell gland are experienced. There is no doubt that vitamin D is involved in these processes inasmuch as vitamin D deficient animals cannot deposit shells around eggs and in fact stop egg production. An examination of the kidney preparations from egg-laying Japanese quail and their male counterparts has revealed that the egg-laying quail have high levels of 25-OH-D_3-1-hydroxylase whereas the male counterparts have almost entirely 25-OH-D_3-24-hydroxylase (Tanaka *et al.*, 1976*b*; Castillo *et al.*, 1977). The administration of estradiol to the mature males brings about a rapid change from the 24-hydroxylase to large amounts of 25-OH-D_3-1-hydroxylase. Thus in egg-laying birds estradiol can stimulate very markedly the 25-OH-D_3-1-hydroxylase. This stimulation has been shown to require the presence of an androgen or proges-terone, and a combination of the estradiol, androgen, and progesterone produces very marked elevations in 25-OH-D_3-1-hydroxylase. The elevation in the 25-OH-D_3-1-hydroxylase occurs after medullary bone has been formed and just prior to its mobilization for the formation of eggshell (L. Castillo, Y. Tanaka, and H. F. DeLuca, unpublished results). Thus it appears that high serum 1,25-$(OH)_2D_3$ levels are found in egg-laying birds at the time medullary bone is being mobilized for the formation of eggshells. A detailed investigation into the egg-laying cycle is now in progress in regard to vitamin D metabolism which will undoubtedly contribute to the elucidation of much of the avian physiology of eggshell formation and of medullary bone formation and utilization. Nevertheless, control by sex hormones of the vitamin D hydroxylases is an important new concept which may provide insight into the postmenopausal osteoporosis syndrome suffered by women, as will be discussed below.

More recently, primarily through the work of Schedl and his collaborators, it has become known that diabetes induced by streptozotocin or alloxan treatment of rats results in poor intestinal calcium absorption, poor bone mineralization, and reduced production of 1,25-$(OH)_2D_3$ (Schneider *et al.*, 1974, 1976). These phenomena can be corrected by the administration of insulin, and furthermore the poor intestinal calcium absorption can be corrected by administration of exogenous 1,25-$(OH)_2D_3$ (Schneider *et al.*, 1974). Thus insulin and diabetes are known to affect 25-OH-D_3-1-hydroxylase, which could provide important new insights into bone disease associated with the diabetic condition (Levin *et al.*, 1976). Undoubtedly, additional interrelationships of the vitamin D endocrine system and other endocrine systems known to affect calcium metabolism will be found. It might therefore appear likely that the vitamin D system is the master calcium- and phosphorus-regulating system and other endocrine systems carry out their functions on calcium and phosphorus metabolism through the vitamin D system. Only additional investigation will provide this kind of insight.

Before leaving the regulation of vitamin D metabolism, it is important to realize that there are a short-range control and a long-range control of calcium metabolism. The parathyroid hormone secretion system reacts within seconds following a hypocalcemic stimulus to secrete parathyroid hormone. The parathyroid hormone has a lifetime of only minutes in the peripheral blood. Furthermore, its action is very rapid in intestine and bone, occurring within minutes, and its lifetime of effectiveness is also measured in minutes. On the other hand, the production of 1,25-$(OH)_2D_3$ or elevation of the 25-OH-D_3-1-hydroxylase by the parathyroid hormone, by phosphate depletion, or by the sex hormones requires many hours (Garabedian *et al.*, 1972; Baxter and DeLuca, 1976; Tanaka *et al.*, 1976*b*). Furthermore, the lifetime of 1,25-$(OH)_2D_3$ in the bloodstream is measured in hours and its effectiveness at the target sites is measured in many hours if not days (Tanaka and DeLuca, 1971). With respect to short-term control of serum calcium concentration, it is clear that the 1,25-$(OH)_2D_3$ system does not operate. Instead, it is the parathyroid system which operates with existing 1,25-$(OH)_2D_3$. In response to a short-term need for calcium there is a stimulation of parathyroid hormone secretion. This 84 amino acid peptide appears at the target sites, where it stimulates renal reabsorption of calcium with existing 1,25-$(OH)_2D_3$, and it also stimulates the mobilization of calcium from bone with existing 1,25-$(OH)_2D_3$. This provides a short-term correction of serum calcium concentration without altering 1,25-$(OH)_2D_3$ levels. If, however, there is a chronic need for calcium, bringing about continued hypocalcemia and continued parathyroid hormone secretion, 1,25-$(OH)_2D_3$ production is stimulated, the 1,25-$(OH)_2D_3$ stimulates the intestine to sequester as much calcium from the environment as possible, and it also increases the sensitivity of kidney and bone to secreted parathyroid hormone. Thus on a long-term basis the animal has a mechanism for increasing the utilization of calcium from the environment, thus protecting the skeleton from loss of calcium in addition to protecting the animal against such acute disorders as hypocalcemic tetany.

It becomes immediately obvious that if there is an impairment of the intestinal mechanism to respond to 1,25-$(OH)_2D_3$ or if there is an impairment of 1,25-$(OH)_2D_3$ production in response to the need for calcium or if there is a lack of dietary calcium or factors which interfere with calcium absorption, the organism will continue to utilize calcium from the skeleton to protect against hypocalcemic tetany. This will bring about a continual erosion of bone, and it is the author's view that this contributes to the disease osteoporosis, which afflicts many adults.

2.12. Mechanism of Action of 1,25-$(OH)_2D_3$

From the above discussion it is clear that 1,25-$(OH)_2D_3$ is likely the metabolically active form in all known systems responsive to vitamin D. Thus in response to the need for calcium, 1,25-$(OH)_2D_3$ stimulates intestinal calcium absorption, bone calcium mobilization, and renal reabsorption of calcium. In response to the need for phosphate, the 1,25-(OH)$_2$D$_3$ stimulates intestinal phosphate transport and the mobilization of phosphate from bone. These mechanisms either singly or in combination bring about an elevation of plasma calcium and phosphorus to levels which prevent rachitic and osteomalacic lesions on one hand and hypocalcemic tetany on the other, and in addition provide for prevention of osteoporosis. This section will deal with what is known concerning the cellular and molecular mechanism of action of 1,25-(OH)$_2$D$_3$ in the target sites.

2.12.1. Intestinal Calcium Absorption

The intestinal calcium absorption response to vitamin D and 1,25-(OH)$_2$D$_3$ has been the most studied, for technical reasons. It is much more difficult to study reactions in such tissues as bone at the biochemical level or even at the physiological and cellular level. Similarly, technical difficulties are experienced in studying the renal mechanisms. Thus it is not surprising that we know much more about the function of the vitamin D compounds in the intestine than in any other target tissue.

Intestinal calcium absorption is an active transport process. It has been adequately demonstrated that calcium is transported from the lumen of intestine to the serosal fluid by a process which requires metabolic energy (Martin and DeLuca, 1969; Wasserman *et al.*, 1961; Schachter, 1963). The transfer occurs against an electrical and concentration gradient. Phosphate is the normal accompanying anion, although it is clear that this transport mechanism does not require the presence of phosphate to operate (Martin and DeLuca, 1969). Magnesium ions are required, and an oxidizable substrate, usually fructose, is provided (Schachter and Rosen, 1959). This transport system can be studied by the everted sac method, which is the least quantitative of all, or it can be studied by elegant techniques such as the Ussing chamber methods using short-circuited electrical systems. Vitamin D improves the transfer of calcium in both directions, although the flux in the mucosal to serosal direction is approximately tenfold higher than the serosal to mucosal flux, thus accounting for the net transfer of calcium against a concentration gradient (Martin and DeLuca, 1969; Wasserman *et al.*, 1961; Schachter, 1963). This system is specific for calcium, and it does not transport magnesium but will transport strontium and to a much less extent barium (Wasserman, 1963). Reports that it is involved in the transfer of other divalent ions lack convincing support, however. Besides metabolic energy, the only other known requirement is sodium ion. Martin and DeLuca (1969) believe that this requirement for sodium is for the expulsion of calcium across the basal-lateral membrane, and in its absence calcium accumulates in the intestinal villus cells charged with the responsibility of transport.

There is disagreement as to whether vitamin D functions only at the brush border membrane surface or whether it functions elsewhere as well (Schachter *et al.*, 1966). Electron micrographic examination of vitamin D deficient intestinal cells shows that calcium perhaps cannot cross the mucosal villus membrane barrier (Sampson *et al.*, 1970). Following vitamin D administration, calcium granules can be seen to appear in vesicles and in mitochondria at the terminal web region (Sampson *et al.*, 1970), and it is believed that these serve to shuttle calcium to the basal-lateral membrane. Most investigators agree that vitamin D must function at the mucosal surface membrane. Schachter *et al.* (1966) believe that vitamin D plays an additional role in the basal-lateral membrane transfer of calcium. So far there is not appreciable evidence to support this position. The current hypothesis of basal-lateral membrane transfer is that a sodium gradient brings about an expulsion of calcium across the basal-lateral membrane providing a calcium-depleted portion of the cell which induces mitochondria to give up their calcium. Thus the mitochondria can be visualized as a shuttle system. Again, evidence for this is primarily microscopic, plus the evidence provided by the abundant literature on calcium transport in isolated mitochondria. An alternative hypothesis is that calcium is packaged in membrane vesicles at the terminal web region which expel their calcium at the basal-lateral membrane under the influence of sodium. These are mere hypotheses and considerable investigation will be required before any are verified (Fig. 20).

Because of the steroid nature of the vitamin D molecule and hence 1,25-$(OH)_2D_3$, it has been visualized that this compound must function in a manner similar to that of the steroid hormones (Chen and DeLuca, 1973b; Brumbaugh and Haussler, 1974, 1975). In fact, recent autoradiographic work in the author's laboratory using physiological amounts of 1,25-$(OH)_2D_3$ has demonstrated a specific localization of the 1,25-$(OH)_2D_3$ in the nuclei of crypt cells and intestinal villus cells (Fig. 21). It had been demonstrated previously that the intestinal nuclear fraction accumulated some 80% of the tritiated 1,25-$(OH)_2D_3$ of the intes-

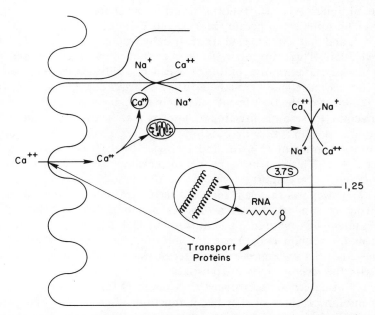

Fig. 20. Proposed mechanism of action of 1,25-$(OH)_2D_3$ in stimulating intestinal calcium transport. 1,25-$(OH)_2D_3$ is believed to act similarly to steroid hormones by binding with a cytosol receptor, which enters the nucleus and causes transcription of messenger RNA, which codes for calcium and presumably phosphorus transport proteins. These function at the brush border surface in an undetermined manner, permitting calcium to enter the absorption cells. Calcium either is packaged near the terminal web into vesicles or is taken up by mitochondria. Calcium is then extruded by a sodium-dependent process.

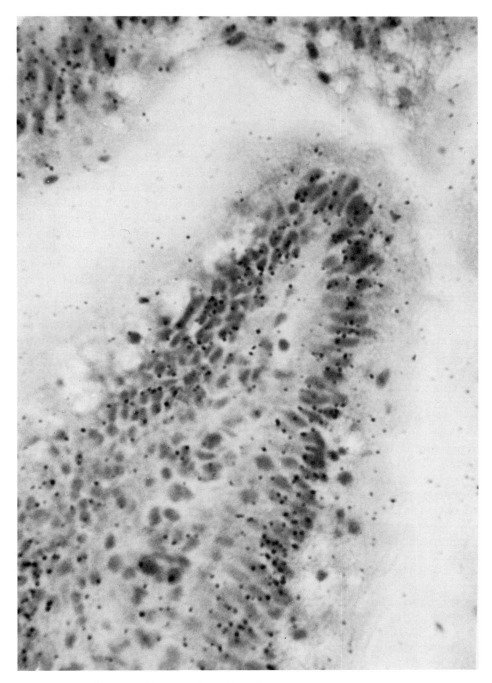

Fig. 21. Autoradiography of frozen sections of intestine obtained from a rachitic chicken given a single injection of 650 pmol of 78 Ci/mmol 1,25-(OH)$_2$D$_3$. This frozen section was stained with hematoxylin and eosin to reveal the cytoplasm and nuclei. Note that the grains of silver resulting from the tritiated 1,25-(OH)$_2$D$_3$ are located almost exclusively over the nuclei. This section is through the villus, but the same has been demonstrated in the case of the crypt cells.

tine (Chen and DeLuca, 1973*b*; Tsai *et al.*, 1972). However, the yield of isolated pure nuclei was so poor that a quantitative assessment of whether all of the tritium was in fact nuclear could not be made (Chen *et al.*, 1970). It is now clear, by means of autoradiography, that the nucleus does accumulate much of the radioactive 1,25-$(OH)_2D_3$.

The steroid hormone dogma holds that the hormone must interact with a cytoplasmic receptor protein which is transferred to the nucleus by a temperature-dependent reaction, bringing about a binding of the receptor protein to the chromatin of nuclei (Brumbaugh and Haussler, 1974). A 3.0–3.7 S cytosolic 1,25-$(OH)_2D_3$ binding protein in the intestine of rachitic chicks was first described by Brumbaugh and Haussler (1974, 1975). More recent work in the author's laboratory has confirmed the existence of this receptor protein and has clearly shown it to be a 3.7 S receptor protein which is easily stabilized with high salt, dithiothreitol, and the absence of proteolytic enzymes (Kream *et al.*, 1976, 1977*a*). Brumbaugh and Haussler (1974) have shown that this receptor molecule plus 1,25-$(OH)_2D_3$ is transferred to nuclear chromatin in a temperature-dependent process, a reaction which has been confirmed by Procsal *et al.* (1975). Although there was a failure by all investigators to demonstrate a similar receptor protein in such animals as rats and mice, more recently, with the chemical synthesis of very high specific activity ^{3}H-1,25-$(OH)_2D_3$, the existence of a 3.2 S receptor protein in the cytosol from vitamin D deficient rats (Kream *et al.*, 1977*b*) and in the bone of embryonic chicks and rats as well as rachitic chicks has been demonstrated (Kream *et al.*, 1977*c*). Thus there is good evidence that 1,25-$(OH)_2D_3$ does interact with a receptor that then transfers to the nuclei to induce synthesis of proteins which play a role in both metabolism of 125-$(OH)_2D_3$ and transport of calcium across the intestine. That vitamin D increases RNA synthesis of the intestine has been demonstrated in a variety of laboratories (Stohs *et al.*, 1967; Hallick and DeLuca, 1969; Zerwekh *et al.*, 1976), and increased template activity of isolated chromatin in response to vitamin D was first demonstrated by Hallick and De-Luca (1969) and more recently by Haussler and his colleagues (Zerwekh *et al.*, 1976). Thus it appears likely that the receptor plus 1,25-$(OH)_2D_3$ does stimulate production of messenger RNA and other RNAs which must play a role in the biogenesis of proteins that carry out calcium transport.

There is considerable question as to what is the nature of the calcium transport proteins or substances which are induced by 1,25-$(OH)_2D_3$. Wasserman and his colleagues in 1966 described a protein in the cytosol of chick intestine which bound calcium specifically and which made its appearance only after administration of vitamin D (Wasserman and Taylor, 1968). This protein, called the "calcium-binding protein," has a molecular weight of 28,000 in the chick (Wasserman *et al.*, 1968) and about 8000–12,000 in mammalian species (Drescher and DeLuca, 1971). There is considerable disagreement as to whether this protein can be considered a transport protein. Emtage and Lawson and their colleagues have provided evidence that following vitamin D administration there is an mRNA which codes for the calcium-binding protein and which can be shown to result in the production of the calcium-binding protein by ribosomes (Emtage *et al.*, 1973). It is not clear from their studies, however, what brings about the appearance of the messenger RNA that codes for the calcium-binding protein. Is it a direct induction by 1,25-$(OH)_2D_3$ and its receptor, or does it occur secondarily to some other event in vitamin D function? There is considerable argument as to whether the calcium-binding protein of Wasserman appears at the correct time (Spencer *et al.*, 1976) and at the right site to bring about transport of calcium (Spencer *et al.*, 1976; Harmeyer and DeLuca, 1969). Wasserman and his colleagues have provided a considerable amount of evidence that the calcium-binding protein level correlates with the calcium transport level under a variety of circumstances. However, the calcium-binding protein does not appear in the intestine in response to 1,25-$(OH)_2D_3$ at the same time that intestinal calcium transport is initiated. Furthermore, when intestinal transport has diminished to predosage level, calcium-binding protein of the intestine remains high (Norman, 1974). Thus there is a search for an additional factor which is responsible for 1,25-$(OH)_2D_3$ induced calcium transport. One possibility is high molecular weight proteins, and others include changes in lipid composition. Thus the mechanism whereby 1,25-$(OH)_2D_3$ stimulates the transfer of calcium across the brush border membrane remains largely unknown.

2.12.2. Mobilization of Calcium from Bone

The mobilization of calcium from bone has been much more difficult to study, although it is possible by means of tissue culture experiments to study the process *in vitro* (Trummel *et al.*, 1969; Raisz *et al.*, 1972; Reynolds *et al.*, 1973). Very early it had been demonstrated that 1,25-$(OH)_2D_3$ is responsible for mobilization of calcium from bone rather than 25-OH-D_3, as described above. Furthermore, Raisz *et al.* (1972) were able to demonstrate that 1,25-$(OH)_2D_3$ was much more effective than 25-OH-D_3 in inducing mobilization of calcium from organ cultures of embryonic bone. Similar experiments were carried out by Reynolds and his colleagues, and in fact Reynolds *et al.* (1973) concluded that 1,25-$(OH)_2D_3$ is likely the major calcium-mobilizing hormone in bone. The mobilization of calcium from bone in response to 1,25-$(OH)_2D_3$ is clearly actinomycin D sensitive (Tanaka and DeLuca, 1971), which indicates that this is also an induction process. Weber *et al.* (1971) have provided evidence by isolation and biochemical methods that the 1,25-$(OH)_2D_3$ appears in the nuclei of bone. So far autoradiography is not yet available on bone to provide for a clear insight into nuclear location of ^{3}H-1,25-$(OH)_2D_3$. Beyond the fact that 1,25-$(OH)_2D_3$ is likely located in the nuclei in bone and beyond the fact that the process is blocked by actinomycin and does occur in tissue culture, little is known concerning the mechanism whereby 1,25-$(OH)_2D_3$ brings about the release of calcium from the bone fluid compartment to the extracellular fluid compartment. In fact, it is not altogether certain whether it is an active transport mechanism that brings about calcium mobilization from bone.

2.12.3. Mechanism Whereby 1,25-$(OH)_2D_3$ Stimulates Intestinal Phosphate Transport

Unfortunately, little is known concerning the ileal and jejunal transport of phosphate in response to vitamin D. Likely this transport of phosphate occurs throughout the entire small intestine, although it is considerably more complicated in the duodenum because of the existence of the massive calcium transport mechanism responsive to 1,25-$(OH)_2D_3$ (Chen *et al.*, 1974). In the jejunum it can be shown that phosphate transport in response to 1,25-$(OH)_2D_3$ requires the presence of sodium ions (Taylor, 1974) and that it is an active transport process (Walling, 1977). Furthermore, it can be shown that this process does not require the presence of calcium in the ambient fluid, which lends considerable support to the contention that this mechanism is independent of the calcium transport mechanism responsive to 1,25-$(OH)_2D_3$. It is also known that sodium ions are required on the mucosal surface rather than on the serosal surface, in contrast to the calcium transport system. Likely much new work can be expected in regard to the mechanism of action of 1,25-$(OH)_2D_3$ at its target sites.

2.13. Analogues of 1,25-$(OH)_2D_3$

Because of the discovery that vitamin D must be converted to a hormonal substance before it can function, there has been a great deal of renewed interest in the synthesis of analogues of vitamin D and its metabolites. This section will deal with the most important analogues and those which have provided new information on the mechanism of vitamin D metabolism and function.

Perhaps the most important of all the analogues known to date are those which can be considered as the vitamin D_2 series. Vitamin D_2 cannot be considered the natural form of the vitamin inasmuch as it is not the form believed to be manufactured in the skin upon ultraviolet irradiation. Instead, the plant sterol ergosterol can be irradiated to produce vitamin D_2 (Askew *et al.*, 1931; Windaus *et al.*, 1932). Vitamin D_2 differs from vitamin D_3 because it has the 24-methyl group in the S configuration and has a double bond in the 22- and 23-positions (Fig. 2). Thus far it is believed that vitamin D_2 in mammals is equally active as vitamin D_3, although some difference in the two might be expected. It is known, for example, that vitamin D_3 is more toxic than vitamin D_2 in rats (Roborgh and DeMan, 1960). Furthermore, it appears that the mobilization of calcium from bone is stimulated more by compounds with the vitamin D_3 side chain than by those with the vitamin D_2 side chain (Roborgh and DeMan, 1960). Thus there may be subtle differences in sensitivity of certain systems to vitamin D_2 vs. vitamin D_3 even in mammals. In the bird it is well known that vitamin D_2 and its analogues are approximately one-tenth as active as vitamin D_3 in all respects (Chen and Bosmann, 1964). Nevertheless, it has been clearly demonstrated that vitamin D_2 is converted to 25-OH-D_2 before it can function (Suda *et al.*, 1969; Drescher *et al.*, 1969) and furthermore that 25-OH-D_2 is converted to 1,25-$(OH)_2D_2$ in the kidney (Jones *et al.*, 1975, 1976*a*). These compounds have all been isolated and chemically identified. Of great interest is the fact that the hydroxylases in liver and kidney which act on vitamin D_3 also act on the vitamin D_2 molecule (Jones *et al.*, 1975, 1976*a*). Of special interest is the fact that the 25-hydroxylase in liver of chickens is just as active on vitamin D_2 as it is on vitamin D_3 (Jones *et al.*, 1976*a*). Furthermore, the 1-hydroxylase in the kidney is likewise equally active on 25-OH-D_2 as it is on 25-OH-D_3. Finally, the 3.7 S cytosol receptor for 1,25-$(OH)_2D_3$ binds just as well to 1,25-$(OH)_2D_2$ (Kream *et al.*, 1977*a*). Nevertheless, birds discriminate against the vitamin D_2 molecule and they do so by rapidly metabolizing vitamin D_2 and its metabolites to water-soluble products which are excreted into the bile and feces. Thus it appears that substitution on the 24-carbon, whether it be hydroxyl or methyl in the bird, in some manner stimulates rapid metabolism and excretion (Imrie *et al.*, 1967). This reduces the lifetime of the vitamin D_2 molecule in the bird and probably accounts for its poor biological activity. $1,25\text{-}(OH)_2D_2$ and $25\text{-}OH\text{-}D_2$ are approximately one-tenth as active as their vitamin D_3 counterparts in the chicken (Drescher *et al.*, 1969; Jones *et al.*, 1976b). Thus the discrimination is applied not only to vitamin D_2 itself but also to its metabolites.

With our understanding of the necessity for 1-hydroxylation of the vitamin D molecule before it can function came the chemical synthesis of $1\alpha\text{-}OH\text{-}D_3$. This compound was chemically synthesized primarily as a chemical exercise to learn how to introduce the hydroxyl group in the 1-position of the vitamin D compounds (Holick *et al.*, 1973b). Thus $1\alpha\text{-}OH\text{-}D_3$ was synthesized in 1972. This interesting analogue has been prepared in radioactive form (Holick *et al.*, 1975a), and it has been demonstrated that prior to its function it is hydroxylated on carbon 25 to form $1,25\text{-}(OH)_2D_3$ (Holick *et al.*, 1976b, 1977). The $1,25\text{-}(OH)_2D_3$ accumulates in the target tissues prior to their response, illustrating that conversion of $1\alpha\text{-}OH\text{-}D_3$ to $1,25\text{-}(OH)_2D_3$ is an initial event in its function. Nevertheless, the $1\alpha\text{-}OH\text{-}D_3$ is an important and interesting compound since it bypasses the necessity for the kidney to insert the 1-hydroxyl for function. Thus patients who fail to make this conversion can be treated with $1\alpha\text{-}OH\text{-}D_3$. This compound in rachitic rats and chicks is approximately 2–3 times more active than vitamin D_3 but is approximately one-half to one-fifth as active as $1,25\text{-}(OH)_2D_3$ (Holick *et al.*, 1975b). Its use in man will likely be restricted to treatment of specific diseases wherein there is insufficient 1-hydroxylation occurring. $1\alpha\text{-}OH\text{-}D_2$ has also been synthesized, and its biological activity in the rat is approximately equal to that of $1\alpha\text{-}OH\text{-}D_3$. In the chick it is about one-tenth as active as $1\alpha\text{-}OH\text{-}D_3$ (Lam *et al.*, 1974b).

Very recently there has been considerable interest in whether all of the hydroxyls of $1,25\text{-}(OH)_2D_3$ are essential for binding to the receptor and for its biological activity. Two groups have synthesized $3\text{-deoxy-}1\alpha\text{-}OH\text{-}D_3$ and another group has synthesized the $3\text{-deoxy-}1,25\text{-}(OH)_2D_3$ (Lam *et al.*, 1974a; Okamura *et al.*, 1974a). Okamura and his colleagues have suggested that because of the absence of the 3-hydroxyl the 1-hydroxyl should remain in a fixed equatorial conformation, which should markedly enhance its activity (Okamura *et al.*, 1974b). However, the results have proved quite the contrary; the 3-deoxy compounds are much less active than their 3-hydroxylated counterparts, providing no support for this concept (Onisko *et al.*, 1977; Stern *et al.*, 1975). Thus the 3-hydroxyl function is important to maximal activity of the $1,25\text{-}(OH)_2D_3$.

The question of a shortened side chain has also been examined in view of the X-ray crystallographic work which shows that $25\text{-}OH\text{-}D_3$ has a fully extended side chain (Trinh-Toan *et al.*, 1977). Shortening of the vitamin D molecule by one carbon in the side chain markedly reduces its biological activity whether it is 1-hydroxylated or not (Holick *et al.*, 1975c); the full length of the side chain appears to be essential for its biological activity.

A considerable number of isomers have been chemically synthesized or in-

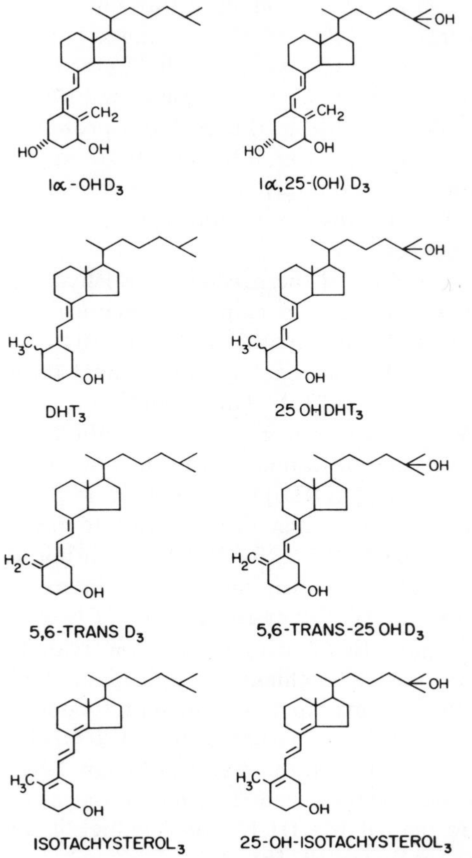

Fig. 22. Analogues of 1,25-(OH)₂D₃ of importance.

advertently synthesized in which the *cis*-triene structure of vitamin D has been considerably altered. Of particular importance are the 5,6-*trans*-vitamin D and 5,6-*trans*-25-OH-D₃ in which ring A is rotated 180° (Holick *et al.*, 1972*c,d*) (Fig. 22). These compounds are considerably more active than vitamin D₃ in stimulating intestinal calcium absorption and bone calcium mobilization in nephrectomized rats. On the other hand, the 5,6-*trans*-vitamin D₃ is considerably less active than its vitamin D₃ counterparts when given to intact animals capable of 1-hydroxylation. 25-OH-Isotachysterol has also been prepared and has been shown to have activity similar to that of 5,6-*trans*-25-OH-D₃ (Holick *et al.*, 1973*c*).

Of considerable importance are the dihydrotachysterols. These compounds were originally prepared as a means for stabilizing vitamin D preparations (von Werder, 1939). Surprisingly, they proved to have remarkable hypercalcemic activity. Dihydrotachysterol₂ has since been marketed in pure form, and dihydrotachysterol₃ has also been prepared (Westerhof and Keverling-Buisman, 1956). These compounds, because of the fact that during reduction the A ring is rotated 180°, have the 3-hydroxyl in the 1-hydroxyl position of 1,25-(OH)₂D₃ and thus can substitute for 1,25-(OH)$_2$D$_3$ (Hallick and DeLuca, 1971). They are therefore much more active in nephrectomized animals than is vitamin D$_3$ or in animals which lack the ability to 1-hydroxylate the vitamin. However, they are much less active than the corresponding D vitamins when given to intact animals (Suda *et al.*, 1970*b*). Thus the *cis*-triene structure is essential for maximal vitamin D activity, although all of the analogues in this respect have not yet been prepared. So far, no true antivitamin D compounds have been prepared despite reports to the contrary.

The large number of analogues now available for study has led to their application to biological activity *in vivo*, and to a testing of their biological effectiveness in isolated cultures of small intestine, isolated cultures of bone, and the receptor protein either by itself or in the intestinal chromatin-binding study. There is remarkable agreement between the activity of the compounds when added to isolated cultures of intestine or bone and their activity in binding to the intestinal cytosol receptor. Figure 23 illustrates the competitive position of a large number of analogues of vitamin D in terms of their competition for the 3.7 S chick intes-

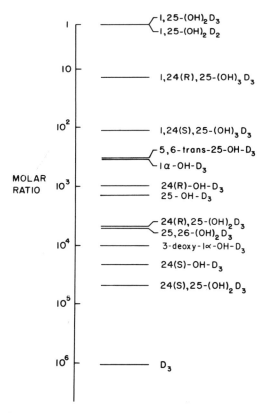

Fig. 23. Relative effectiveness of vitamin D and its metabolites to displace 1,25-(OH)$_2$D$_3$ from the 3.7 S intestinal cytosol receptor obtained from chicks. Graphically indicated is the ratio of the molar concentration of analogue to the molar concentration of radioactive 1,25-(OH)$_2$D$_3$ required to bring about a 50% displacement of the radioactive 1,25-(OH)$_2$D$_3$ from the receptor protein.

tinal cytosol receptor. Of particular importance is the fact that $1,25\text{-}(OH)_2D_3$ is bound highly specifically to this protein. However, it can be displaced with a ten- to twenty-fold excess of $1,24(R),25\text{-}(OH)_3D_3$, but a 500-fold excess is required for $1\alpha\text{-}OH\text{-}D_3$ to provide a displacement and even more is required for $25\text{-}OH\text{-}D_3$. It is important to note, however, that these compounds can substitute on the receptor for $1,25\text{-}(OH)_2D_3$ when provided in large enough concentrations (Kream *et al.*, 1977*a*). This important fact may provide the answer as to how vitamin D can be toxic when its conversion to the active forms is under such elaborate control. In any case, the structure–function relationships have illustrated that very few changes can be made in the vitamin D molecule without diminishing its biological activity or its activity in binding to the receptor proteins.

2.14. Toxicity of Vitamin D

Vitamin D is one of the two vitamins which when taken in large amounts can be toxic. The symptoms of vitamin D toxicity are nausea, lack of appetite, vomiting, itchiness of skin, polyuria, and ultimately death. The pathological reasons for vitamin D toxicity are hypercalcemia, nephrocalcinosis, aortic and heart calcification, subcutaneous calcification, intestinal calcification, and arrhythmia of the heart. Thus it appears that the basic functions of vitamin D are drastically exaggerated by giving large amounts of the vitamin. A large amount of vitamin D will cause a marked demineralization of bone at the same time that it is causing hypercalcemia and nephrocalcinosis (Kramer and Knof, 1954; Nicolaysen and Eeg-Larsen, 1953). Vitamin D intake is normally approximately 10 μg per day or less. Exactly where the danger area is for man is not entirely decided, although there is a considerable safety factor between the 400 units per day recommended daily intake and the toxicity level (Kramer and Knof, 1954). Although there have been reports of toxicity with as little as 2000 units per day, this likely represents an abnormal pathological condition. There are diseases in which there is a sensitivity to vitamin D, giving rise to hypercalcemia and toxicity. Such diseases are sarcoidosis and idiopathic hypercalcemia. In any case, because of the danger of vitamin D toxicity, it is recommended that no one take more than 400 units per day of any form of vitamin D unless it is specifically prescribed by a physician who is carefully monitoring serum calcium concentration and other evidences of vitamin D toxicity.

Vitamin D_3, as discussed above, is stored in the adipose tissues, and during the course of many months large amounts of vitamin D can be stored in the fat depots, making it very difficult to rapidly withdraw vitamin D from use by the patient. So when a patient experiences hypervitaminosis D, not only must the physician withdraw vitamin D but also the serum calcium must be rapidly reduced, usually by means of a furosamide diuretic agent and removal of calcium from the diet, and if hypercalcemia still persists the physician must treat the patient with some glucocorticoid, usually prednisone. Exactly how prednisone counteracts the effects of excess vitamin D is not known, although there has been the suggestion that it alters vitamin D metabolism (Carre *et al.*, 1974). It has also been suggested that prednisone directly affects calcium transport in the small intestine and may well also affect bone. In any case, it is currently the final method of choice for treatment of hypercalcemia of hypervitaminosis D.

With the elaborate control mechanisms on vitamin D metabolism as described above, it is difficult to understand how vitamin D toxicity can result. It has already been described in the previous section that large doses of vitamin D can overcome control at the 25-hydroxylation site. Thus the administration of large and toxic amounts of vitamin D will bring about very high levels of 25-OH-D_3 on the order of 1000 ng/ml as compared to the usual levels of approximately 15–25 ng/ml (Haddad and Stamp, 1974). It is indeed possible that even higher concentrations of 25-OH-D_3 will be found in the plasma. As illustrated in the previous section, the receptor proteins can interact with 25-OH-D_3 when present in large amounts. So it is possible that 25-OH-D_3 may be the compound which is bringing about the toxic reaction by interacting with or substituting for 1,25-$(OH)_2D_3$ on the receptors of intestine and bone, bringing excessive amounts of calcium into the system and exceeding the ability of the mechanisms to prevent metastatic calcification. Alternatively, contaminating 5,6-*trans*-vitamin D_3 may be responsible since it acts in the absence of kidney control mechanisms. Since the liver will rapidly hydroxylate 5,6-*trans*-vitamin D_3 to the 25-hydroxy derivative, this is not an unlikely possibility. Only continued investigation into this problem will provide the answer, although it already seems likely by direct analysis in the author's laboratory that hypervitaminosis D does not bring about increased levels of 1,25-$(OH)_2D_3$. In fact, it appears that hypervitaminosis D causes below-normal levels of 1,25-$(OH)_2D_3$.

2.15. Vitamin D Metabolism and Disease

With the discovery that vitamin D must be converted to metabolically active forms before it can function and that these reactions are markedly regulated in true hormonal fashion has come the possibility that a variety of pathological states may in part result from disturbed vitamin D metabolism. So far, there are some very clear disease states in which this is quite obviously the case. This section will describe some of those diseases and the methods now available for treatment.

2.15.1. Hypoparathyroidism and Pseudohypoparathyroidism

Patients suffering from either idiopathic or surgically induced hypoparathyroidism have been traditionally difficult to manage. Hypoparathyroid patients, have, of course, lost the ability to sense hypocalcemia and respond appropriately. As a result, serum calcium levels drop to very low values while serum inorganic phosphorus may rise. Obviously the primary difficulty which arises is failure of the neuromuscular conjunction, giving rise to tetany and convulsions. These patients have low urinary cyclic AMP and undetectable levels of immunoreactive parathyroid hormone in their plasma. Previously they have been treated by administering large amounts of vitamin D or the analogue dihydrotachysterol$_2$ described above. In most cases, this treatment has been successful with the maintenance of serum calcium in the approximate normal range. However, the major difficulty is that the patients either become refractory to the dosage of vitamin D used or suddenly become hypercalcemic. A significant number become refractory to almost all forms of vitamin D. The episodes of hypercalcemia and danger of nephrocalcinosis are difficult to deal with, as described above. Obviously these patients have lost their abilty to synthesize $1,25\text{-}(OH)_2D_3$ in response to the need for calcium, and as shown in Table IV their serum values of $1,25\text{-}(OH)_2D_3$ are below the normal range. Since $1,25\text{-}(OH)_2D_3$ can initiate intestinal calcium absorption without the parathyroid hormone, it is obvious that treatment can be carried out with $1,25\text{-}(OH)_2D_3$ or an analogue plus sufficient amounts of dietary calcium to support the serum calcium concentration (Kooh *et al.*, 1975; Neer *et al.*, 1975; Russell *et al.*, 1974). Without parathyroid hormone, the utilization of bone for maintenance of serum calcium seems less fruitful. The administration of 1–2 μg/day of $1,25\text{-}(OH)_2D_3$ or 2–4 μg/day of $1\alpha\text{-}OH\text{-}D_3$ plus dietary calcium, giving 800 mg elemental calcium per day, is an effective treatment of these patients. Even the extremely resistant patients have responded satisfactorily to this treatment, with one or two possible exceptions. Pseudohypoparathyroid patients differ by secreting parathyroid hormone in response to hypocalcemia, but the kidney and bone fail to respond to their own secreted parathyroid hormone. Cases differ also with respect to whether or not the cyclic AMP response of kidney to parathyroid hormone is found. In any case, the patients are hypocalcemic in the face of large circulating levels of parathyroid hormone. These patients are also satisfactorily treated by means of oral $1,25\text{-}(OH)_2D_3$ or $1\alpha\text{-}OH\text{-}D_3$ plus oral calcium (Kooh *et al.*, 1975; Neer *et al.*, 1975; Russell *et al.*, 1974). $25\text{-}OH\text{-}D_3$ in pharmacological amounts has also been successful in the treatment of most patients, although it would appear that the 1-hydroxylated vitamins are the treatment of choice.

2.15.2. Renal Osteodystrophy

One of the disease states which has received a great deal of attention has been the bone disease associated with chronic renal failure (DeLuca and Avioli, 1977). As shown in Fig. 24, there is a diminished if not absent mechanism for produc-

Table IV. Plasma Levels of 25-OH-D₃ and 1,25-(OH)₂D₃

Subjects	25-OH-D₃ (ng/ml)	1,25-(OH)₂D₃ (pg/ml)
Children (under 15 yr old)	25	80
Adults (50–60 yr old)	26	35
Hypoparathyroid adults	25	18
Hyperparathyroid adults	27	110
Anephric patients	27	0
Chronic renal failure patients (GFR <40)	25	0
Vitamin D dependency rickets patients (treated with vitamin D)	300	20
Vitamin D dependency rickets patients [treated with 1,25-(OH)₂D₃]	—	85
Osteoporotics	27	26

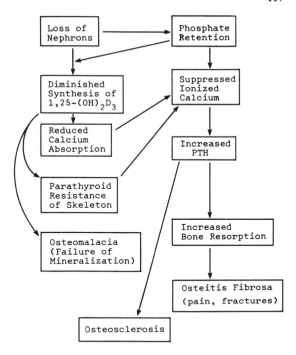

Fig. 24. Factors involved in the genesis of renal osteodystrophy. The primary factor is the loss of nephrons, which diminishes biosynthesis of 1,25-$(OH)_2D_3$. Phosphate retention suppresses ionized calcium, and all these factors bring about massively increased parathyroid hormone secretion and hypertrophy. Lack of 1,25-$(OH)_2D_3$ and the large circulating levels of parathyroid hormone bring about osteomalacia and osteitis fibrosa.

tion of 1,25-$(OH)_2D_3$ in these patients either because of loss of renal mass or because of accumulation of inorganic phosphorus in the cells which biosynthesize 1,25-$(OH)_2D_3$. Nevertheless, as shown in Table IV, serum levels of 1,25-$(OH)_2D_3$ in nephrectomized or renal failure patients are essentially undetectable (Eisman *et al.*, 1976; Haussler *et al.*, 1976). Thus there is no doubt that they have lost the ability to make this important calcium-mobilizing hormone. This disease is complicated by the fact that the kidney no longer excretes phosphate, which means that phosphate retention occurs, causing a secondary hyperparathyroidism (Bricker *et al.*, 1969). The lack of calcium absorption because of lack of 1,25-$(OH)_2D_3$ also aggravates the secondary hyperparathyroidism (Bricker *et al.*, 1969). Finally, the absence of adequate amounts of 1,25-$(OH)_2D_3$ may eventually lead to bone resistance to parathyroid hormone (Massry *et al.*, 1973). During the course of the entire development, excessively secreted parathyroid hormone causes a marked bone calcium mobilization and erosion, giving rise to osteitis fibrosa (DeLuca and Avioli, 1977). The lack of 1,25-$(OH)_2D_3$ may also result in a failure of mineralization, resulting in osteomalacia (DeLuca and Avioli, 1977; Stanbury, 1967). These patients have been successfully treated in 90% of the cases with 1,25-$(OH)_2D_3$, 25-OH-D_3 in pharmacological amounts, and 1α-OH-D_3 (Silverberg *et al.*, 1975; Chan *et al.*, 1975; Brickman *et al.*, 1974; Teitelbaum *et al.*, 1976). The dosage level found satisfactory for 1,25-$(OH)_2D_3$ is on the order of 0.5–1 μg/day, for 1α-OH-D_3 about 2 μg/day, and for 25-OH-D_3 in the range of 20–50 μg/day. Before treatment of this bone disease with 1,25-$(OH)_2D_3$ or any form of vitamin D_3 can be initiated, serum inorganic phosphorus levels must be adjusted to the normal range; serum phosphorus cannot be excessive because of danger to soft tissue calcification, nor can it be very low, because this would interfere with healing of osteomalacia (Pierides *et al.*, 1976). Once the serum inorganic phosphorus level is adjusted, either by phosphate binders or simply by phosphate deprivation, then 1,25-$(OH)_2D_3$ or another metabolite can be administered, provided that the patient is not extremely hyperparathyroid. If the parathyroids have become autonomous, partial parathyroidectomy may be necessary before initiation of treatment. Following these adjustments, treatment with the vitamin D metabolites has usually been successful in reversing all forms of the bone disease. It is perhaps much more important, however, to prevent development of renal osteodystrophy by the administration of one of the vitamin D metabolites very early during the development of chronic renal failure. Thus when the glomerular filtration rate falls appreciably below 80, it may be of considerable help to begin administration of the missing vitamin D metabolite.

2.15.3. *Corticoid-Induced Osteoporosis*

A major problem involved in the use of glucocorticoids in a wide variety of diseases has been the thinning of the bones or osteoporosis. The exact mechanism of this is not understood, although an interference with intestinal calcium absorption has been suggested (Carre *et al.*, 1974; Favus *et al.*, 1973; Harrison and Harrison, 1960). In the absence of sufficient calcium absorption, the skeleton is used to support serum calcium concentration, bringing about an erosion of the skeleton and giving rise to osteoporosis. Whether or not the glucocorticoids have any influence on vitamin D metabolism is a matter of debate. Carre *et al.* (1974) have provided evidence that glucocorticoids do cause the conversion of 1,25-$(OH)_2D_3$ to a biologically inactive metabolite. On the other hand, Favus *et al.* (1973) have reported that vitamin D metabolism is not affected by glucocorticoid treatment and that the glucocorticoid acts directly on the calcium transport mechanism. Whatever the mechanism, Hahn and his colleagues have clearly demonstrated that 25-OH-D_3 in sufficient quantities on the order of 10–50 μg/day will correct the osteoporosis (T. J. Hahn, personal communication). More recent results with 1,25-$(OH)_2D_3$ at the Mayo Clinic have provided a similar conclusion, namely that the osteoporosis can be corrected by the vitamin D metabolites (B. L. Riggs, personal communication).

2.15.4. *Postmenopausal and Senile Osteoporosis*

It is at this time unclear whether postmenopausal and senile osteoporosis involve an abnormality in vitamin D metabolism. It is well known that intestinal calcium absorption and serum levels of 1,25-$(OH)_2D_3$ diminish with age (Riggs *et al.*, 1977; Heaney, 1977). Furthermore, osteoporotic patients lose the ability to increase intestinal calcium absorption as the calcium needs of the body arise (Riggs *et al.*, 1977). Preliminary results have suggested that osteoporotic patients may have somewhat reduced levels of 1,25-$(OH)_2D_3$ in their plasma as compared to nonosteoporotic age-matched controls (Gallagher *et al.*, 1976). Furthermore, the intestinal calcium absorption of these patients can be enhanced by 1,25-$(OH)_2D_3$ but not by vitamin D_3 itself. Thus the inability of the intestine to adapt to low calcium intake suggests that such patients cannot produce the required amount of 1,25-$(OH)_2D_3$ to meet their needs. It is possible, therefore, that a component of the disease is inability to make sufficient amounts of this active metabolite of vitamin D, and treatment with it or an analogue may be beneficial. This is perhaps underscored by the discovery that the sex hormones play an important role in the biogenesis of 1,25-$(OH)_2D_3$ in egg-laying birds (Tanaka *et al.*, 1976*b*; Castillo *et al.*, 1977). Possibly this control might also be found in mammals, and thus an inability to make 1,25-$(OH)_2D_3$ may be aggravated by the lack of sex hormones. Trials are currently under way and the answer to whether the active forms of vitamin D might be useful in the treatment of this disease should be known within 2 or 3 years. It is also possible that the active forms of vitamin D may not benefit patients who have serious osteoporosis, but may be useful in preventing onset of the disease if taken as replacement therapy for lack of the required amount of 1,25-$(OH)_2D_3$ to enhance absorption of calcium as is needed.

2.15.5. Hepatic Disorders

So far no clear-cut hepatic disorder has been reported in which there is severe bone disease related to a defect in vitamin D metabolism. Certainly rickets is associated with biliary atresia. Much of this is failure to absorb vitamin D adequately from the intestine, although there may also be a failure in production of the 25-hydroxy derivative. Tyrosinemia may carry with it a defect in 25-hydroxylation. Liver cirrhosis does not appear to cause markedly reduced 25-OH-D_3 levels in the blood, although some bone disease has been ascribed to cirrhotic livers. It seems possible that patients having hepatic disorders might be benefited by administration of 25-hydroxylated forms of the vitamin.

2.15.6. Osteomalacia Induced by Phenobarbital and Dilantin

There has been considerable controversy as to the exact incidence of osteomalacia among patients who are on anticonvulsant therapy with phenobarbital and dilantin (Kruse, 1968; Christiansen *et al.*, 1973; Stamp *et al.*, 1972). There does appear, however, to be as high as 33% incidence of osteomalacia or rickets among such patients. Often, however, these are institutionalized patients with complications insofar as dietary management is concerned. Nevertheless, it has been demonstrated by Hahn and his colleagues that administration of 4000 units of ordinary vitamin D daily will correct the bone disease; alternatively, it can be corrected with metabolites of vitamin D at much lower dosage levels (Stamp *et al.*, 1972).

2.15.7. Neonatal Hypocalcemia

Premature infants often develop severe hypocalcemia because of inadequate development of parathyroid glands and of the vitamin D endocrine system. This hypocalcemia can be very nicely corrected with exogenous administration of 1,25-$(OH)_2D_3$ (Kooh *et al.*, 1976).

2.15.8. *Vitamin D Dependency Rickets*

Vitamin D dependency rickets is an autosomal recessively inherited disease in which children present with rickets in the face of normal intakes of vitamin D (Prader *et al.*, 1961). This disease is characterized by early onset and by amino-aciduria and is completely corrected by the administration of large amounts of vitamin D, on the order of 50,000–100,000 units per day (Prader *et al.*, 1961; Fraser *et al.*, 1973). Alternatively, patients can be treated with 10,000–20,000 units per day of 25-OH-D$_3$, still a pharmacological dose (Fraser *et al.*, 1973). As little as 1 μg/day of 1,25-(OH)$_2$D$_3$ will completely correct the disease and initiate mineralization of the rachitic skeleton (Fraser *et al.*, 1973; Reade *et al.*, 1975). One microgram of 1,25-(OH)$_2$D$_3$ on a weight basis might be considered equivalent to 50 units of ordinary vitamin D$_3$. It is therefore obvious that a dose approximating the physiological range of 1,25-(OH)$_2$D$_3$ will completely correct the disease, which has led to the argument that this disease is a defect in 1-hydroxylation of the 25-OH-D$_3$. Recently it has been confirmed, as shown in Table IV, that the serum circulating levels of 1,25-(OH)$_2$D$_3$ are low, especially when compared to those of age-matched controls (H. F. DeLuca, A. J. Hamstra, and D. Fraser, unpublished results). The cure with large amounts of vitamin D is probably the result of high circulating levels of 25-OH-D$_3$, which will interact with the receptor directly. Thus treatment with large amounts of vitamin D$_3$ does not appreciably increase the serum circulating level of 1,25-(OH)$_2$D$_3$ yet it does cure the disease (Fig. 25).

2.15.9. *Vitamin D Resistant Hypophosphatemic Rickets*

Vitamin D resistant hypophosphatemic rickets is an X-linked dominant disease and is the common vitamin D resistant rickets described in the literature

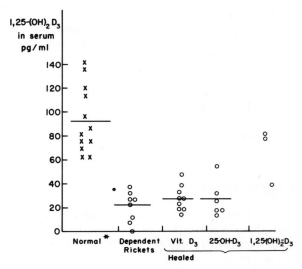

Fig. 25. Serum levels of 1,25-(OH)$_2$D$_3$ in normal children less than 12 yr of age, in children suffering from vitamin D dependency rickets, and in vitamin D dependency rachitic children healed with either vitamin D$_3$, 25-OH-D$_3$, or 1,25-(OH)$_2$D$_3$.

(Williams *et al.*, 1966; Winters *et al.*, 1958). Children manifest this disease later in life, with low blood phosphorus, severe rickets, approximately normal serum calcium concentration, and either normal or reduced intestinal calcium absorption. Recently a strain of mice has been developed which has a disease with the same genetic and phenotypic characteristics as the human disease (Eicher *et al.*, 1976). Using these mice it has been possible to demonstrate that there is a generalized defect in phosphate transport reactions, especially in the kidney and intestine (Eicher *et al.*, 1976; O'Doherty *et al.*, 1976). These phosphate transport reactions do not respond to $1,25\text{-}(OH)_2D_3$ (O'Doherty *et al.*, 1976). Treatment of patients is primarily by administration of large amounts of phosphate to compensate for the large losses in the urine. This, however, will bring about secondary hyperparathyroidism, which can be corrected by the counter-administration of $1,25\text{-}(OH)_2D_3$ or $1\alpha\text{-}OH\text{-}D_3$. Blood levels of $1,25\text{-}(OH)_2D_3$ in these patients have been reported to be normal (Haussler *et al.*, 1976), although more recent results indicate that they might be low, which would account for the reduced intestinal calcium absorption (H. Rasmussen and H. F. DeLuca, unpublished results). It is possible that the phosphate transport defect in the renal tubule cells might result in high cellular phosphate concentration, which would suppress $25\text{-}OH\text{-}D_3\text{-}1$-hydroxylation. This in turn would give low circulating levels of the hormone and low rates of intestinal calcium absorption.

Undoubtedly other applications of the vitamin D metabolites will be found in human disease. However, it is already clear that these compounds will be of great use in the treatment of a variety of metabolic bone diseases.

2.16. Conclusion

The understanding of vitamin D, its functions, and its metabolism has been advanced dramatically in the past one and one-half decades. We now realize that vitamin D is the precursor of at least one hormone which functions in the regulation of calcium and phosphorus metabolism. This regulation is brought about by the parathyroid hormone on one hand and serum inorganic phosphorus on the other. It is also brought about by sex hormones and probably directly or indirectly by insulin. The mechanism of action of the active form of vitamin D on intestinal calcium transport, bone calcium mobilization, and renal reabsorption mechanisms remains largely unknown, although it seems likely that there is a receptor specific for $1,25\text{-}(OH)_2D_3$ which appears then to interact with the nucleus to bring about the induction of proteins which function in calcium and phosphorus transport. A number of analogues of $1,25\text{-}(OH)_2D_3$ have been made, and the application of the vitamin D metabolites and their analogues to the treatment of a variety of metabolic bone diseases is clear. Thus vitamin D is unique among the vitamins inasmuch as it must now be considered a prohormone.

ACKNOWLEDGMENTS

Some of the original investigations reported in this chapter were supported by Grant AM-14881 from the United States Public Health Service and by the Harry Steenbock Research Fund.

2.17. References

Anderson, H. C., 1969, Vesicles associated with calcification in the matrix of epiphyseal cartilage, *J. Cell Biol.* **41**:59.

Anderson, H. C., and Reynolds, J. J., 1973, Pyrophosphate stimulation of initial mineralization in cultured embryonic bones: Fine structure of matrix vesicles and their role in mineralization, *Dev. Biol.* **34**:211.

Askew, F. A., Bourdillon, R. B., Bruce, H. M., Jenkins, R. G. C., and Webster, T. A., 1931, The distillation of vitamin D, *Proc. R. Soc. London Ser. B* **107**:76.

Barnes, M. M., Constable, B. J., Morton, L. F., and Kodicek, E., 1973, Bone collagen metabolism in vitamin D deficiency, *Biochem. J.* **132**:113.

Baylink, D., Wergedal, J., and Stauffer, M., 1971, Formation, mineralization, and resorption of bone in hypophosphatemic rats, *J. Clin. Invest.* **50**:2519.

Baxter, L. A., and DeLuca, H. F., 1976, Stimulation of 25-hydroxyvitamin D_3-1α-hydroxylase by phosphate depletion, *J. Biol. Chem.* **251**:3158.

Bekemeier, H., 1958, Versuche zur maxmalen antirachitischen UV-Aktivierung isolierter menschlicher Haut, *Acta Biol. Med. Ger.* **1**:756.

Bell, D. J., and Freeman, B. M. (eds.), 1971, *Physiology and Biochemistry of the Domestic Fowl*, Vol. 3, Academic Press, London.

Bhattacharyya, M. H., and DeLuca, H. F., 1973, The regulation of rat liver calciferol-25-hydroxylase, *J. Biol. Chem.* **248**:2969.

Bhattacharyya, M. H., and DeLuca, H. F., 1974, Subcellular location of rat liver calciferol-25-hydroxylase, *Arch. Biochem. Biophys.* **160**:58.

Bills, C. E., 1954, Vitamin D chemistry, in: *The Vitamins*, 1st ed., Vol. 2 (W. H. Sebrell, Jr., and R. S. Harris, eds.), pp. 132–203, Academic Press, New York.

Blondin, G. A., Kulkarni, B. D., and Nes, W. R., 1967, A study of the origin of vitamin-D from 7-dehydrocholesterol in fish, *Comp. Biochem. Physiol.* **20**:379.

Blunt, J. W., and DeLuca, H. F., 1969, The synthesis of 25-hydroxycholecalciferol: A biologically active metabolite of vitamin D_3, *Biochemistry* **8**:671.

Blunt, J. W., DeLuca, H. F., and Schnoes, H. K., 1968, 25-Hydroxycholecalciferol: A biologically active metabolite of vitamin D_3, *Biochemistry* **7**:3317.

Bonjour, J.-P., and Fleisch, H., 1977, Effects of vitamin D and its metabolites on the renal handling of phosphate, in: *Vitamin D: Biochemical, Chemical and Clinical Aspects Related to Calcium Metabolism* (A. W. Norman, K. Schaefer, J. W. Coburn, H. F. DeLuca, D. Fraser, H. G. Grigoleit, and D. von Herrath, eds.), pp. 419–431, de Gruyter, Berlin.

Boyle, I. T., Gray, R. W., and DeLuca, H. F., 1971, Regulation by calcium of *in vivo* synthesis of 1,25-dihydroxycholecalciferol and 21,25-dihydroxycholecalciferol, *Proc. Natl. Acad. Sci. USA* **68**:2131.

Boyle, I. T., Miravet, L., Gray, R. W., Holick, M. F., and DeLuca, H. F., 1972*a*, The response of intestinal calcium transport of 25-hydroxy and 1,25-dihydroxy vitamin D in nephrectomized rats, *Endocrinology* **90**:605.

Boyle, I. T., Gray, R. W., Omdahl, J. L., and DeLuca, H. F., 1972*b*, Calcium control of the *in vivo* biosynthesis of 1,25-dihydroxyvitamin D_3: Nicolaysen's endogenous factor, in: *Endocrinology 1971* (S. Taylor, ed.), pp. 468–476, Heinemann Medical Books, London.

Boyle, I. T., Omdahl, J. L., Gray, R. W., and DeLuca, H. F., 1973, The biological activity and metabolism of 24,25-dihydroxyvitamin D_3, *J. Biol. Chem.* **248**:4174.

Bricker, N. S., Slatopolsky, E., Reiss, E., and Avioli, L. V., 1969, Calcium phosphorus and bone in renal disease and transplantation, *Arch. Intern. Med.* **123**:543.

Brickman, A. S., Sherrard, D. J., Jowsey, J., Singer, F. R., Baylink, D. J., Maloney, N., Massry, S. G., Norman, A. W., and Coburn, J. W., 1974, 1,25-Dihydroxycholecalciferol: Effect on skeletal lesions and plasma parathyroid hormone levels in uremic osteodystrophy, *Arch. Intern. Med.* **134**:883.

Brumbaugh, P. F., and Haussler, M. R., 1974, 1α,25-Dihydroxycholecalciferol receptors in intestine. II. Temperature-dependent transfer of the hormone to chromatin via a specific cytosol receptor, *J. Biol. Chem.* **249**:1258.

Brumbaugh, P. F., and Haussler, M. R., 1975, Nuclear and cytoplasmic binding components for vitamin D metabolites, *Life Sci.* **16**:353.

Carlsson, A., 1952, Tracer experiments on the effect of vitamin D on the skeletal metabolism of calcium and phosphorus, *Acta Physiol. Scand.* **26**:212.

Carlsson, A., and Hollunger, G., 1954, Effect of vitamin D on the absorption of inorganic phosphate, *Acta Physiol. Scand.* **31**:301.

Carre, M., Ayigbede, O., Miravet, L., and Rasmussen, H., 1974, The effect of prednisolone upon the metabolism and action of 25-hydroxy and 1,25-dihydroxyvitamin D_3, *Proc. Natl. Acad. Sci. USA* **71**:2996.

Castillo, L., Tanaka, Y., and DeLuca, H. F., 1975, The mobilization of bone mineral by 1,25-dihydroxyvitamin D_3 in hypophosphatemic rats, *Endocrinology* **97**:995.

Castillo, L., Tanaka, Y., DeLuca, H. F., and Sunde, M. L., 1977, The stimulation of 25-hydroxyvitamin D_3-1α-hydroxylase by estrogen, *Arch. Biochem. Biophys.* **179**:211.

Chan, J. C. M., Oldham, S. B., Holick, M. F., and DeLuca, H. F., 1975, 1α-Hydroxyvitamin D_3 in chronic renal failure: A potent analogue of the kidney hormone, 1,25-dihydroxycholecalciferol, *J. Am. Med. Assoc.* **234**:47.

Chen, P. S., Jr., and Bosmann, H. B., 1964, Effect of vitamin D_2 and D_3 on serum calcium and phosphorus in rachitic chicks, *J. Nutr.* **83**:133.

Chen, T. C., and DeLuca, H. F., 1973a, Stimulation of [^{3}H]uridine incorporation into nuclear RNA of rat kidney by vitamin D metabolites, *Arch. Biochem. Biophys.* **156**:321.

Chen, T. C., and DeLuca, H. F., 1973b, Receptors of 1,25-dihydroxycholecalciferol in rat intestine, *J. Biol. Chem.* **248**:4890.

Chen, T. C., Weber, J. C., and DeLuca, H. F., 1970, On the subcellular location of vitamin D metabolites in intestine, *J. Biol. Chem.* **245**:3776.

Chen, T. C., Castillo, L., Korycka-Dahl, M., and DeLuca, H. F., 1974, Role of vitamin D metabolites in phosphate transport of rat intestine, *J. Nutr.* **104**:1056.

Christiansen, C., Rødbro, P., and Lund, M., 1973, Effect of vitamin D on bone mineral mass in normal subjects and in epileptic patients on anticonvulsants: A controlled therapeutic trial, *Brt. Med. J.* **5**:208.

Daniels, F., Jr., 1964, Man and radiant energy: Solar radiation, in: *Handbook of Physiology*, Section 4: *Adaptation to Environment* (F. Field, Jr., ed.), pp. 969–987, American Physiologic Society, Williams and Wilkins, Baltimore.

DeLuca, H. F., 1967, Mechanism of action and metabolic fate of vitamin D, *Vitamins Hormones* **25**:315.

DeLuca, H. F., 1974, Vitamin D: The vitamin and the hormone, *Fed. Proc.* **33**:2211.

DeLuca, H. F., 1977, The vitamin D endocrine system, in: *Proceedings of the Josiah Macy Foundation Conference on Renal Function*, February 7–9, 1977, Charleston, S.C., in press.

DeLuca, H. F., and Avioli, L. V., 1977, Renal osteodystrophy, in: *Renal Disease*, 4th ed. (D. Black, ed.), Blackwell Scientific Publications, Oxford.

DeLuca, H. F., and Schnoes, H. K., 1976, Metabolism and mechanism of action of vitamin D, *Annu. Rev. Biochem.* **45**:631.

DeLuca, H. F., Zile, M. H., and Neville, P. F., 1969, Chromatography of vitamins A and D, in: *Lipid Chromatographic Analysis*, Vol. 2 (G. V. Marinetti, ed.), pp. 345–457, Dekker, New York.

DeLuca, H. F., Blunt, J. W., and Rikkers, H., 1971, VII. Biogenesis, in: *The Vitamins*, Vol. 3 (W. H. Sebrell, Jr., and R. S. Harris, eds.), pp. 213–230, Academic Press, New York.

Dent, C. E., 1977, Rickets: Clinical syndromes and classification, in: *Inborn Errors of Calcium and Bone Metabolism* (H. Bickel and J. Stern, eds.), Livingstone, England.

Dent, C. E., and Smith, R., 1969, Nutritional osteomalacia, *Q. J. Med.* **38**:195.

Drescher, D., and DeLuca, H. F., 1971, Vitamin D stimulated calcium binding protein from rat intestinal mucosa: Purification and some properties, *Biochemistry* **10**:2302.

Drescher, D., DeLuca, H. F., and Imrie, M. H., 1969, On the site of discrimination of chicks against vitamin D_2, *Arch. Biochem. Biophys.* **130**:657.

Eicher, E. M., Southard, J. L., Scriver, C. R., and Glorieux, F. H., 1976, Hypophosphatemia: Mouse model for human familial hypophosphatemic (vitamin D-resistant) rickets, *Proc. Natl. Acad. Sci. USA* **73**:4667.

Eisman, J. A., Hamstra, A. J., Kream, B. E., and DeLuca, H. F., 1976, 1,25-Dihydroxyvitamin D in biological fluids: A simplified and sensitive assay, *Science* **193**:1021.

Emtage, J. S., Lawson, D. E. M., and Kodicek, E., 1973, Vitamin D-induced synthesis of mRNA for calcium-binding protein, *Nature (London)* **246**:100.

Esvelt, R. P., DeLuca, H. F., and Schnoes, H. K., 1977, The production of vitamin D_3 in ultraviolet irradiated rat skin, in: *Proceedings of the 6th Parathyroid Conference* (abstr.), June 12–17, 1977, in press.

Favus, M. J., Kimberg, D. V., Millar, G. N., and Gershon, E., 1973, Effects of cortisone administration on the metabolism and localization of 25-hydroxycholecalciferol in the rat, *J. Clin. Invest.* **52**:1328.

Favus, M. J., Walling, M. W. and Kimberg, D. V., 1974, Effects of dietary calcium restriction and chronic thyroparathyroidectomy on the metabolism of [^{3}H]-25-hydroxyvitamin D_3 and the active transport of calcium by rat intestine, *J. Clin. Invest.* **53**:1139.

Fieser, L. F., and Fieser, M., 1959, Vitamin D, in: *Steroids*, pp. 90–168, Reinhold, New York.

Forte, L. R., Nickols, G. A., and Anast, C. S., 1976, Renal adenylate cyclase and the interrelationship between parathyroid hormone and vitamin D in the regulation of urinary phosphate and adenosine cyclic 3′, 5′-monophosphate excretion, *J. Clin. Invest.* **57**:559.

Fraser, D. R., and Kodicek, E., 1965, Vitamin D esters: Their isolation and identification in rat tissues, *Biochem. J.* **96**:59P.

Fraser, D. R., and Kodicek, E., 1970, Unique biosynthesis by kidney of a biologically active vitamin D metabolite, *Nature (London)* **228**:764.

Fraser, D. R., and Kodicek, E., 1973, Regulation of 25-hydroxycholecalciferol-1-hydroxylase activity in kidney by parathyroid hormone, *Nature (London) New Biol.* **241**:163.

Fraser, D., Kooh, S. W., and Scriver, C. R., 1967, Hyperparathyroidism as the cause of hyperaminoaciduria and phosphaturia in human vitamin D deficiency, *Pediatr. Res.* **1**:425.

Fraser, D., Kooh, S. W., Kind, H. P., Holick, M. F., Tanaka, Y., and DeLuca, H. F., 1973, Pathogenesis of hereditary and vitamin D dependent rickets: An inborn error of vitamin D metabolism involving defective conversion of 25-hydroxyvitamin D to 1α,25-dihydroxyvitamin D, *New Engl. J. Med.* **289**:817.

Frolik, C. A., and DeLuca, H. F., 1971, 1,25-Dihydroxycholecalciferol: The metabolite of vitamin D responsible for increased intestinal calcium transport, *Arch. Biochem. Biophys.* **147**:143.

Frolik, C. A., and DeLuca, H. F., 1972, Metabolism of 1,25-dihydroxycholecalciferol in the rat, *J. Clin. Invest.* **51**:2900.

Gallagher, C., Riggs, L., Eisman, J., Arnaud, S., and DeLuca, H. F., 1976, Impaired intestinal calcium absorption in postmenopausal osteoporosis: Possible role of vitamin D metabolites and PTH, *Clin. Res.* **24**:360A.

Garabedian, M., Holick, M. F., DeLuca, H. F., and Boyle, I. T., 1972, Control of 25-hydroxycholecalciferol metabolism by the parathyroid glands, *Proc. Natl. Acad. Sci. USA* **69**:1673.

Garabedian, M., Tanaka, Y., Holick, M. F., and DeLuca, H. F., 1974, Response of intestinal calcium transport and bone calcium mobilization of 1,25-dihydroxyvitamin D_3 in thyroparathyroidectomized rats, *Endocrinology* **94**:1022.

Ghazarian, J. G., and DeLuca, H. F., 1974, 25-Hydroxycholecalciferol-1-hydroxylase: A specific requirement for NADPH and a hemoprotein component in chick kidney mitochondria, *Arch. Biochem. Biophys.* **160**:63.

Ghazarian, J. G., Schnoes, H. K., and DeLuca, H. F., 1973, Mechanism of 25-hydroxycholecalciferol 1α-hydroxylation: Incorporation of oxygen-18 into the 1α position of 25-hydroxycholecalciferol, *Biochemistry* **12**:2555.

Ghazarian, J. G., Jefcoate, C. R., Knutson, J. C., Orme-Johnson, W. H., and DeLuca, H. F., 1974, Mitochondiral cytochrome P_{450}: A component of chick kidney 25-hydroxycholecalciferol-1α-hydroxylase; *J. Biol. Chem.* **249**:3026.

Glimcher, M. J., 1959, Macromolecular aggregation stage and reactivity of collagen in calcification, *Soc. Gen. Physiol.* **6**:53.

Goldblatt, H., and Soames, K. M., 1923, Studies on the fat soluble growth-promoting factor (I) storage (II) synthesis, *Biochem. J.* **17**:446.

Gonnerman, W. A., Toverud, S. U., Ramp, W. K., and Mechanic, G. L., 1976, Effects of dietary vitamin D and calcium on lysyl oxidase activity in chick bone metaphyses, *Proc. Soc. Exp. Biol. Med.* **151**:453.

Gray, R., Boyle, I., and DeLuca, H. F., 1971, Vitamin D metabolism: The role of kidney tissue, *Science* **172**:1232.

Gray, R. W., Omdahl, J. L., Ghazarian, J. G., and DeLuca, H. F., 1972, 25-Hydroxycholecalciferol-1-hydroxylase: Subcellular location and properties, *J. Biol. Chem.* **247**:7528.

Greaves, J. D., and Schmidt, C. L. A., 1933, The role played by bile in the absorption of vitamin D in the rat, *J. Biol. Chem.* **102**:101.

Haddad, J. G., and Stamp, T. C. B., 1974, Circulating 25-hydroxyvitamin D in man, *Am. J. Med.* **57**:57.

Hahn, T. J., Birge, S. J., Scharp, C. R., and Avioli, L. V., 1972a, Phenobarbital-induced alterations in vitamin D metabolism, *J. Clin. Invest.* **51**:741.

Hahn, T. J., Hendin, B. A., Scharp, C. R., and Haddad, J. G., Jr., 1972b, Effect of chronic anticonvulsant therapy on serum 25-hydroxycholecalciferol levels in adults, *New Engl. J. Med.* **287**:900.

Hallick, R. B., and DeLuca, H. F., 1969, Vitamin D_3-stimulated template activity of chromatin from rat intestine, *Proc. Natl. Acad. Sci. USA* **63**:528.

Hallick, R. B., and DeLuca, H. F., 1971, 25-Hydroxydihydrotachysterol₃-biosynthesis *in vivo* and *in vitro*, *J. Biol. Chem.* **246**:5733.

Harmeyer, J., and DeLuca, H. F., 1969, Calcium-binding protein and calcium absorption after vitamin D administration, *Arch. Biochem. Biophys.* **133**:247.

Harnden, D., Kumar, R., Holick, M. F., and DeLuca, H. F., 1976, Side chain metabolism of 25-hydroxy-[26,27-^{14}C]vitamin D_3 and 1,25-dihydroxy-[26,27-^{14}C]vitamin D_3 *in vivo*, *Science* **193**:493.

Harris, L. J., 1956, Vitamin D and bone, in: *The Biochemistry and Physiology of Bone* (G. H. Bourne, ed.), pp. 581–622, Academic Press, New York.

Harrison, H. C., Harrison, H. E., and Park, E. A., 1958, Vitamin D and citrate metabolism: Effect of vitamin D in rats fed diets adequate in both calcium and phosphorus, *Am. J. Physiol.* **192**:432.

Harrison, H. E., and Harrison, H. C., 1941, The renal excretion of inorganic phosphate in relation to the action of vitamin D and parathyroid hormone, *J. Clin. Invest.* **20**:47.

Harrison, H. E., and Harrison, H. C., 1960, Transfer of Ca^{45} across intestinal wall *in vitro* in relation to action of vitamin D and cortisol, *Am. J. Physiol.* **199**:265.

Harrison, H. E., and Harrison, H. C., 1961, Intestinal transport of phosphate: Action of vitamin D, calcium, and potassium, *Am. J. Physiol.* **201**:1007.

Haussler, M. R., Baylink, D. J., Hughes, M. R., Brumbaugh, P. F., Wergedal, J. E., Shen, F. H., Nielsen, R. L., Counts, S. J., Bursac, K. M., and McCain, T. A., 1976, The assay of 1α,25-dihydroxyvitamin D_3: Physiologic and pathologic modulation of circulating hormone levels, *Clin. Endocrinol.* **5**:151s.

Havinga, E., 1973, Vitamin D, example and challenge, *Experientia* **29**:1181.

Heaney, R. P., 1977, Vitamin D and osteoporosis, in: *Vitamin D: Biochemical, Chemical and Clinical Aspects Related to Calcium Metabolism* (A. W. Norman, K. Schaefer, J. W. Coburn, H. F. DeLuca, D. Fraser, H. G. Grigoleit, and D. von Herrath, eds.), pp. 627–633, de Gruyter, Berlin.

Henry, H. L., Midgett, R. J., and Norman, A. W., 1974, Regulation of 25-hydroxyvitamin D_3-1-hydroxylase *in vivo*, *J. Biol. Chem.* **249**:7584.

Hess, A. F., Weinstock, M., and Helman, F. D., 1925, The antirachitic value of irradiated phytosterol and cholesterol. I., *J. Biol. Chem.* **63**:305.

Holick, M. F., and DeLuca, H. F., 1974, Chemistry and biological activity of vitamin D, its metabolites and analogs, in: *Advances in Steroid Biochemistry and Pharmacology*, Vol. 4 (M. H. Briggs and G. A. Christie, eds.), pp. 111–155, Academic Press, New York.

Holick, M. F., Schnoes, H. K., DeLuca, H. F., Suda, T., and Cousins, R. J., 1971a, Isolation and identification of 1,25-dihydroxycholecalciferol: A metabolite of vitamin D active in intestine, *Biochemistry* **10**:2799.

Holick, M. F., Schnoes, H. K., and DeLuca, H. F., 1971b, Identification of 1,25-dihydroxycholecalciferol, a form of vitamin D_3 metabolically active in the intestine, *Proc. Natl. Acad. Sci. USA* **68**:803.

Holick, M. F., Garabedian, M., and DeLuca, H. F., 1972a, 1,25-Dihydroxycholecalciferol: Metabolite of vitamin D_3 active on bone in anephric rats, *Science* **176**:1146.

Holick, M. F., Schnoes, H. K., DeLuca, H. F., Gray, R. W., Boyle, I. T., and Suda, T., 1972b, Isolation and identification of 24,25-dihydroxycholecalciferol: A metabolite of vitamin D_3 made in the kidney, *Biochemistry* **11**:4251.

Holick, M. F., Garabedian, M., and DeLuca, H. F., 1972c, 5,6-*trans* isomers of cholecalciferol and 25-hydroxycholecalciferol: Substitutes for 1,25-dihydroxycholecalciferol in anephric animals, *Biochemistry* **11**:2715.

Holick, M. F., Garabedian, M., and DeLuca, H. F., 1972*d*, 5,6-*trans*-25-Hydroxycholecalciferol: Vitamin D analog effective on intestine of anephric rats, *Science* **176**:1247.

Holick, M. F., Kleiner-Bossaller, A., Schnoes, H. K., Kasten, P. M., Boyle, I. T., and DeLuca, H. F., 1973*a*, 1,24,25-Trihydroxyvitamin D₃: A metabolite of vitamin D₃ effective on intestine, *J. Biol. Chem.* **248**:6691.

Holick, M. F., Semmler, E. J., Schnoes, H. K., and DeLuca, H. F., 1973*b*, 1α-Hydroxy derivative of vitamin D₃: A highly potent analog of 1α,25-dihydroxyvitamin D₃, *Science* **180**:190.

Holick, M. F., DeLuca, H. F., Kasten, P. M., and Korycka, M. B., 1973*c*, Isotachysterol₃ and 25-hydroxyisotachysterol₃: Analogs of 1,25-dihydroxyvitamin D₃, *Science* **180**:964.

Holick, M. F., Holick, S. A., Tavela, T., Gallagher, B., Schnoes, H. K., and DeLuca, H. F., 1975*a*, Synthesis of [6-³H]-1α-hydroxyvitamin D₃ and its metabolism *in vivo* to [³H]-1α,25-dihydroxy-vitamin D₃, *Science* **190**:576.

Holick, M. F., Kasten-Schraufrogel, P., Tavela, T., and DeLuca, H. F., 1975*b*, Biological activity of 1α-hydroxyvitamin D₃ in the rat, *Arch. Biochem. Biophys.* **166**:63.

Holick, M. F., Garabedian, M., Schnoes, H. K., and DeLuca, H. F., 1975*c*, Relationship of 25-hydroxyvitamin D₃ side chain structure to biological activity, *J. Biol. Chem.* **250**:226.

Holick, M. F., Baxter, L. A., Schraufrogel, P. K., Tavela, T. E., and DeLuca, H. F., 1976*a*, Metabolism and biological activity of 24,25-dihydroxyvitamin D₃ in the chick, *J. Biol. Chem.* **251**:397.

Holick, M. F., Tavela, T. E., Holick, S. A., Schnoes, H. K., DeLuca, H. F., and Gallagher, B. M., 1976*b*, Synthesis of 1α-hydroxy[6-³H]vitamin D₃ and its metabolism to 1α,25-dihydroxy[6-³H]-vitamin D₃ in the rat, *J. Biol. Chem.* **251**:1020.

Holick, M. F., deBlanco, M. C., Clark, M. B., Henley, J. W., Neer, R. M., DeLuca, H. F., and Potts, J. T., Jr., 1977, The metabolism of [6-³H]1α-hydroxycholecalciferol to [6-³H]1α,25-dihydroxy-cholecalciferol in man, *J. Clin. Endocrinol. Metab.* **44**:595.

Horsting, M., 1970, The enzymatic conversion of cholecalciferol to 25-hydroxycholecalciferol, Ph.D. thesis, University of Wisconsin-Madison.

Horsting, M., and DeLuca, H. F., 1969, *In vitro* production of 25-hydroxycholecalciferol, *Biochem. Biophys. Res. Commun.* **36**:251.

Howland, J., and Kramer, B., 1921, Calcium and phosphorus in the serum in relation to rickets, *Am. J. Dis. Child.* **22**:105.

Huldshinsky, K., 1919, Heilung von Rachitis durch künstliche Höhensonne, *Deutsch Med. Wochenschr.* **45**:712.

Idler, D. R., and Baumann, C. A., 1952, Skin sterols. II. Isolation of Δ⁷-cholestenol, *J. Biol. Chem.* **195**:623.

Imrie, M. H., Neville, P. F., Snellgrove, A. W., and DeLuca, H. F., 1967, Metabolism of vitamin D₂ and vitamin D₃ in the rachitic chick, *Arch. Biochem. Biophys.* **120**:525.

Jones, G., and DeLuca, H. F., 1975, High-pressure liquid chromatography: Separation of the metab-olites of vitmains D₂ and D₃ on small-particle silica columns, *J. Lipid Res.* **16**:448.

Jones, G., Schnoes, H. K., and DeLuca, H. F., 1975, Isolation and identification of 1,25-dihydroxyvita-min D₂, *Biochemistry* **14**:1250.

Jones, G., Schnoes, H. K., and DeLuca, H. F., 1976*a*, An *in vitro* study of vitamin D₂ hydroxylases in the chick, *J. Biol. Chem.* **251**:24.

Jones, G., Baxter, L. A., DeLuca, H. F., and Schnoes, H. K., 1976*b*, Biological activity of 1,25-dihydroxyvitamin D₂ in the chick, *Biochemistry* **15**:713.

Kimberg, D. V., Schachter, D., and Schenker, H., 1961, Active transport of calcium by intestine: Effects of dietary calcium, *Am. J. Physiol.* **200**:1256.

Kleeman, C. R., Bernstein, D., Rockney, R., Dowling, J. T., and Maxwell, M. H., 1961, Studies on the renal clearance of diffusible calcium and the role of the parathyroid glands in its regulation, in: *The Parathyroids* (R. O. Greep and R. V. Talmage, eds.), pp. 353–387, Thomas, Springfield, Ill.

Knudson, A., and Benford, F., 1938, Quantitative studies of the effectiveness of ultraviolet radiation of various wavelengths in rickets, *J. Biol. Chem.* **124**:287.

Knutson, J. C., and DeLuca, H. F., 1974, 25-Hydroxyvitamin D₃-24-hydroxylase: Subcellular location and properties, *Biochemistry* **13**:1543.

Kodicek, E., 1956, Metabolic studies on vitamin D, in: *Ciba Foundation Symposium on Bone Struc-ture and Metabolism* (G. W. E. Wolstenholme and C. M. O'Connor, eds.), pp. 161–174, Little, Brown, Boston.

Kooh, S. W., Fraser, D., DeLuca, H. F., Holick, M. F., Belsey, R. E., Clark, M. B., and Murray, T. M., 1975, Treatment of hypoparathyroidism and pseudohypoparathyroidism with metabolites of vitamin D: Evidence for impaired conversion of 25-hydroxyvitamin D to 1α,25-dihydroxyvitamin D, *New Engl. J. Med.* **293**:840.

Kooh, S. W., Fraser, D., Toon, R., and DeLuca, H. F., 1976, Response of protracted neonatal hypocalcemia to 1α,25-dihydroxyvitamin D_3, *Lancet* **2**:1105.

Kramer, B., and Knof, A., 1954, Chemical pathology and pharmacology, in: *The Vitamins*, 1st ed., Vol. 2 (W. H. Sebrell, Jr., and R. S. Harris, eds.), pp. 248–253, Academic Press, New York.

Kream, B. E., Reynolds, R. D., Knutson, J. C., Eisman, J. A., and DeLuca, H. F., 1976, Intestinal cytosol binders of 1,25-dihydroxyvitamin D_3 and 25-hydroxyvitamin D_3, *Arch. Biochem. Biophys.* **176**:779.

Kream, B. E., Jose, M. J. L., and DeLuca, H. F., 1977a, The chick intestinal cytosol binding protein for 1,25-dihydroxyvitamin D_3: A study of analog binding, *Arch. Biochem. Biophys.* **179**:462.

Kream, B. E., Yamada, S., Schnoes, H. K., and DeLuca, H. F., 1977b, A specific cytosol binding protein for 1,25-dihydroxyvitamin D_3 in rat intestine, *J. Biol. Chem.* **252**:4501.

Kream, B. E., Eisman, J. A., and DeLuca, H. F., 1977c, Intestinal cytosol binders for 1,25-dihydroxyvitamin D_3: Use in a competitive binding protein assay, in: *Vitamin D: Biochemical, Chemical and Clinical Aspects Related to Calcium Metabolism* (A. W. Norman, K. Schaefer, J. W. Coburn, H. F. DeLuca, D. Fraser, H. G. Grigoleit, and D. von Herrath, eds.), pp. 501–510, de Gruyter, Berlin.

Krieger, C. H., and Steenbock, H., 1940, Cereals and rickets. XII. The effect of calcium and vitamin D on the availability of phosphorus, *J. Nutr.* **20**:125.

Kruse, R., 1968, Osteopathien bei antiepileptischer Langzeittherapie, *Monatsschr. Kinderheilk.* **116**:378.

Kumar, R., Harnden, D., and DeLuca, H. F., 1976, Metabolism of 1,25-dihydroxyvitamin D_3: Evidence for side-chain oxidation, *Biochemistry* **15**:2420.

Lam, H. Y., Onisko, B. L., Schnoes, H. K., and DeLuca, H. F., 1974a, Synthesis and biological activity of 3-deoxy-1α-hydroxyvitamin D_3, *Biochem. Biophys. Res. Commun.* **59**:845.

Lam, H.-Y. P., Schnoes, H. K., and DeLuca, H. F., 1974b, 1α-Hydroxyvitamin D_2: A potent synthetic analog of vitamin D_2, *Science* **186**:1038.

Lam, H.-Y., Schnoes, H. K., and DeLuca, H. F., 1975, Synthesis and biological activity of 25ξ,26-dihydroxycholecalciferol, *Steroids* **25**:247.

Larkins, R. G., MacAuley, S. J., and MacIntyre, I., 1974, Feedback control of vitamin D metabolism by nuclear action of 1,25-dihydroxyvitamin D_3 on the kidney, *Nature (London)* **252**:412.

Levin, M. E., Boisseau, V. C., and Avioli, L. V., 1976, Effects of diabetes mellitus on bone mass in juvenile and adult-onset diabetes, *New Engl. J. Med.* **294**:241.

Lowe, J. T., and Steenbock, H., 1936, Cereals and rickets. VII. The role of inorganic phosphorus in calcification on cereal diets, *Biochem. J.* **30**:1126.

Lund, J., and DeLuca, H. F., 1966, Biologically active metabolite of vitamin D_3 from bone, liver, and blood serum, *J. Lipid Res.* **7**:739.

Lund, J., DeLuca, H. F., and Horsting, M., 1967, Formation of vitamin D esters *in vivo*, *Arch. Biochem. Biophys.* **120**:513.

Madhok, T. C., Schnoes, H. K., and DeLuca, H. F., 1977, Mechanism of 25-hydroxyvitamin D_3-24-hydroxylation: Incorporation of oxygen-18 into the 24 position of 25-hydroxyvitamin D_3, *Biochemistry* **16**:2142.

Martin, D. L., and DeLuca, H. F., 1969, Calcium transport and the role of vitamin D, *Arch. Biochem. Biophys.* **134**:139.

Massry, S. G., Coburn, J. W., Lee, D. B. N., Jowsey, J., and Kleeman, C. R., 1973, Skeletal resistance to parathyroid hormone in renal failure, *Ann. Intern. Med.* **78**:357.

Mawer, E. B., 1977, Aspects of the control of vitamin D metabolism in man, in: *Vitamin D: Biochemical, Chemical and Clinical Aspects Related to Calcium Metabolism* (A. W. Norman, K. Schaefer, J. W. Coburn, H. F. DeLuca, D. Fraser, H. G. Grigoleit, and D. von Herrath, eds.), pp. 165–173, de Gruyter, Berlin.

Maximow, A. A., and Bloom, W., 1957, Bone, in: *A Textbook of Histology*, 7th ed., Saunders, Philadelphia.

McCollum, E. V., and Davis, M., 1913, The necessity of certain lipins in the diet during growth, *J. Biol. Chem.* **15**:167.

McCollum, E. V., Simmonds, N., and Pitz, W., 1916, The relation of the unidentified dietary factors, the fat-soluble A, and water-soluble B, of the diet to the growth-promoting properties of milk, *J. Biol. Chem.* **27**:33.

McCollum, E. V., Simmonds, N., Becker, J. E., and Shipley, P. G., 1922a, Studies on experimental rickets. XXI. An experimental demonstration of the existence of a vitamin which promotes calcium deposition, *J. Biol. Chem.* **53**:293.

McCollum, E. V., Simmonds, N., Shipley, P. G., and Park, E. A., 1922b, Studies on experimental rickets. XV. The effect of starvation on the healing of rickets, *Bull. Johns Hopkins Hosp.* **33**:31.

McCollum, E. V., Simmonds, N., Becker, J. E., and Shipley, P. G., 1925, Studies on experimental rickets. XXVI. A diet composed principally of purified foodstuffs for use with the "line test" for vitamin D studies, *J. Biol. Chem.* **65**:97.

Mellanby, E., 1919a, An experimental investigation on rickets, *Lancet* **1**:407.

Mellanby, E., 1919b, A further determination of the part played by accessory food factors in the aetiology of rickets, **52**:1iii.

Morrissey, R. L., and Wasserman, R. H., 1971, Calcium absorption and calcium-binding protein in chicks on differing calcium and phosphorus intakes, *Am. J. Physiol.* **220**:1509.

Napoli, J. L., Reeve, L. E., Eisman, J. A., Schnoes, H. K., and DeLuca, H. F., 1977, *Solanum glaucophyllum* as a source of 1,25-dihydroxyvitamin D_3, *J. Biol. Chem.* **252**:2580.

Neer, R. M., Holick, M. F., DeLuca, H. F., and Potts, J. T., Jr., 1975, Effects of 1α-hydroxyvitamin D_3 and 1,25-dihydroxyvitamin D_3 on calcium and phosphorus metabolism in hypoparathyroidism, *Metabolism* **24**:1403.

Neuman, W. F., and Neuman, M. W., 1958, *The Chemical Dynamics of Bone*, University of Chicago Press, Chicago.

Neville, P. F., and DeLuca, H. F., 1966, The synthesis of $[1,2{-}^3H]$vitamin D_3 and the tissue localization of a 0.25 μg (10 IU) dose per rat, *Biochemistry* **5**:2201.

Nicolaysen, R., 1937a, Studies upon the mode of action of vitamin D. III. The influence of vitamin D on the absorption of calcium and phosphorus in the rat, *Biochem. J.* **31**:122.

Nicolaysen, R., 1937b, Studies upon the mode of action of vitamin D. II. The influence of vitamin D on the faecal output of endogenous calcium and phosphorus in the rat, *Biochem. J.* **31**:107.

Nicolaysen, R., and Eeg-Larsen, N., 1953, The biochemistry and physiology of vitamin D, *Vitamins Hormones* **11**:29.

Nicolaysen, R., Eeg-Larsen, N., and Malm, O. J., 1953, Physiology of calcium metabolism, *Physiol. Rev.* **33**:424.

Norman, A. W., 1974, 1,25-Dihydroxyvitamin D_3: A kidney-produced steroid hormone essential to calcium homeostasis, *Am. J. Med.* **57**:21.

Norman, A. W., and DeLuca, H. F., 1963a, The preparation of H^3-vitamins D_2 and D_3 and their localization in the rat, *Biochemistry* **2**:1160.

Norman, A. W., and DeLuca, H. F., 1963b, Chromatographic separation of mixtures of vitamin D_2, ergosterol, and tachysterol$_2$, *Anal. Chem.* **35**:1247.

Norman, A. W., Lund, J., and DeLuca, H. F., 1964, Biologically active forms of vitamin D_3 in kidney and intestine, *Arch. Biochem. Biophys.* **108**:12.

O'Doherty, P. J. A., DeLuca, H. F., and Eicher, E. M., 1976, Intestinal calcium and phosphate transport in genetic hypophosphatemic mice, *Biochem. Biophys. Res. Commun.* **71**:617.

Okamura, W. H., Mitra, M. N., Wing, R. M., and Norman, A. W., 1974a, Chemical synthesis and biological activity of 3-deoxy-1α-hydroxyvitamin D_3 an analog of 1α,25-dihydroxyvitamin D_3, the active form of vitamin D_3, *Biochem. Biophys. Res. Commun.* **60**:179.

Okamura, W. H., Norman, A. W., and Wing, R. M., 1974b, Vitamin D: Concerning the relationship between molecular topology and biological function, *Proc. Natl. Acad. Sci. USA* **71**:4194.

Olson, E. B., Jr., Knutson, J. C., Bhattacharyya, M. H., and DeLuca, H. F., 1976, The effect of hepatectomy on the synthesis of 25-hydroxyvitamin D_3, *J. Clin. Invest.* **57**:1213.

Omdahl, J. L., and DeLuca, H. F., 1971, Strontium induced rickets: Metabolic basis, *Science* **174**:949.

Omdahl, J. L., and DeLuca, H. F., 1972, Rachitogenic activity of dietary strontium. I. Inhibition of intestinal calcium absorption and 1,25-dihydroxycholecalciferol synthesis, *J. Biol. Chem.* **247**:5520.

Omdahl, J. L., and DeLuca, H. F., 1973, Regulation of vitamin D metabolism and function, *Physiol. Rev.* **53**:327.

Omdahl, J. L., Gray, R. W., Boyle, I. T., Knutson, J., and DeLuca, H. F., 1972, Regulation of metabolism of 25-hydroxycholecalciferol by kidney tissue *in vitro* in dietary calcium, *Nature (London) New Biol.* **237**:63.

Onisko, B. L., Lam, H.-Y., Reeve, L. E., Schnoes, H. K., and DeLuca, H. F., 1977, Synthesis and bioassay of 3-deoxy-lα-hydroxyvitamin D_3, an active analog of lα,25-dihydroxyvitamin D_3, *Bioorg. Chem.* **6**:203.

Orr, W. J., Holt, L. E., Jr., Wilkins, L., and Boone, F. H., 1923, The calcium and phosphorus metabolism in rickets, with special reference to ultraviolet ray therapy, *Am. J. Dis. Child.* **26**:362.

Osborne, T. B., and Mendel, L. B., 1917, The role of vitamins in the diet, *J. Biol. Chem.* **31**:149.

Partridge, J. J., Shiuey, S.-J., Baggiolini, E. G., Hennessy, B., and Uskokovic, M. R., 1977, A stereoselective synthesis of lα,24R,25-trihydroxycholecalciferol, a metabolite of vitamin D_3, in: *Vitamin D: Biochemical, Chemical and Clinical Aspects Related to Calcium Metabolism* (A. W. Norman, K. Schaefer, J. W. Coburn, H. F. DeLuca, D. Fraser, H. K. Grigoleit, and D. von Herrath, eds.), pp. 47–55, de Gruyter, Berlin.

Pedersen, J. I., Ghazarian, J. G., Orme-Johnson, N. R., and DeLuca, H. F., 1976, Isolation of chick renal mitochondrial ferredoxin active in the 25-hydroxyvitamin D_3-lα-Hydroxylase system, *J. Biol. Chem.* **251**:3933.

Pierides, A. M., Kerr, D. N. S., Ellis, H. A., Peart, K. M., O'Riordan, J. L. H., and DeLuca, H. F., 1976, lα-Hydroxycholecalciferol in hemodialysis renal osteodystrophy: Adverse effects of anticonvulsant therapy, *Clin. Nephrol.* **5**:189.

Pileggi, V. J., DeLuca, H. F., and Steenbock, H., 1955, The role of vitamin D in the prevention of rickets in rats on cereal diets, *Arch. Biochem. Biophys.* **58**:194.

Pileggi, V. J., DeLuca, H. F., Cramer, J. W., and Steenbock, H., 1956, Citrate in the prevention of rickets in rats, *Arch. Biochem. Biophys.* **60**:52.

Ponchon, G., and DeLuca, H. F., 1969, The role of the liver in the metabolism of vitamin D, *J. Clin. Invest.* **48**:1273.

Prader, A., Illig, R., and Heierli, E., 1961, Eine besondere Form der primären vitamin D-resistenten Rachitis mit Hypocalcämie und autosomaldominantem Erbgang: die hereditäre Pseudomangelrachitis, *Helv. Paediat. Acta* **16**:452.

Procsal, D. A., Okamura, W. H., and Norman, A. W., 1975, Structural requirements for the interaction of lα,25-(OH)$_2$-vitamin D_3 with its chick intestinal receptor system, *J. Biol. Chem.* **250**:8382.

Puschett, J. B., Beck, W. S., Jr., and Jelonek, A., 1975, Parathyroid hormone and 25-hydroxy vitamin D_3: Synergistic and antagonistic effects on renal phosphate transport, *Science* **190**:473.

Raisz, L. G., Trummel, C. L., Holick, M. F., and DeLuca, H. F., 1972, 1,25-Dihydroxycholecalciferol: A potent stimulator of bone resorption in tissue culture, *Science* **175**:768.

Rasmussen, H., DeLuca, H., Arnaud, C., Hawker, C., and von Stedingk, M., 1963, The relationship between vitamin D and parathyroid hormone, *J. Clin. Invest.* **42**:1940.

Reade, T. M., Scriver, C. R., Glorieux, F. H., Nogrady, B., Delvin, E., Poirier, R., Holick, M. F., and DeLuca, H. F., 1975, Response to crystalline lα-hydroxyvitamin D_3 in vitamin D dependency, *Pediatr. Res.* **9**:593.

Recommended Dietary Allowances, 8th ed., 1974, National Academcy of Sciences, Washington, D.C.

Reed, C. I., Struck, H. C., and Steck, I. E., 1939, *Vitamin D*, University of Chicago Press, Chicago.

Reynolds, J. J., Holick, M. F., and DeLuca, H. F., 1973, The role of vitamin D metabolites in bone resorption, *Calc. Tiss. Res.* **12**:295.

Ribovich, M. L., and DeLuca, H. F., 1975, The influence of dietary calcium and phosphorus on intestinal calcium transport in rats given vitamin D metabolites, *Arch. Biochem. Biophys.* **170**:529.

Ribovich, M. L., and DeLuca, H. F., 1976, Intestinal calcium transport: Parathyroid hormone and adaptation to dietary calcium, *Arch. Biochem. Biophys.* **175**:256.

Riggs, B. L., and Gallagher, J. C., 1977, Evidence for bihormonal deficiency state (estrogen and 1,25-dihydroxyvitamin D) in patients with post-menopausal osteoporosis, in: *Vitamin D: Biochemical, Chemical and Clinical Aspects Related to Calcium Metabolism* (A. W. Norman, K. Schaefer, J. W. Coburn, H. F. DeLuca, D. Fraser, H. G. Grigoleit, and D. von Herrath, eds.), pp. 639–648, de Gruyter, Berlin.

Robison, R., and Soames, K. M., 1924, The possible significance of hexose phosphoric esters in ossification. Part II. The phosphoric esterase of ossifying cartilage. *Biochem. J.* **18**:740.

Roborgh, J. R., and DeMan, T. J., 1960, The hypercalcemic activity of dihydrotachysterol and dihydrotachysterol₃ and of the vitamins D₂ and D₃ after intravenous injection of aqueous preparations, *Biochem. Pharmacol.* **3**:277.

Rosenstreich, S. J., Rich, C., and Volwiler, W., 1971, Deposition in and release of vitamin D₃ from body fat: Evidence for a storage site in the rat, *J. Clin. Invest.* **50**:679.

Russell, R. G., Smith, R., Walton, R. J., Preston, C., Basson, R., Henderson, R. G., and Norman, A. W., 1974, 1,25-Dihydroxyvitamin D₃ and 1α-hydroxyvitamin D₃ in hypoparathyroidism, *Lancet* **2**:14.

Sampson, H. W., Matthews, J. L., Martin, J. H., and Kunin, A. S., 1970, An electron microscopic localization of calcium in the small intestine of normal, rachitic, and vitamin D-treated rats, *Calc. Tiss. Res.* **5**:305.

Schachter, D., 1963, Vitamin D and the active transport of calcium by the small intestine, in: *The Transfer of Calcium and Strontium Across Biological Membranes* (R. H. Wasserman, ed.) pp. 197–210, Academic Press, New York.

Schachter, D., and Rosen, S. M., 1959, Active transport of Ca⁴⁵ by the small intestine and its dependence on vitamin D, *Am. J. Physiol.* **196**:357.

Schachter, D., Finkelstein, J. D., and Kowarski, S., 1964, Metabolism of vitamin D. I. Preparation of radioactive vitamin D and its intestinal absorption in the rat, *J. Clin. Invest.* **43**:787.

Schachter, D., Kowarski, S., Finkelstein, J. D., and Wang Ma, R., 1966, Tissue concentration differences during active transport of calcium by intestine, *Am. J. Physiol.* **211**:1131.

Schneider, L. E., Wilson, H. D., and Schedl, H. P., 1974, Effects of alloxan diabetes on duodenal calcium-binding protein in the rat, *Am. J. Physiol.* **227**:832.

Schneider, L. E., Omdahl, J., and Schedl, H. P., 1976, Effects of vitamin D and its metabolites on calcium transport in the diabetic rat, *Endocrinology* **99**:793.

Sebrell, W. H., Jr., and Harris, R. S. (eds.), 1954, Vitamin D group, in: *The Vitamins*, Vol. 2, pp. 131–266, Academic Press, New York.

Semmler, E. J., Holick, M. F., Schnoes, H. K., and DeLuca, H. F., 1972, The synthesis of 1α,25-dihydroxycholecalciferol—a metabolically active form of vitamin D₃, *Tetrahedron Lett.* **40**:4147.

Shipley, P. G., Kramer, B., and Howland, J., 1925, Calcification of rachitic bones *in vitro*, *Am. J. Dis. Child.* **30**:37.

Silverberg, D. S., Bettcher, K. B., Dossetor, J. B., Overton, T. R., Holick, M. F., and DeLuca, H. F., 1975, Effect of 1,25-dihydroxycholecalciferol in renal osteodystrophy, *Can. Med. Assoc. J.* **112**:190.

Smerdon, G. T., 1950, Daniel Whistler and the English disease: A translation and biographical note, *J. Hist. Med.* **5**:397.

Spencer, R., Charman, M., Wilson, P., and Lawson, E., 1976, Vitamin D-stimulated intestinal calcium absorption may not involve calcium binding protein directly, *Nature (London)* **263**:161.

Stamp, T. C. B., Round, J. M., Rowe, D. J. F., and Haddad, J. G., 1972, Plasma levels and therapeutic effect of 25-hydroxycholecalciferol in epileptic patients taking anticonvulsant drugs, *Br. Med. J.* **4**:9.

Stanbury, S. W., 1967, Bony complications of renal disease, in: *Renal Disease* (D. A. K. Black, ed.), pp. 665–713, Davis, Philadelphia.

Steele, T. H., and DeLuca, H. F., 1976, Influence of dietary phosphorus on renal phosphate reabsorption in the parathyroidectomized rat, *J. Clin. Invest.* **57**:867.

Steele, T. H., Engle, J. E., Tanaka, Y., Lorenc, R. S., Dudgeon, K. L., and DeLuca, H. F., 1975, Phosphatemic action of 1,25-dihydroxyvitamin D₃, *Am. J. Physiol.* **229**:489.

Steenbock, H., 1924, The induction of growth promoting and calcifying properties in a ration by exposure to light, *Science* **60**:224.

Steenbock, H., and Black, A., 1924, Fat-soluble vitamins. XVII. The induction of growth-promoting and calcifying properties in a ration by exposure to ultraviolet light, *J. Biol. Chem.* **61**:405.

Steenbock, H., and Black, A., 1925, Fat-soluble vitamins. XXIII. The induction of growth-promoting and calcifying properties in fats and their unsaponifiable constituents by exposure to light, *J. Biol. Chem.* **64**:263.

Steenbock, H., and Herting, D. C., 1955, Vitamin D and growth, *J. Nutr.* **57**:449.

Steenbock, H., Kletzien, S. W. F., and Halpin, J. G., 1932, The reaction of the chicken to irradiated ergosterol and irradiated yeast as contrasted with the natural vitamin D in fish liver oils, *J. Biol. Chem.* **97:**249.

Stern, P. H., Trummel, C. L., Schnoes, H. K., and DeLuca, H. F., 1975, Bone resorbing activity of vitamin D metabolites and congeners *in vitro:* Influence of hydroxyl substituents in the A ring, *Endocrinology* **97:**1552.

Stohs, S. J., Zull, J. E., and DeLuca, H. F., 1967, Vitamin D stimulation of [^{3}H]orotic acid incorporation into ribonucleic acid of rat intestinal mucosa, *Biochemistry* **6:**1304.

Suda, T., DeLuca, H. F., Schnoes, H. K., and Blunt, J. W., 1969, The isolation and identification of 25-hydroxyergocalciferol, *Biochemistry* **8:**3515.

Suda, T., DeLuca, H. F., Schnoes, H. K., Tanaka, Y., and Holick, M. F., 1970a, 25,26-Dihydroxycholecalciferol, a metabolite of vitamin D$_3$ with intestinal calcium transport activity, *Biochemistry* **9:**4776.

Suda, T., Hallick, R. B., DeLuca, H. F., and Schnoes, H. K., 1970b, 25-Hydroxydihydrotachysterol$_3$: Synthesis and biological activity, *Biochemistry* **9:**1651.

Tanaka, Y., and DeLuca, H. F., 1971, Bone mineral mobilization activity of 1,25-dihydroxycholecalciferol, a metabolite of vitamin D, *Arch. Biochem. Biophys.* **146:**574.

Tanaka, Y., and DeLuca, H. F., 1973, The control of 25-hydroxyvitamin D metabolism by inorganic phosphorus, *Arch. Biochem. Biophys.* **154:**566.

Tanaka, Y., and DeLuca, H. F., 1974a, Role of 1,25-dihydroxyvitamin D$_3$ in maintaining serum phosphorus and curing rickets, *Proc. Natl. Acad. Sci. USA* **71:**1040.

Tanaka, Y., and DeLuca, H. F., 1974b, Stimulation of 24,25-dihydroxyvitamin D$_3$ production by 1,25-dihydroxyvitamin D$_3$, *Science* **183:**1198.

Tanaka, Y., DeLuca, H. F., Omdahl, J., and Holick, M. F., 1971, Mechanism of action of 1,25-dihydroxycholecalciferol on intestinal calcium transport, *Proc. Natl. Acad. Sci. USA* **68:**1286.

Tanaka, Y., Frank, H., and DeLuca, H. F., 1973a, Biological activity of 1,25-dihydroxyvitamin D$_3$ in the rat, *Endocrinology* **92:**417.

Tanaka, Y., Frank, H., and DeLuca, H. F., 1973b, Intestinal calcium transport: Stimulation by low phosphorus diets, *Science* **181:**564.

Tanaka, Y., DeLuca, H. F., Ikekawa, N., Morisaka, M., and Koizumi, N., 1975a, Determination of stereochemical configuration of the 24-hydroxyl group of 24,25-dihydroxyvitamin D$_3$ and its biological importance, *Arch. Biochem. Biophys.* **170:**620.

Tanaka, Y., Lorenc, R. S., and DeLuca, H. F., 1975b, The role of 1,25-dihydroxyvitamin D$_3$ and parathyroid hormone in the regulation of chick renal 25-hydroxyvitamin D$_3$-24-hydroxylase, *Arch. Biochem. Biophys.* **171:**521.

Tanaka, Y., DeLuca, H. F., Akaiwa, A., Morisaki, M., and Ikekawa, N., 1976a, Synthesis of 24S and 24R-hydroxy-[24-^{3}H]vitamin D$_3$ and their metabolism in rachitic rats, *Arch. Biochem. Biophys.* **177:**615.

Tanaka, Y., Castillo, L., and DeLuca, H. F., 1976b, Control of the renal vitamin D hydroxylases in birds by the sex hormones, *Proc. Natl. Acad. Sci. USA* **73:**2701.

Tanaka, Y., Castillo, L., DeLuca, H. F., and Ikekawa, N., 1977, The 24-hydroxylation of 1,25-dihydroxyvitamin D$_3$, *J. Biol. Chem.* **252:**1421.

Taylor, A. N., 1974, *In vitro* phosphate transport in chick ileum: Effect of cholecalciferol, calcium, sodium and metabolic inhibitors, *J. Nutr.* **104:**489.

Teitelbaum, S. L., Bone, J. M., Stein, P. M., Gilden, J. J., Bates, M., Boisseau, V. C., and Avioli, L. V., 1976, Calcifediol in chronic renal insufficiency, *J. Am. Med. Assoc.* **235:**164.

Trinh-Toan, Ryan, R. C., Simon, G. L., Calabrese, J. C., Dahl, L. F., and DeLuca, H. F., 1977, Crystal structure of 25-hydroxyvitamin D$_3$ monohydrate: A stereochemical analysis of vitamin D molecules, *J. Chem. Soc. Perkin Trans.* **2:**393.

Trummel, C. L., Raisz, L. G., Blunt, J. W., and DeLuca, H. F., 1969, 25-Hydroxycholecalciferol: Stimulation of bone resorption in tissue culture, *Science* **163:**1450.

Tsai, H. C., Wong, R. G., and Norman, A. W., 1972, Studies on calciferol metabolism. IV. Subcellular localization of 1,25-dihydroxy-vitamin D$_3$ in intestinal mucosa and correlation with increased calcium transport, *J. Biol. Chem.* **247:**5511.

Velluz, L., and Amiard, G., 1949a, Chimie organique-le précalciférol, *Compt. Rend.* **228:**692.

Velluz, L., and Amiard, G., 1949b, Chimie organique-équilibre de reaction entre précalciférol et calciférol, *Compt. Rend.* **228:**853.

Velluz, L., and Amiard, G., 1949c, Chimie organique-nouveau précurseur de la vitamine D_3, *Compt. Rend.* **228**:1037.

Velluz, L., Amiard, G., and Petit, A., 1949, Le précalciférol: Ses relations d'équilibre avec le calciférol, *Bull. Soc. Chim. France* **16**:501.

Vida, J. A., 1971, II. Chemistry, in: *The Vitamins*, 2nd ed. (W. H. Sebrell, Jr., and R. S. Harris, eds.), pp. 180–203, Academic Press, New York.

von Werder, F., 1939, Über Dihydro-tachysterin, *Hoppe-Seyler's Z. Physiol. Chem.* **260**:119.

Waddell, J., 1934, The provitamin D of cholesterol. I. The antirachitic efficacy of irradiated cholesterol, *J. Biol. Chem.* **105**:711.

Walling, M. W., 1977, Effects of 1,25-dihydroxyvitamin D_3 on active intestinal inorganic phosphate absorption, in: *Vitamin D: Biochemical, Chemical and Clinical Aspects Related to Calcium Metabolism* (A. W. Norman, K. Schaefer, J. W. Coburn, H. F. DeLuca, D. Fraser, H. G. Grigoleit, and D. von Herrath, eds.), pp. 321–330, de Gruyter, Berlin.

Wasserman, R. H., 1963, *The Transfer of Calcium and Strontium Across Biological Membranes*, Academic Press, New York.

Wasserman, R. H., 1975, Active vitamin D-like substances in solanum malacoxylon and other calcinogenic plants, *Nutr. Rev.* **33**:1.

Wasserman, R. H., and Taylor, A. N., 1968, Vitamin D-dependent calcium-binding protein: Response to some physiological and nutritional variables, *J. Biol. Chem.* **243**:3987.

Wasserman, R. H., and Taylor, A. N., 1973, Intestinal absorption of phosphate in the chick: Effect of vitamin D_3 and other parameters, *J. Nutr.* **103**:586.

Wasserman, R. H., Kallfelz, F. A., and Comar, C. L., 1961, Active transport of calcium by rat duodenum *in vivo*, *Science* **133**:883.

Wasserman, R. H., Corradino, R. A., and Taylor, A. N., 1968, Vitamin D-dependent calcium-binding protein: Purification and some properties, *J. Biol. Chem.* **243**:3978.

Wasserman, R. H., Henion, J. D., Haussler, M. R., and McCain, T. A., 1976, Calcinogenic factor in *Solanum malacoxylon:* Evidence that it is 1,25-dihydroxyvitamin D_3-glycoside, *Science* **194**:853.

Weber, J. C., Pons, V., and Kodicek, E., 1971, The localization of 1,25-dihydroxycholecalciferol in bone cell nuclei of rachitic chicks, *Biochem. J.* **125**:147.

Westerhof, P., and Keverling Buisman, J. A., 1956, Investigations on sterols. VI. The preparation of dihydrotachysterol, *Rec. Trav. Chem.* **75**:453.

Williams, T. F., Winter, R. W., and Burnett, C. H., 1966, Familial (hereditary) vitamin D-resistant rickets with hypophosphatemia, in: *The Metabolic Basis of Inherited Disease*, 2nd ed. (J. B. Stanbury, J. B. Wyngaarden, and D. S. Fredrickson, eds.), pp. 1179–1204, McGraw-Hill, New York.

Windaus, A., and Trautman, G., 1937, Crystalline vitamin D_4, *Hoppe-Seyler's Z. Physiol. Chem.* **247**:185.

Windaus, A., Linsert, O., Lüttringhaus, A., and Weidlich, G., 1932, Crystalline-vitamin D_2, *Ann.* **492**:226.

Windaus, A., Lettre, H., and Schenck, F., 1935, 7-Dehydrocholesterol, *Ann.* **520**:98.

Windaus, A., Schenck, F., and von Werder, F., 1936, Über das antirachitisch wirksame Bestrahlungsprodukt aus 7-dehydro-Cholesterin, *Hoppe-Seyler's Z. Physiol. Chem.* **241**:100.

Winters, R. W., Graham, J. B., Williams, T. F., Falls, V. W., and Burnett, C. W., 1958, A genetic study of familial hypophosphatemia and vitamin D-resistant rickets with a review of the literature, *Medicine* **37**:97.

Zerwekh, J. E., Lindell, T. J., and Haussler, M. R., 1976, Increased intestinal chromatin template activity: Influence of 1α,25-dihydroxyvitamin D_3 and hormone receptor complexes: *J. Biol. Chem.* **251**:2388.

Chapter 3

Vitamin E

Milton Leonard Scott

3.1. Introduction

"Vitamin E" is the generic term for a group of lipid-soluble tocol and tocotrienol derivatives possessing varying degrees of vitamin activity, the most active of these compounds being α-tocopherol. The tocopherols and tocotrienols are produced in green plants. The tocopherols are stored in highest concentration in the lipid portion of the germ plasm while the tocotrienols occur primarily in the outer coating of the seed. An essential role for vitamin E in seed germination, photosynthetic processes, or other functions of plant growth has not been established.

In animals, vitamin E, mainly as α-tocopherol, is concentrated in the phospholipids of mitochondria, endoplasmic reticulum, and plasma membranes of cells.

Vitamin E is nature's best fat-soluble antioxidant. Whether its function in plants and animals lies solely in its antioxidant properties or is mediated through some entirely different mechanism still remains to be elucidated.

Under certain circumstances, vitamin E appears to be critically required for maintenance of the health and integrity of every cell of the animal body; under other circumstances, no dietary need for vitamin E can be shown.

Many factors markedly alter the vitamin E requirements of animals. Prooxidants such as oxidizing agents, ozone, peroxidizing polyunsaturated fatty acids, and catalytic trace elements dramatically increase the dietary requirement for vitamin E. In the opposite direction, addition to the diet of synthetic fat-soluble antioxidants, selenium, and sulfur amino acids may appear to completely obviate a dietary requirement for vitamin E. Since those factors which tend to increase peroxidation in the body cause an increased dietary requirement for vitamin E, and those factors which decrease lipid peroxidation or destroy peroxides markedly reduce the vitamin E requirement, it has been almost impossible to establish a recommended dietary allowance for this vitamin that is applicable to the wide variety of dietary conditions existing throughout the world.

Over the more than 50 years since vitamin E was discovered, research on the

Milton Leonard Scott • Department of Poultry Science and Division of Nutritional Sciences, Cornell University, Ithaca, New York 14853.

many interrelationships existing between this vitamin and other dietary factors has led to such confusion that a number of researchers and clinicians have tended to relegate vitamin E to a position of little importance.

Recent research on the interrelationships of vitamin E and synthetic antioxidants, selenium, and sulfur amino acids has led to a much better understanding of the separate roles of each of these dietary factors and has provided a means of assessing the dietary vitamin E requirement necessary for its optimum functioning at the subcellular level. This work has demonstrated a fundamental role of vitamin E of great importance in preserving the health and integrity of the vital cells of the body.

A multiplicity of symptoms occur in animals suffering from vitamin E deficiency. It is clear from studies of these deficiency diseases that vitamin E is involved in maintaining the health of the brain, the vascular system, the erythrocytes, the skeletal muscles, the liver, the heart, and the gonads and in preventing yellow fat and ceroid deposition in adipose and other tissues.

Some but not all of these deficiency diseases are prevented by synthetic antioxidants, some are prevented by either vitamin E or selenium, and still others are prevented by either vitamin E or sulfur amino acids. It is apparent that vitamin E functions in at least two different metabolic roles: (1) as a fat-soluble antioxidant and (2) in one or more specific roles interrelated with the metabolism of selenium and sulfur amino acids.

This chapter will describe the chemistry and biochemistry of vitamin E, current knowledge concerning nutritional requirements for this vitamin, and its role in prevention of deficiency diseases.

3.2. History

The year was 1922. Remarkable progress had been made in the knowledge of nutrition during the previous decade. The century-old nutritional factors—the saccharine principle, the oily principle, and the albuminous principle proposed by William Prout (1827/1828)—were now known as carbohydrates, fats, and proteins. Studies of Osborne and Mendel (1914) at Yale had shown that certain of the protein constituent amino acids were essential while others apparently could be synthesized in the animal body.

Eijkman (1897) demonstrated the nutritional basis of the disease beriberi, while Grijns (1901) concluded that an essential nutrient in rice hulls prevented this disease. Both Funk (1912) and Hopkins (1906) postulated the existence of essential nutrients, while Funk must be credited with coining the term "vitamins." In 1913, independent investigations of McCollum and Davis and of Osborne and Mendel had demonstrated the existence in certain fats of a required fat-soluble vitamin which McCollum termed "fat-soluble vitamin A," and thus Funk's "water-soluble vitamine" became known as "water-soluble vitamin B." Guinea pig studies by Holst and Frølich during the early 1900s had led to the recognition of vitamin C as the antiscorbutic vitamin. McCollum and co-workers had just separated the activities of vitamin A and the antirachitic factor by bubbling oxygen through warm cod liver oil for several hours. The resulting oil lost its effectiveness for curing vitamin A deficiency but still possessed antirachitic activity. The antirachitic vitamin later was named "vitamin D" by McCollum (1925).

3.2.1. Discovery of Vitamin E

It was in this exciting era that H. M. Evans of the University of California undertook studies "on the relations between fertility and nutrition." His co-worker in these studies was Katherine S. Bishop. In a series of papers (Evans and Bishop, 1922), it was reported that reproduction failed in pregnant female rats fed diets containing all of the then known nutrients. Although estrus, mating, and every detectable phase of the beginning of a normal pregnancy had ensued, the fetuses soon died and were resorbed unless the diet was supplemented with small amounts of fresh lettuce, wheat germ, or dried alfalfa leaves. Evans and Bishop showed that the unknown fat-soluble reproductive factor was none of the known nutrients. Daily single drops of wheat germ oil completely prevented this gestation resorption, whereas the well-known rich source of vitamins A and D, cod liver oil, not only failed to prevent the disease but also appeared to increase its severity.

By 1924, Barnett Sure of the University of Arkansas, in independent studies, concluded that this fat-soluble reproductive factor was a new vitamin, and he named it "vitamin E."

3.2.2. A Variety of Vitamin E Deficiency Diseases in Different Species

Throughout the 1920s, vitamin E was recognized only as an unidentified fat-soluble factor required for reproduction in rats. Vitamin E was found to be required not only for normal gestation in female rats but also for prevention of sterility in male rats (Mason, 1933) and chickens (Adamstone and Card, 1934). After the classical experiments of Pappenheimer and Goettsch (1931) and Goettsch and Pappenheimer (1931) showing that vitamin E is required also for prevention of encephalomalacia in chicks and nutritional muscular dystrophy in rabbits and guinea pigs, studies were made with a wide variety of experimental animals which in time led to recognition of vitamin E deficiency as the cause of many different pathological conditions.

The report that vitamin E prevented a nutritional muscular dystrophy in animals was heralded with much hope and expectation among clinicians working with incurable muscular dystrophy diseases in humans, such as progressive muscular dystrophy, myasthenia gravis, and Duchenne's disease. Imagine their disappointment and frustration when it was found that dietary vitamin E, even at high levels, had no significant, lasting effect on the course of these human muscular dystrophies (Milhorat, 1954; Berneske *et al.*, 1960)!

Nutritional muscular dystrophy was observed in the striated muscles of vitamin E deficient rats (Olcott, 1938; Pappenheimer, 1940) and ducklings (Pappenheimer and Goettsch, 1933), and in the gizzard musculature of turkeys (Jungherr and Pappenheimer, 1937). Nutritional muscular dystrophy was found in chicks only when the diet was deficient in both vitamin E and sulfur amino acids (Dam *et al.*, 1952). Muscular dystrophy had been described in lambs by Metzger and Hagen, who reported in the spring of 1927 that many lambs born in upstate New

York suffered from degeneration of the skeletal muscles, followed by an increase in the sarcolemma and connective tissue. They believed it to be due to a nutritional deficiency and called it a "stiff lamb" disease of unknown etiology. Whiting *et al.* (1949) reported that this disease was due to a vitamin E deficiency.

Muscular dystrophy was produced experimentally in sheep (Madsen *et al.*, 1935) and goats (Davis *et al.*, 1938), but only when the diets naturally low in vitamin E contained considerable amounts of cod liver oil or other fish oils. Although iron treatment of diets, which destroyed all measurable vitamin E, produced muscular dystrophy in guinea pigs and rabbits, it produced no muscle lesions or reproductive disturbances in goats (Thomas and Cannon, 1937).

These results, impossible to understand in the late 1930s, are more readily explainable today. Selenium is now known to prevent nutritional muscular dystrophy in goats and lambs but not in rabbits or guinea pigs, while vitamin E is effective for rabbits and guinea pigs and has little or no effect on goats or sheep in the absence of an adequate level of dietary selenium.

In 1939, Dam and Glavind reported that chicks receiving certain types of vitamin E deficient diets showed no signs of encephalomalacia but instead developed a severe edematous condition characterized by accumulation of a plasmalike fluid in the subcutaneous tissues. This disorder was named "exudative diathesis."

Dam and Granados in 1945 showed the presence of peroxides in adipose tissues of rats and chicks fed diets deficient in vitamin E and rich in unsaturated fatty acids. This condition was characterized by a yellow-brown discoloration of the adipose tissue. It later was found that steatitis, or yellow fat disease, due to vitamin E deficiency was responsible for serious problems in vitamin E deficient minks (Stowe and Whitehair, 1963) and pigs (Robinson and Coey, 1951). The development of ceroid pigments in vitamin E deficient animals and its possible relationship to lipofuscin age pigments are subjects of intense current interest and investigation (Tappel, 1970, 1975; Pryor, 1971).

Prevention of erythrocyte hemolysis by vitamin E was discovered by György and Rose in 1949. These workers found that erythrocytes from vitamin E deficient rats were hemolyzed by hydrogen peroxide solutions or by dialuric acid, whereas those of rats receiving vitamin E were very resistant to hemolysis. The peroxidative hemolysis test has been used widely as an indicator of vitamin E status.

3.2.3. Characterization, Identification, and Synthesis of Vitamin E

In 1927, Evans *et al.* reported that a potent concentrate of vitamin E could be obtained by saponification of wheat germ oil and removal of the sterols from the unsaponifiable fraction which contained vitamin E. Olcott and Mattill (1931) concentrated the lipids of lettuce, isolated a fraction having high vitamin E activity, and showed that this fraction had excellent antioxidant properties. In 1936, Evans *et al.* prepared three allophanic acid esters of three alcohols which they had isolated from the nonsaponifiable fraction of wheat germ oil. One derivative was found to have very high vitamin E activity. In 1938 Erhard Fernholz undertook thermal degradation of the vitamin, showed that it contained a phytyl chain and a hydroquinone moiety, and suggested a structure of the vitamin later proven to be true.

Following its isolation in pure form, Evans felt that the new vitamin needed a name descriptive of its metabolic function. For this task he enlisted the help of George M. Calhoun, Professor of Greek at the University of California. Since the vitamin permits an animal to bear offspring, Professor Calhoun suggested the name *tokos*, which in Greek means "offspring," followed by the verb *pherein*, the Greek term for "to bear." They decided to add the ending *ol* denoting that the compound was an alcohol (Evans, 1962). Thus Evans *et al.* (1936) named their three compounds "α-, β-, and γ-tocopherols."

α-Tocopherol was synthesized by Karrer *et al.* in 1938, and by Smith *et al.* later the same year. Bergel *et al.* (1938) also synthesized α-tocopherol and three dimethyltocopherols by condensing phytol with the corresponding hydroquinones. Walter John (1937) paved the way for this work by purifying and determining the structure of β-tocopherol. δ-Tocotrienol, the last of the eight natural vitamin E compounds, was isolated and chemically characterized in 1966 by Whittle *et al.* of the University of Liverpool.

The history of vitamin E has been reviewed by Mason (1944), Dam (1957), McCollum (1957), Evans (1962), and Morton (1968). Historically, our knowledge of vitamin E developed in quite distinct phases.

During the 1920s attention was devoted chiefly to studies of the effects of vitamin E deficiency on resorption gestation in the female and testicular degeneration in the male rat which produced sterility. Interest was also concentrated on the relation of vitamin E deficiency to late lactation paralysis in young animals and to growth retardation in rats. Based on the ability of the vitamin to prevent fetal resorption in female rats, a bioassay for vitamin E was developed and the distribution of vitamin E was determined in a wide variety of plant and animal tissues. Potent concentrates of the vitamin were prepared from wheat germ oil and other vegetable oils. The report of Pappenheimer and Goettsch in 1931 that nutritional muscular dystrophy occurred in rabbits and guinea pigs fed vitamin E deficient diets stimulated numerous studies with a wide variety of animals. Vitamin E deficiency produced many different signs and symptoms in the various animal species, and in a single species, the chick, three distinct vitamin E deficiency diseases were found to occur: exudative diathesis, encephalomalacia, and muscular dystrophy.

3.2.4. Vitamin E Interrelationships with Other Factors

These observations raised many questions. Research on vitamin E was in a state of total confusion. In 1944, Dam reported that the symptoms in the chick could be enhanced or suppressed by dietary changes unrelated to the vitamin E content of the diet. For example, he showed that when diets low in vitamin E were fed to chicks in the presence of cod liver oil, lard, or linseed oil, exudative diathesis occurred as the main symptom, whereas the same diets in which fatty acids from hog liver replaced cod liver oil favored the production of encephalomalacia. Vitamin E corrected both symptoms. However, he also found that inositol counteracted both symptoms. Lipocaic acid prevented exudative diathesis but not encephalomalacia. Cholesterol hastened exudative diathesis when the diet contained 5% cod liver oil and a low level of salt, but prevented encephalomalacia when the diet contained 30% lard.

3.2.5. Research That Brought Some Order Out of the Confusion

In 1953, Singsen *et al.* found that the addition of the synthetic antioxidant diphenylparaphenylene diamine (DPPD) to a diet low in vitamin E and containing 2% of cod liver oil prevented encephalomalacia in chicks as effectively as vitamin E.

During the early 1950s Klaus Schwarz of the National Institutes of Health, studying dietary liver necrosis in rats receiving a diet low in vitamin E, showed that dried brewers yeast, which contains no vitamin E, was as effective as vitamin E in preventing necrosis. He believed that brewers yeast and certain other natural materials contained an unknown water-soluble factor, in no way chemically related to vitamin E, which was able to replace vitamin E in metabolism, at least insofar as the prevention of hepatic necrosis was concerned. Schwarz called this unknown substance "Factor 3" (Schwarz, 1951). In the meantime, in the author's laboratory it was found that dried brewers yeast would replace vitamin E in the prevention of a severe leg disorder in young turkeys produced experimentally by the addition of cod liver oil to the diet (Scott, 1953). In view of Schwarz's work on Factor 3, research was initiated aimed at characterizing the separate effects of vitamin E and the factor in dried brewers yeast. Chicks fed a basal diet showed a 100% incidence of exudative diathesis and died within approximately 3 weeks. Addition of either vitamin E or brewers yeast completely prevented exudative diathesis and promoted good growth (Scott *et al.*, 1955).

Following the discovery by Schwarz and Foltz (1957) that potent concentrates of Factor 3 contained selenium, which was far more effective than vitamin E in preventing dietary liver necrosis, studies by Scott *et al.* (1957) and Stokstad *et al.* (1957) showed that selenium also is as effective as vitamin E in preventing exudative diathesis in chicks.

These results brought researchers to the realization that vitamin E functions in at least two metabolic roles: (1) as a fat-soluble antioxidant and (2) in a more specific role interrelated with the metabolism of selenium and sulfur amino acids.

Noguchi *et al.* (1973) and Scott *et al.* (1974) provided evidence concerning the mechanisms of action of both vitamin E and selenium in the prevention of exudative diathesis in chicks. These studies demonstrated that the metabolic actions of vitamin E and selenium are completely dissimilar. In the prevention of exudative diathesis and of *in vitro* peroxidative damage to hepatic microsomes, vitamin E appears to act in a mechanism differing from the "antioxidant" effect which is involved in prevention of encephalomalacia in chickens. The Cornell studies confirmed and extended the original discovery by Rotruck *et al.* (1972, 1973) that one important function of selenium is its role as an integral part of glutathione peroxidase. This selenoenzyme protects cellular and subcellular membranes from peroxidative damage, apparently by destroying fatty acid hydroperoxides before they can undergo chain reactions and cause malfunctions in the membranes. In the chick, plasma glutathione peroxidase appears to be particularly important for protection of the plasma membranes of the capillary cells, thereby preventing exudative diathesis. These studies provide evidence in confirmation of the hypothesis of Tappel (1972) that vitamin E reacts or functions as a chain-breaking antioxidant, neutralizing free radicals and preventing peroxidation of lipids within the membranes.

Thus it now appears that vitamin E in cellular and subcellular membranes is the first line of defense against peroxidation of vital phospholipids. Even with adequate vitamin E, however, some peroxides are formed. Selenium, in glutathione peroxidase, is a second line of defense which destroys these peroxides before they have an opportunity to cause damage to the membranes.

3.3. Chemistry of Vitamin E

α-Tocopherol (Fig. 1), the most active form of vitamin E, is a 6-hydroxychroman derivative with methyl groups in positions 2, 5, 7, and 8 and a phytyl side chain

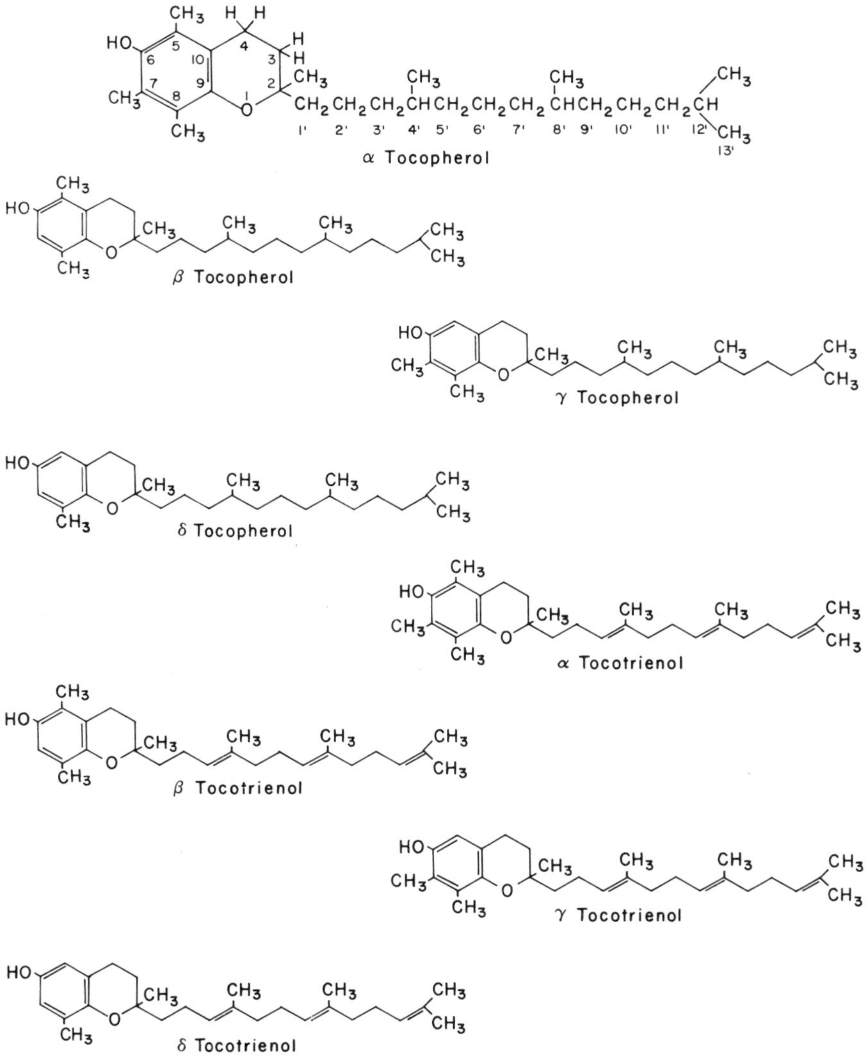

Fig. 1. Chemical structures of naturally occurring tocopherols and tocotrienols.

attached at carbon 2. Eight forms of vitamin E are known to exist in nature: α-, β-, γ-, and δ-tocopherols contain saturated phytol side chains, while α-, β-, γ-, and δ-tocotrienols possess three double bonds in the side chain (Pennock *et al.*, 1964). According to Morton (1968), it appears that δ-tocotrienol is the earliest member of the group to be formed by biosynthetic processes in plants. Methylation in the plant leads to the successive formation of γ-, β-, and α-trienols; hydrogenation of these produces the respective tocopherols. The tocopherols possess three asymmetrical carbon atoms located at C-2, C-4', and C-8' (Fig. 1).

Using the specifications for asymmetrical configuration proposed by Cahn *et al.* (1956), natural α-tocopherol, termed "*d*-α-tocopherol," may be described chemically as 2*R*-(4'*R*,8'*R*)-5,7,8-trimethyltocol, where the term "tocol" is the accepted name for the two-ring structure basic to all vitamin E compounds. Thus natural β-tocopherol is 2*R*-(4'*R*,8'*R*)-5,8-dimethyltocol, γ-tocopherol is 2*R*-(4'*R*,8'*R*)-7,8-dimethyltocol, and δ-tocopherol is 2*R*-(4'*R*,8'*R*)-8-methyltocol. The α-, β-, γ-, and δ-tocotrienols possess methyl groups in the same positions present in the corresponding tocopherols.

3.3.1. Chemical Synthesis of dl-α-Tocopherol

The first total synthesis of *dl*-α-tocopherol or (2*RS*-(4'*R*,8'*R*)-5,7,8-trimethyltocol was achieved by condensation of trimethyl hydroquinone with phytyl bromide (Karrer *et al.*, 1938). A similar synthesis was performed by Smith *et al.* (1938). Others showed that phytyl bromide could be replaced by natural phytol (Karrer and Isler, 1938; Bergel *et al*, 1938; Smith *et al*, 1939*b*; Fieser *et al.*, 1940) or by isophytol (Karrer and Isler, 1941).

Isler *et al.* (1962) described the preparation of 2-*l*-α-tocopherol [2*S*-(4'*R*,8'*R*)-5,7,8-trimethyltocol] by treatment of natural *d*-α-tocopherol with ferric chloride to produce α-tocopheryl quinone. Recyclization of the quinone in the presence of zinc chloride produced 75% *l*-, 25% *d*-α-tocopherol. The composition of this mixture was determined by measurement of the rotatory power of the potassium ferricyanide oxidation product according to the method of Nelan and Robeson (1962). The *p*-phenylazobenzoate of the *l*-α-tocopherol epimer was recrystallized, saponified, and chromatographed, yielding pure *l*-α-tocopherol (Robeson and Nelan, 1962).

Two principal sources of vitamin E are in commercial use: *d*-α-tocopherol and *dl*-α-tocopherol, plus acetate and succinate esters of these compounds. The acetate esters are prepared chemically by reaction of the alcohol forms with acetic anhydride; they do not exist in nature. *d*-α-Tocopherol is largely obtained from natural sources by molecular distillation. However, some of the *d*-α-tocopherol is prepared by further methylation of β-, γ-, and δ-tocopherols, or by hydrogenation of α-tocotrienol.

Synthetic *dl*-α-tocopherol and its esters are prepared from isophytol. This synthesis yields a mixture of eight isomers. Although the synthetic *dl*-α-tocopherol racemic only at carbon 2 has been prepared from natural phytol, and samples of this compound may be available for research, it is not a commercial source of vitamin E.

Details concerning these syntheses and those of many other tocopherol deriva-

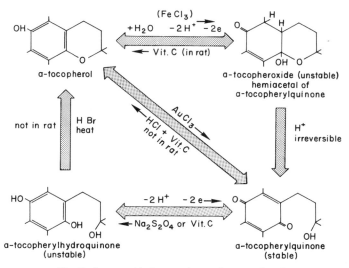

Fig. 2. Some oxidation products of α-tocopherol.

tives are presented by Isler *et al.* (1962). Commercial production of vitamin E is described by Rubel (1969).

The eight isomers resulting from the synthesis of *dl*-α-tocopherol prepared from isophytol have been separated into two components of four isomers each (Mayer *et al.*, 1963; Mayer and Isler, 1971). One mixture of four isomers of α-tocopherol contains $2R,4'RS,8'RS$-α-tocopherol (termed the *d*-epimers of tocopherol) while the other mixture contains $2S,4'RS,8'RS$- and represents the *l*-epimers of synthetic α-tocopherol. No *l*-isomers are commercially available.

Liebermann's nitroso reaction is a characteristic reaction of phenols in which either the *o*- or *p*- position is unsubstituted; only α-tocopherol and α-tocotrienol have substitutions at the *o*- or *p*-position. Thus all tocols except α-tocopherol and α-tocotrienol react with nitrous acid to form nitroso derivatives that are yellow. The difference between the total tocopherols as measured by the FeC1$_3$ method and the non-α forms as determined by the nitroso reaction gives the α-tocopherol plus α-tocotrienol.

3.3.2. Metabolic Degradation Products of Tocopherol

Some oxidation products of α-tocopherol are shown in Fig. 2. Of these, only α-tocopheroxide has been shown to have vitamin E activity, and this compound is quickly and irreversibly converted to inactive α-tocopherol quinone.

Csallany *et al.* (1962) reported isolation of tocopherol-*p*-quinone from liver of rats fed *d*-α-[^{14}C]tocopherol. Csallany and Draper (1963) also reported the existence of a dimeric metabolite of *d*-α-tocopherol in rat liver. The structure of this compound was established by Csallany (1971).

Simon *et al.* (1955) isolated a glucuronic acid derivative from the urine of humans ingesting large quantities of α-tocopherol. They tentatively identified the metabolite as the lactone of 2-(3-hydroxy-3-methyl-5-carboxypentyl)-3-5-6-

trimethylbenzoquinone. Schwarz *et al.* (1958) reported that this compound, both as the lactone and as the free acid, prevented respiratory decline (i.e., a failure of respiration of liver slices of vitamin E deficient rats), whereas *dl*-α-tocopherol had no significant influence on respiration under similar conditions. Since ubiquinone is now known to be required for normal activity of the respiratory chain of enzymes, it appears possible that this vitamin E metabolite may be capable of substituting, in part, for ubiquinone.

Skinner and Parkhurst (1971) have reviewed the reaction products of tocopherol, beginning with the early work of Evans *et al.* (1927), who showed that the biological activity of vitamin E was destroyed by bromination but not by hydrogenation. Smith *et al.* (1939*a*) showed that the red oxidation product produced by treatment of *d*-α-tocopherol with silver nitrate was a red orthoquinone which he termed "α-tocored." John and Emte (1941) produced a hydroxy-*p*-quinone by oxidation of α-tocopherol with silver nitrate in boiling ethanol. This compound, purple in base and yellow in acid, was termed "tocopurple" and also could be produced by ferric chloride oxidation or by treatment of tocored with hydrochloric acid. Frampton *et al.* (1954) produced tocored by oxidation of α-tocopherol with ferric chloride in methanol at 50°C. The possibility of a free-radical-initiated oxidation of vitamin E in the body led Inglett and Mattill (1955) to study the effect of oxidation of α- and γ-tocopherols with benzoyl peroxide at 30°C. A rapid oxidation of α-tocopherol occurred at this temperature, yielding α-tocopheryl quinone. Using azobisisobutyrolnitrile as the free radical initiator, Skinner (1964) found no α-tocopherol quinone, but instead identified the dihydroxy dimer of α-tocopherol as the main oxidation product. Skinner and Parkhurst (1971) also discussed the reduction products, halogenation reactions, phosphate derivatives, and alkylation leading to several derivatives of tocopherols substituted at the 5-methyl group. The *in vitro* antioxidant activities of vitamin E and numerous derivatives substituted in the 5-position and/or for the phytyl side chain were compared with their relative antihemolytic activities by Skinner *et al.* (1971) using the *in vitro* hemolysis test based on addition of dialuric acid to erythrocytes from vitamin E deficient rats. All antioxidants were not equally effective in preventing the dialuric acid induced hemolysis. None of the compounds evaluated was as effective as α-tocopherol. The most active of the nontocopherol antioxidants were pyrogallol, *N*,*N*′-diphenyl-*p*-diamine (DPPD), butylated hydroxytoluene (BHT), 1,2-dihydroxy-6-ethoxy-2,2,4-trimethylquinoline (ethoxyquin), and 4,4′-thiobis (6-*tert*-butyl-*ortho*-cresol), as well as the 5-hydroxymethyl and 5-methoxymethyl derivatives of tocopherol. Orthoquinone and 5-aldehyde were better antihemolytic agents than the antioxidant results indicated. Gloor *et al.* (1966) found α-tocopheramine (6-amino-5,7,8-trimethyltocol) to be as active as α-tocopherol in the *in vitro* erythrocyte hemolysis test, while γ-tocopherol had only 10% of the activity of α-tocopherol. These activities correlate well with the biological activities of the compounds in prevention of other deficiency diseases (Bieri, 1969). However, biological activity and *in vitro* antihemolytic activity do not always correlate since not all tocopherol derivatives are equally well absorbed and retained in the animal body.

Ascorbic acid, an excellent water-soluble antioxidant in certain systems, is inactive in preventing hemolysis.

3.3.3. Nomenclature

Since some important vitamin E preparations are mixtures of diastereoisomers of α-tocopherol, this has presented great difficulties to the committees on nomenclature of the International Union of Pure and Applied Chemists (IUPAC), the International Union of Nutrition Sciences (IUNS), and the American Institute of Nutrition (AIN). The Commission on Biochemical Nomenclature (CPN) in 1972 attempted to prepare a system of designation of the various stereoisomeric tocopherols and their mixtures. However, complete agreement on nomenclature could not be arrived at by the various committees of AIN, IUNS, and IUPAC. Pending further definite recommendations and adoption of revised nomenclature by AIN, therefore, the *Journal of Nutrition* [Vol. 105(1), 1975] proposed nomenclature consistent with current usage and with most recommendations of the above committees.

The term "vitamin E" should be used as the generic descriptor for all toco- and tocotrienol derivatives qualitatively exhibiting the biological activity of *d*-α-tocopherol. Thus phrases such as "vitamin E activity," "vitamin E deficiency," and "vitamin E in the form of . . ." represent preferred usage.

The term "tocopherol" should be used as the generic descriptor for all methyl tocols. Since the tocotrienols also possess vitamin E activity, "tocopherol" is not synonymous with "vitamin E." Esters of the tocopherols and tocotrienols should be designated "tocopheryl esters" and "tocotrienyl esters," respectively. Table I presents currently accepted nomenclature for the various forms of vitamin E.

A compound from tobacco, solanochromene, is related to *d*-α-tocotrienol but contains eight unsaturated isoprene units in the side chain and a double bond between carbons 3 and 4 in the oxygen-containing ring. (It is the latter double bond that distinguishes chromenes from chromans.)

3.3.4. Properties of the Vitamins E

Both *d*-α-tocopherol and *dl*-α-tocopherol are practically insoluble in water but are almost completely soluble in oils, fats, acetone, alcohol, chloroform, ether, benzene, and other fat solvents. All tocopherols are stable to heat and alkalis in the absence of oxygen and are unaffected by acids up to $100°C$; they are slowly oxidized by atmospheric oxygen, a process which is rapidly increased by heat and catalyzed by ferric or silver salts. On exposure to light, the tocopherols gradually darken. They are not precipitated by digitonin. The commercial forms of vitamin E are *d*-α-tocopheryl acetate and *dl*-α-tocopheryl acetate.

Pure *d*-α-tocopherol is a viscous, pale yellow liquid; molecular weight 430.69; melting point $2.5–3.5°C$; boiling point at 0.1 atm (used in molecular distillation) $200–220°C$; density 0.95 at $25°C$ in reference to water at $4°C$; specific rotation $[\alpha]_{5461}^{25} = 3°$ (benzene), $+0.32°$ (alcohol); absorption maximum 294 nm; $E_{1\,cm}^{1\%} = 71$. The refractive index in sodium light spectrum at $20°C$ is 1.5045.

β-Tocopherol is a viscous, pale yellow oil; absorption maximum 297 nm; $E_{1\,cm}^{1\%} = 86.4$.

The absorption maximum for γ-tocopherol is 298 nm; $E_{1\,cm}^{1\%} = 92.8$; boiling

Table I. Nomenclature of the Epimers and Isomers of Vitamin E

Trivial names	Chemical names	Biopotencies (IU/mg)	Sources
d-α-Tocopherol	2R-(4'R,8'R)-5,7,8-Trimethyltocol[a]	1.49	Wheat germ and other vegetable oils (some synthetic)
d-α-Tocopheryl acetate	2R-(4'R,8'R)-5,7,8-Trimethyltocol acetate	1.36	Chemical esterification
l-α-Tocopherol	2S-(4'RS,8'RS)-5,7,8-Trimethyltocol	0.36	Synthetic
2-l-α-Tocopherol (2-*epi*-α-tocopherol)	2S-(4'R,8'R)-5,7,8-Trimethyltocol	0.36	Synthetic
dl-α-Tocopherol (all-*rac*-α-tocopherol)	2RS-(4'RS,8'RS)-5,7,8-Trimethyltocol	1.1	Synthetic
dl-α-Tocopheryl	2RS-(4'RS,8'RS)-5,7,8-trimethyltocol acetate	1.0	Synthetic
2-dl-α-Tocopherol	2RS-(4'R,8'R)-5,7,8-Trimethyltocol	1.1	Synthetic
2-dl-α-Tocopheryl acetate	2RS-(4'R,8'R)-5,7,8-Trimethyltocol acetate	1.0	Synthetic
d-β-Tocopherol	2R-(4'R,8'R)-5,8-Dimethyltocol	0.12	Wheat germ and other vegetable oils
d-γ-Tocopherol	2R-(4'R,8'R)-7,8-Dimethyltocol	0.05	Corn oil
d-δ-Tocopherol	2R-(4'R,8'R)-8-Methyltocol		Soybean oil
d-δ₂-Tocopherol	2R-(4'R,8'R)-5,7-Dimethyltocol		Rice oil
d-α-Tocotrienol	*trans*-2R-5,7,8-Trimethyl tocotrienol[b]	0.32	Wheat oil
d-β-Tocotrienol	*trans*-2R,5,8-Dimethyl-tocotrienol	0.05	Plant oils
d-γ-Tocotrienol	*trans*-2R,7,8-Dimethyl-tocotrienol		Plant oils
d-δ-Tocotrienol	*trans*-2R,8-Methyl-tocotrienol		Plant oils

[a]In other chemical terminology, this compound has been referred to as 2,5,7,8-tetramethyl-2-(4',8',12'-trimethyl-tridecyl)-6-chromanol.
[b]Also has been referred to as 2,5,7,8-tetramethyl-2-(4',8',12'-trimethyl-trideca-3',7',11'-trienyl)-6-chromanol.

point at 0.1 atm 200–210°C; specific rotation at 20°C in alcohol −2.4°. It crystallizes at 30°C.

δ-Tocopherol has an absorption maximum of 298 nm; $E_{1\,cm}^{1\%} = 91.2$; specific rotation at 25°C and 5460 Å +1.1°.

α-Tocotrienol has an absorption maximum in ethanol of 292.5 nm; $E_{1\,cm}^{1\%} = 91$.

β-Tocotrienol has an absorption maximum in ethanol of 296 nm; $E_{1\,cm}^{1\%} = 87$.

δ₂-Tocopherol has an absorption maximum in ethanol of 292 nm; $E_{1\,cm}^{1\%} = 83$. It crystallizes at −4°C.

3.3.5. Properties of Derivatives

d-α-Tocopherol forms an allophanate ($C_{31}H_{52}N_2O_4$), a crystalline material with a melting point of 172–173°C, and a p-nitrophenylurethane derivative ($C_{36}H_{54}N_2O_5$) with a melting point of 130–131°C.

d-α-Tocopheryl succinate is a white powder with a melting point of 76–77°C. Biological activity of this compound is 1.21 IU/mg.

α-Tocopheryl succinate polyethylene glycol esters have been made. These are viscous liquids soluble in water up to 25%, depending on the amount of polyethylene glycol incorporated in the formation of the compound.

δ_2-Tocopherol gives a 4-phenyl azobenzoate derivative ($C_{41}H_{56}N_2O_3$) which crystallizes from dilute 2-propanol and melts at 65°C.

β-Tocotrienol forms a 4-phenyl azobenzoate ($C_{41}H_{50}N_2O_3$) as orange crystals from isopropanol, melting point 70–71°C.

3.4. Deficiency Diseases

Most vitamin E deficiency diseases are now quite well defined. Over the 50 years since the discovery of vitamin E, this fat-soluble factor has been shown to be concerned in maintaining the health of the brain, the vascular system, the red blood cells, the skeletal muscles, the liver, the heart, the gonads, and the developing fetus, and in preventing yellow fat and ceroid in adipose and other tissues. Vitamin E appears to have at least two metabolic functions, one as a fat-soluble antioxidant and another in a more specific role interrelated with the function of selenium and the metabolism of sulfur amino acids. Vitamin E requirements are influenced by dietary levels of polyunsaturated fatty acids, other antioxidants, sulfur amino acids, and selenium. A summary of the vitamin E deficiency diseases of animals and the effects of the factors interrelated with vitamin E nutrition is presented in Table II.

Several "vitamin E deficiency diseases" respond also to synthetic antioxidants. Thus in prevention of these diseases vitamin E appears to act as an antioxidant. Other diseases have a more specific requirement for vitamin E, some require selenium as the primary preventive factor, several may respond to either vitamin E or selenium, and others may require both vitamin E and selenium. The "vitamin E responsive diseases" are embryonic degeneration in female and sterility in male rats and erythrocyte hemolysis, plasma protein loss, anemia, and nutritional muscular dystrophy in rabbits, guinea pigs, dogs, minks, and monkeys. The "selenium and vitamin E responsive diseases" are necrotic liver degeneration in rats and pigs; muscular dystrophies in lambs, calves, pigs, horses, and turkeys; infertility in ewes; unthriftiness in cattle and sheep; poor hair or feather development in pigs, horses, and chickens; and exudative diathesis in chicks.

"Antioxidant and vitamin E responsive diseases" are encephalomalacia, steatitis, and depigmentation of rat incisors. The "cystine and vitamin E responsive disease" is nutritional muscular dystrophy in chicks.

Table II. Vitamin E Deficiency Diseases

Disease	Experimental animal	Tissue affected	PUFA influence	Prevented by			
				Vitamin E	Se	Antioxidant	Sulfur amino acids
1. Reproductive failure							
Embryonic degeneration							
Type A	Rat, hamster, mouse, hen, turkey	Vascular system of embryo	X	X		X	
Type B	Cow, ewe			—[a]	X[b]		
Sterility (male)	Rat, guinea pig, hamster, dog, cock, rabbit, monkey	Male gonads		X			
2. Liver, blood, brain, capillaries, pancreas							
Liver necrosis	Rat, pig	Liver		X	X		
Fibrosis	Chick, mouse	Pancreas		X	X		
Erythrocyte hemolysis	Rat, chick, man (premature infant)	Erythrocytes	X	X		X	
Plasma protein loss	Chick, turkey	Serum albumin		X	X		
Anemia	Monkey	Bone marrow		X		X	
Encephalomalacia	Chick	Cerebellum	X	X		X	
Exudative diathesis	Chick, turkey	Vascular system		X	X		
Kidney degeneration	Rat, mouse, monkey, mink	Kidney tubular epithelium	X	X	X		
Steatitis (ceroid)	Mink, pig, chick	Adipose tissue	X	X		X	
Depigmentation	Rat	Incisors	X	X		X	
3. Nutritional myopathies							
Type A (nutritional muscular dystrophy)	Rabbit, guinea pig, monkey, duck, mouse, mink	Skeletal muscle		X		?	
Type B (white muscle disease)	Lamb, calf, kid	Skeletal and heart muscles	—[a]		X[b]		
Type C	Turkey	Gizzard, heart	—[a]		X		
Type D	Chicken	Skeletal muscle	—[c]	X			X

[a] Not effective in diets severely deficient in selenium.
[b] When added to diets containing low levels of vitamin E.
[c] A low level (0.5%) of linoleic acid was necessary to produce dystrophy; higher levels did not increase vitamin E required for prevention.

3.4.1. Deficiency Diseases in Mammals

3.4.1.1. Reproductive Failure in the Female

A widely used biological assay for vitamin E has been a determination of the amount of an unknown material required to prevent fetal death and resorption in rats, compared with the amount of vitamin E required when added to a specific basal diet. The vitamin E level which prevents fetal resorption under these conditions is so critical that this assay method is the standard for assessing the relative biopotencies of various forms of vitamin E. In determining the vitamin E content of an unknown material, the linoleic acid and antioxidant in the test material must be eliminated or appropriate controls must be used. The primary metabolic defect which causes death and subsequent resorption of the embryo is unknown. Degenerative changes occur in the uterus. The vascular system of the embryo also undergoes degeneration and the embryo suffers from a severe anemia.

Resorption gestation in vitamin E deficient female rats has sometimes been represented as a simple antioxidant deficiency. Søndergaard (1967) concluded that "Antioxidants cannot protect against fetal resorption. They can, however, improve the reproductive capacity by protecting suboptimal amounts of vitamin E present in the diet or in the various organs of the rat." Draper *et al.* (1964), however, showed that the antioxidant diphenyl-*p*-phenylenediamine (DPPD) prevents this deficiency disease in rats fed diets very low in vitamin E over a period of three generations, and results in normal gestation when added to diets which had caused fetal death and resorption in the previous gestation period. Ethoxyquin, which readily prevents encephalomalacia, does not prevent resorption gestation, a finding which adds to the difficulty of determining the possible mode of action of vitamin E in prevention of this deficiency disease.

Although certain forms of reproductive failure in cattle and sheep have been reported in New Zealand which respond primarily to dietary or injected selenium (Andrews *et al.*, 1968), the classical reproductive failure characterized by death and resorption of the fetus in rats fed vitamin E deficient diets appears to be primarily a vitamin E responsive deficiency disease. Selenium compounds have no effect on the resorption gestation syndrome in rats (Harris *et al.*, 1958; Christensen *et al.*, 1958).

3.4.1.2. Testicular Degeneration in Vitamin E Deficient Male Animals

According to Mason (1954), irreversible testicular degeneration in rats occurs with vitamin E deficient diets whether or not they are supplemented with polyunsaturated fatty acids. This disease is not responsive to antioxidant action.

Vitamin E deficient rabbits, hamsters, dogs, monkeys, and pigs show progressive degeneration but require a long period of chronic deficiency for severe degeneration to occur. There is a gradual loss in testes weight and depletion of germinal epithelium with pronounced accumulation of acid-fast pigment in the Sertoli syncytium and interstitial tissues. Unlike the irreversibility of testicular injury in the rat, α-tocopherol therapy in the hamster gradually restored germinal epithelium in many but not all seminiferous tubules and also caused a striking influx of new spermatozoa into epididymal ducts which prior to therapy were completely empty.

Subcutaneous injections of subtoxic doses of cadmium chloride will produce injury to the testis of the rat and other mammalian species (Parizek and Zahor, 1956). This injury is mediated through an effect of cadmium on the permeability of the capillary bed of the testis parenchyma, leading to ischemic necrosis of the seminiferous tubules. Although Parizek (1957) indicated that this cadmium injury is prevented by zinc, Mason and Young (1967) have shown that selenium dioxide is approximately 100 times as effective as zinc acetate and that selenium is effective when injected simultaneously with cadmium chloride, whereas zinc acetate prevents testis injury only when given at least 8 hr before the cadmium chloride injection. Similar interrelationships have been discovered concerning the mutual effects of selenium, cadmium, mercury, thallium, and arsenic (Levander and Argrett, 1969). These effects appear to be concerned with the influence of one element on the rate and mode of excretion of one or more of the others. The separate roles of vitamin E, selenium, and zinc in maintaining the health and integrity of the testes remain to be determined.

3.4.1.3. Nutritional Muscular Dystrophy in Rats, Mice, Rabbits, Guinea Pigs, Dogs, Minks, and Monkeys

Two types of nutritional muscular dystrophy appear to exist in mammals. One responds primarily to vitamin E, the other primarily to selenium. Nutritional muscular dystrophy in rabbits, guinea pigs, dogs, and monkeys appears to be a vitamin E responsive disease. In these animals it is not prevented by selenium.

Vitamin E responsive muscular dystrophy appears not to be due to a simple need for an antioxidant. This disease does not respond to levels of dietary synthetic antioxidants several times as high as those needed to prevent encephalomalacia in chicks, yet the disease is invariably prevented or cured by supplementing the diet with normal levels of vitamin E.

Grossly, nutritional muscular dystrophy in these species is observed as a progressive muscular weakness. Histopathologically, the disease is characterized as a hyaline, waxy, or Zenker's degeneration involving loss of striations, multiplication and irregular distribution of sarcolemma nuclei, and swelling of the sarcoplasm, which becomes structureless and vacuolated. Irregularly scattered fibers are affected.

Edema and inflammatory reactions appear in the interstitial tissues. Reparative changes near the intact regions of the sarcolemma often are seen, while in other areas complete atrophy and replacement by connective tissue occur simultaneously.

According to Mason (1944), an acute type of dystrophy occurs in the striated musculature of rats only during the period of early adolescence and rapid growth. This disease frequently shows spontaneous remissions and rapid repair of the lesions, poor response to therapy, and sudden death. In contrast, the nutritional muscular dystrophy in guinea pigs and rabbits is easily produced in mature animals, is gradual in onset, rarely shows spontaneous remissions, and readily responds to vitamin E therapy by structural and metabolic repair, unless severely advanced stages have been reached.

Inclusion of torula yeast in a vitamin E deficient diet delayed the onset of muscular dystrophy in the guinea pig but did not prevent the disease (Seidel and Harper, 1960). Dietary addition or subcutaneous injection of coenzyme Q_{10} isolated from Torula yeast (1–4 mg per animal per day) did not alleviate muscular dystrophy. Their results with selenium were similar to those obtained previously with rabbits by Draper (1957) and by Hove *et al.* (1958), who showed that muscular dystrophy of vitamin E deficiency in rabbits is not affected by sodium selenite or selenocystine. Bonetti and Stirpe (1963) reported that sodium selenite (0.15 mg/kg of diet) reduced the severity of muscular dystrophy in rats but not in guinea pigs fed vitamin E deficient diets. Nafstad and Tollersrud (1970) have suggested that nutritional muscular dystrophy in swine also may be predominantly the result of selenium deficiency.

Mice appear to be quite resistant to nutritional muscular dystrophy of vitamin E deficiency.

3.4.1.4. Muscular Dystrophy and "Mulberry Heart" in Young Pigs

Dystrophy of the skeletal and heart muscles occurs in young vitamin E deficient pigs. The skeletal muscles have a grayish color, interspersed with streaks of white due to calcified foci. Intermuscular and interstitial edema is common, the edematous fluid often tinged with heme. This condition is prevented by vitamin E but not by selenium. "Mulberry heart" is the name given to a vitamin E deficiency syndrome of pigs characterized by hydropericardium, hydrothorax, ascites, and venous engorgement with congestion of the liver, kidneys, and the great veins. Coagulated blood is usually present in both ventricles. Subepicardial hemorrhages, most evident over the right ventricle and atrium, occur together with severe subendocardial and myocardial hemorrhages. Changes in the right ventricle include varying degrees of myocardial injury, ranging from swollen fibers to complete necrosis and calcification in some areas (Nafstad and Tollersrud, 1970).

Stowe and Whitehair (1963) reported nutritional muscular dystrophy in vitamin E deficient minks, characterized by swollen, differentially stained fibers, vacuole degeneration, sarcolemmal and myoblastic proliferation, and calcification of the nonphagocytized myofibrils. The myocarditis was focal, calcified, and necrotic, and the fatty hepatic infiltration was accompanied by centrolobular hepatic hemorrhage. Either vitamin E or selenium was completely effective in preventing all signs of nutritional muscular dystrophy in the minks.

Hayes *et al.* (1969) reported nutritional muscular dystrophy in 32 male beagle puppies fed vitamin E deficient diets with various levels of safflower oil. Neuro-axonal dystrophy and myodegeneration were found in all vitamin E deficient dogs. The requirement for vitamin E was directly related to the polyunsaturated fatty acids (PUFA) consumption.

According to Dinning and Day (1957) and Fitch and Dinning (1963), rhesus monkeys (*Macaca mulatta*) receiving vitamin E deficient diets stop growing and begin to lose weight as an early manifestation of the deficiency syndrome. Despite progressive debility, the monkeys continue to eat well. As the muscular weakness develops, the ratio of urinary creatine to creatinine increases rapidly. When the ratio exceeds 1.5:1, the monkeys have reached a severe deficiency state which if not corrected will lead to death in a few days.

Degeneration of the muscle fibers is associated with increased numbers of sarcolemma nuclei and an abnormal alignment of the nuclei within the muscle fibers. The absence of dietary fat does not appreciably alter the onset of muscular dystrophy or the histological appearance of the muscles. The creatinuria responds promptly to α-tocopherol therapy, usually within 48–72 hr.

Creatinuria also has been reported to be associated with nutritional muscular dystrophy in vitamin E deficient rats (Century and Horwitt, 1960; Gerber *et al.*, 1964) and rabbits (Bowman and Dyer, 1960).

3.4.1.5. Creatinuria

Response of dystrophic rats, rabbits, guinea pigs, and monkeys to vitamin E therapy is immediately reflected in a decrease in creatine excretion and a return of the creatine/creatinine ratio to normal. Gerber *et al.* (1964) found that synthesis of creatine from glycocyamine in perfused livers was unchanged by vitamin E deficiency. They postulated that the inability of skeletal muscle to utilize creatine, rather than an increased synthesis of creatine by vitamin E deficient animals, is primarily responsible for the creatinuria observed in vitamin E deficiency. This was supported by studies of Fitch and Dinning (1964) with vitamin E deficient monkeys using [1-^{14}C]creatine. The deficient monkeys retained creatine poorly; this points to a defect in the ability of skeletal muscle to retain creatine as the explanation of the creatinuria and of the low concentration of creatine in the muscles of vitamin E deficient animals.

3.4.1.6. Nutritional Muscular Dystrophy in Lambs and Calves

A "stiff lamb" disease was first described as a myopathy of unknown etiology by Metzger and Hagen in 1927. The disease was produced experimentally by feeding pregnant ewes a diet consisting of hay and cull beans obtained from certain areas of New York state. Early studies (Madsen *et al.*, 1935; Davis *et al.*, 1938) showed that muscular dystrophy could be produced in sheep and goats only when the natural diets contained considerable amounts of cod liver oil or other fish oils. In 1953, Blaxter *et al.* reported eight experiments with calves in which muscular dystrophy did not occur, even when the intake of vitamin E was reduced below 10 mg/kg body weight, unless cod liver oil was added to the diet. Addition of 0.7 g cod liver oil per day per kilogram of body weight increased the vitamin E requirement to 4 mg/kg body weight per day for prevention of muscular dystrophy. Wright and Bell (1964) fed 48 pregnant ewes a purified diet deficient in both vitamin E and selenium but containing relatively little polyunsaturated lipid. Vitamin E deficiency increased the serum glutamic oxaloacetic transaminase (SGOT) in the young ewes. Selenium delayed this increase but did not prevent it. No signs of severe muscular dystrophy occurred in either the ewes or the lambs. Thus muscular dystrophy in ruminants is not easily produced in the absence of dietary sources of unsaturated fatty acids but can be elicited readily by addition of fat fractions with a high degree of unsaturation. Blaxter (1955) observed that the polyunsaturated fatty acids did not cause destruction of tocopherol in the diet, reduce absorption from the gut, or induce oxidation of the vitamin to its quinone. The tocopherol content of dystrophic muscles usually was higher than that of normal muscle, and the livers of dystrophic calves contained more tocopherol than those of normal calves. Thus acceleration of *in vivo* oxidation of tocopherol by unsaturated fatty acids could not account for the dystrophy resulting from the polyunsaturated oils.

After the reports on the effect of inorganic selenium compounds in preventing dietary liver necrosis in rats and exudative diathesis in chicks, studies were initiated almost simultaneously by Proctor *et al.* (1958) and by Muth *et al.* (1958) in which complete prevention of nutritional myopathy in newborn lambs was achieved by addition to the diet of 0.1–1.0 mg Se/kg total diet fed to the ewes. Oldfield *et al.* (1960) found that selenium given prenatally *per os* to ewes or postnatally by injection to the lambs prevents myopathy and results in increased growth of lambs. The growth response to selenium fed prenatally is greater than that obtained with selenium injected into the lambs. Massive oral doses of vitamin E (2000 IU/lamb) at birth protect the lambs from white muscle disease but do not improve growth.

Similar results were reported by Andrews *et al.* (1968) in New Zealand. These and many other subsequent studies have shown nutritional muscular dystrophy in lambs and calves results primarily from selenium deficiency.

3.4.1.7. Erythrocyte Hemolysis

In vitro dialuric acid induced hemolysis or erythrocytes from vitamin E deficient rats, first discovered by Rose and György (1950, 1952), was further studied by many researchers including Bunyan *et al.* (1960*a,b*), Friedman *et al.* (1958), Bieri and Poukka (1970), and Brin *et al.* (1974), who showed that erythrocytes from vitamin E depleted rats were highly susceptible to dialuric acid peroxidative hemolysis while those from rabbits were not. The rabbit red blood cells became susceptible, and muscular dystrophy appeared following supplementation of the diet with arachidonic acid. Fragilographs of red blood cells from arachidonic-fed rabbits were biphasic, suggesting two populations of red blood cells, one old and one young. Thus the feeding of polyunsaturated fatty acids (as arachidonate or to a lesser extent as cod liver oil or safflower oil) initiated the onset of erythrocyte hemolysis and of muscular dystrophy in the rabbits. Barker *et al.* (1973) have reported a rapid micromethod using the fragilograph for the functional evaluation of vitamin E status in rats.

An increased tendency of the erythrocytes to hemolyze has become recognized as a sensitive measure of vitamin E deficiency in several species. The test is carried out *in vitro* using isotonic dialuric acid (Rose and György, 1950, 1952), hydrogen peroxide (György *et al.*, 1952), or oxygen-containing isotonic saline (Christensen *et al.*, 1956) to induce the hemolysis. Dam (1957, 1962) has shown that erthrocyte hemolysis is prevented by both vitamin E and synthetic antioxidants. The addition of tocopherols *in vitro* to erythrocytes from vitamin E deficient animals also decreases the tendency of the cells to hemolyze. α-Tocopherol has been found both *in vivo* and *in vitro* to be more effective than β-, γ-, or δ-tocopherol even though α-tocopherol is generally a weaker *in vitro* antioxidant than the other tocopherols (Rose and György, 1952).

Christensen *et al.* (1958) reported that complete *in vitro* hemolysis of erythrocytes from rats fed vitamin E deficient diets occurred whether or not the diet was supplemented with selenite. Gitler *et al.* (1958) also reported that selenium did not alter the rate of hemolysis in rats or in chicks.

Krishnamurthy and Bieri (1961) found a marked initial inhibition of the hemolysis in erythrocytes from vitamin E deficient rats given dietary selenite (0.5 mg Se/kg diet) which disappeared after about 20 min; after 1 hr of incubation, hemolysis of the basal- and selenite-fed groups was similar. Vitamin E supplementation provides good protection against erythrocyte hemolysis. Since selenium is an integral part of glutathione peroxidase, which is present in erythrocytes in high amounts (Rotruck *et al.*, 1972), selenium via this enzyme may aid vitamin E in protecting erythrocytes from peroxidation and hemolysis.

For uniformity in the erythrocyte hemolysis test, the 25% hydrogen peroxide solution employed should be made up and stored in a Pyrex bottle for several days prior to use (Horwitt, 1962).

The catalase content of the blood apparently is of little consequence in the hemolysis test, since its activity is limited to the first few minutes when hydrogen peroxide concentrations are relatively high. Glutathione peroxidase is an important factor in addition to vitamin E for complete protection of erythrocytes from peroxide-induced hemolysis (Cohen and Hochstein, 1962; Rotruck *et al.*, 1972).

3.4.1.8. *Dietary Liver Necrosis*

Necrotic degeneration of the liver, described as a vitamin E deficiency disease in the rat by Schwarz (1944), has been shown to occur in the pig (Obel, 1953) and the mouse (DeWitt and Schwarz, 1958). In the mouse a multiple necrotic degeneration involved the liver, heart, kidneys, and the peripheral muscles, and there were severe atrophy of the pancreas and degeneration of the testes. The heart lesion was seen several weeks before the other gross changes were observed. The overall incidence of heart necrosis was 91%, of liver necrosis 54%.

In the rat and the pig the liver appears to be the site predisposed to necrosis. However, the kidney also is affected, showing intercortical medullary degeneration.

Necrotic liver degeneration in vitamin E and selenium deficient rats occurs abruptly during the terminal stage of the disease. The gross necrotic changes develop shortly before death, survival time ranging from 21 to 45 days. The necrosis is massive. Infarctlike changes occurring in some areas have an irregular serpentine pattern corresponding to the branching of efferent veins. Some areas appear to be hemorrhagic due to an accumulation and congestion of blood within the sinusoids, suggesting a blockage of the efferent vascular system. Necrosis is primarily centrolobular but is not confirmed to a discrete central zone. A narrow layer of intact cells usually surrounds the portal veins. This zone is extremely narrow in the most severely necrotic areas of the liver.

In describing the pathology of dietary liver necrosis, Fite (1954) observed a diffuse change in the nucleus of the cells, which showed karyolysis and karyorrhexis, and granules in the cytoplasm, which appeared to be calcified.

Although earlier studies indicated that dietary liver necrosis in rats could be prevented by vitamin E, selenium, or cystine, the effect of cystine in preventing liver necrosis has been shown to result from contamination of cystine with traces of selenium (Schwarz *et al.*, 1959). In rats receiving a torula yeast diet, several antioxidants including ethoxyquin and DPPD were effective in preventing dietary liver necrosis (Schwarz, 1958). However, this disease appears to be prevented by either vitamin E or selenium; the antioxidants may act by protecting vitamin E and/or selenium, allowing the low levels in the basal diet to be adequate for prevention of the disease.

Liver necrosis in the pig, also called "toxic liver dystrophy," is now known to be a factor in a long-recognized disease entity of pigs termed "hepatosis dietetica." The dominant pathological finding is centrolobular necrosis. This disease usually is but one portion of the overall vitamin E deficiency syndrome of piglets, which also includes muscular dystrophy, "mulberry heart," and steatitis.

3.4.1.9. Depigmentation of Rat Incisors

Vitamin E deficient rats were reported to lose the normal yellow-orange color of the incisor teeth (Davies and Moore, 1941). Depigmentation did not occur with fat-free diets, but the addition of either cod liver oil or lard produced complete depigmentation (Granados and Dam, 1945). According to Søndergaard (1967), the depigmentation is due to loss from the incisor of the ferric ion responsible for pigmentation of the enamel. Christensen *et al.* (1958) and Aterman (1959) found that whereas vitamin E and selenium both prevented dietary liver necrosis in rats, selenium had no effect on incisor pigmentation.

Irving (1959) reported that selenium at 0.9 mg/kg of diet was effective in curing vitamin E deficiency in rat incisors, but 0.3 mg/kg of diet produced only a temporary recovery of the pigment. A possible explanation of these discrepancies lies in the finding of Seward *et al.* (1966) that selenium (0.1 mg/kg of diet) was almost completely ineffective in counteracting depigmentation of the incisors in rats fed a low-protein diet containing 1% cod liver oil but did prevent depigmentation in rats fed the diet without cod liver oil. This may have been due to a sparing effect of selenium on traces of vitamin E in the diet or the animal body.

3.4.1.10. Vitamin E Protection against Hepatotoxic Effects of Carbon Tetrachloride

Hove (1948) showed that addition of vitamin E to low-protein diets resulted in a marked protection against fatal carbon tetrachloride poisoning in rats. Recknagel and Ghoshal (1965) demonstrated that CCl_4 administration caused accelerated lipoperoxidation in the liver. Recknagel and Ghoshal (1966) suggested that carbon tetrachloride may initiate a chain reaction of peroxidation in the lipids of hepatic microsomes and mitochondria. Green *et al.* (1969), however, were unable to demonstrate any *in vivo* production of hepatic peroxides by administration of an intraperitoneal dose of CCl_4 (2 ml/kg body weight); they found an increased level of α-tocopherol in the livers of the rats treated with CCl_4. These workers suggested, therefore, that CCl_4 does not increase *in vivo* lipid peroxidation.

Gallagher (1961), Seward *et al.* (1966), and Fodor and Kemény (1965) showed that selenium (0.2 mg/kg diet) is effective in preventing the lethal effects of CCl_4 in vitamin E deficient rats. Muth (1960) reported that selenium also is effective in counteracting CCl_4 poisoning in sheep.

Since one of the metabolic roles of selenium is as an integral part of glutathione peroxidase, the failure of Green *et al.* to observe *in vivo* peroxides in their CCl₄-treated rats may have resulted from hepatic glutathione peroxidase, which would destroy any peroxides that may have been formed.

3.4.1.11. Yellow Fat, Brown Fat, Ceroid, and Lipofuscins (Steatitis)

Peroxidation of polyunsaturated fats and formation of yellow to brown pigments occur in adipose and in lipid membranes of other tissues in animals fed vitamin E deficient diets containing cod liver oil or other sources of PUFA. Female rats receiving a vitamin E deficient diet often show an accumulation of brown ceroid in the uterus. Kittens receiving certain canned foods have developed "yellow fat disease." This was reported by Coffin and Holzworth (1954) and by Cordy (1954) to be very similar to the vitamin E deficiency steatitis in minks (Mason and Hartsough, 1951). Yellow fat disease is a constant finding in vitamin E deficient fur-bearing animals, but also occurs in calves, lambs, pigs, dogs, chickens, and turkeys. The color is a mixture of two pigments: one, a yellow fat-soluble pigment showing yellowish-green fluorescence; the other, dark brown and insoluble in both fat solvents and water. The insoluble brown pigment, named "ceroid" by Lillie *et al.* (1942), apparently results from a polymerization of peroxidation products of PUFA with lipoproteins. Gedigk (1959) found the lipid components of ceroid and lipofuscin to be identical, with some differences in their protein components. The lipofuscin age pigments may represent the presence of ceroid pigments within the lysosomes of cells where, because of their insolubility and indigestibility by the lysosomal enzymes, they remain and accumulate throughout the life of the animal (Tappel, 1968, 1975). The extent to which antioxidants, particularly vitamin E, can slow the formation of lipofuscins and perhaps delay aging remains to be determined. Attempts to show the existence of peroxides in phospholipid membranes *in vivo* have failed.

Vitamin E and antioxidants inhibit peroxidation of body lipids and production of ceroid pigment (Dam, 1957, 1962). However, selenite (1.0 mg Se/kg diet) in the absence of vitamin E was completely ineffective in preventing either the yellow fat of the adipose tissue or the brown discoloration of the female rat uterus (Chistensen *et al.*, 1958).

3.4.1.12. Degeneration of the Kidneys

Rats receiving vitamin E deficient diets containing a high level of PUFA over prolonged periods show a cloudy swelling in the kidneys caused by degeneration of the tubulae cortorti (Martin and Moore, 1939; Moore *et al.*, 1958; Emmel, 1957). Addition of vitamin E, methylene blue, or DPPD to the diet prevents the kidney changes but selenium does not (Søndergaard, 1967).

3.4.2. Anemia in Monkeys and Pigs

3.4.2.1. Monkeys

Fitch (1968) reviewed the research showing that severe anemia is responsible for much of the morbidity in the vitamin E deficient monkey. The hemoglobin concentration of vitamin E deficient monkeys remains at almost normal levels for considerable lengths of time, then drops abruptly. Thirty-two monkeys were fed a purified diet deficient in vitamin E, and 30 control monkeys received the same diet supplemented with the vitamin. The vitamin E deficient monkeys began to show signs at 5 months but some did not succumb until 30 months. The first sign was an abrupt weight loss, then creatinuria and anemia followed in the absence of treatment by death. Concurrent with the weight loss were muscle weakness and wasting similar to the nutritional muscular dystrophies observed in other animal species.

Although anemia might be expected to occur as a result of severe erythrocyte hemolysis in vitamin E deficient animals, the work of Fitch and others indicates that the anemia in monkeys stems from a lack of hematopoiesis in the bone marrow rather than from excessive red blood cell destruction. Murty *et al.* (1970) reported evidence that vitamin E may be concerned in heme synthesis. This has not been confirmed by others.

Both the anemia and the muscular dystrophy were quickly cured by vitamin E therapy. Some anemic, vitamin E deficient monkeys also responded to coenzyme Q_{10} but the response was limited to a reticulocytosis. Vitamin E was vastly superior to coenzyme Q in bringing about remission. Fitch suggested that these compounds may have a sparing effect on the basic, specific need for vitamin E.

3.4.2.2. Young Pigs

The effects of parenteral administration of vitamin E to piglets born of mothers fed with and without vitamin E supplementation were studied by Baustad and Nafstad (1972). Weekly injections of 100 mg dl-α-tocopheryl acetate resulted in increased values for hemoglobin and packed cell volume, total red cell counts, and reticulocyte response after iron injections when compared with the values for piglets not receiving vitamin E. Morphological abnormalities of blood and bone marrow cells charactertistic of vitamin E deficiency in monkeys were observed in the vitamin E deficient piglets.

3.4.3. Vitamin E Deficiency Diseases of Poultry

Chickens show three distinct vitamin E deficiency diseases—encephalomalacia, exudative diathesis, and nutritional muscular dystrophy. Severe selenium deficiency produces pancreatic degeneration and fibrosis which are not prevented or cured by dietary vitamin E.

Although turkeys receiving vitamin E and selenium deficient diets may show exudative diathesis to a mild degree, the main symptoms are nutritional muscular dystrophy of the gizzard and heart muscles. Vitamin E deficient ducks develop a nutritional muscular dystrophy very similar to that of rabbits and guinea pigs.

3.4.3.1. Encephalomalacia in Chicks

The degenerative changes of the cerebellum in the young growing chick, termed "encephalomalacia," respond to vitamin E or certain fat-soluble antioxidants but not to dietary selenium or sulfur amino acids. Encephalomalacia occurs only in chicks fed a vitamin E deficient diet containing polyunsaturated fatty acids. The incidence and severity of the disease are markedly increased with increasing levels of dietary linoleic acid. According to Adamstone (1947), encephalomalacia is a nervous derangement characterized by ataxia, backward or downward retractions of the head (sometimes with lateral twisting) (Fig. 3), forced movements, increasing incoordination, rapid contraction and relaxation of the legs, and finally complete prostration and death. Complete paralysis of the wings or legs is not observed.

The cerebellum, the striatal hemispheres, the medulla oblongata, and the mesencephalon are affected most commonly in the order named. When examined soon after the appearance of the signs of encephalomalacia, the cerebellum is softened and swollen, and the meninges are edematous. Minute hemorrhages often are visible on the surface of the cerebellum. The convolutions are flattened. Within a day or two after signs of encephalomalacia appear, greenish-yellow opaque necrotic areas can be found in the cerebellum. Damage to other parts of the brain usually is apparent on microscopic examination. Degenerative neuronal changes occur particularly in the Purkinje cells and in the large motor nuclei. The cells are shrunken and hyperchromatic.

In the prevention of encephalomalacia, there is little doubt that vitamin E functions as an antioxidant and that the disease can be prevented by any antioxidant which can prevent oxidation of linoleic acid. However, the primary metabolic defect responsible for the cerebellar degeneration and hemorrhage is not known.

In 1970, Yoshida *et al.* reported that nutritional encephalomalacia could be produced in young chicks receiving diets containing low levels of vitamin E by the addition of dilauryl succinate to the diet. Encephalomalacia induced by dilauryl succinate could be prevented by further addition of *dl*-α-tocopheryl acetate to the diet (Yoshida and Hoshii, 1972).

Fig. 3. Encephalomalacia.

Fig. 4. Nutritional muscular dystrophy. Middle chicken received vitamin E.

The possibility that encephalomalacia is produced by the toxic action of certain compounds on the cerebellar vascular system is not consistent with the finding that DPPD effectively prevents encephalomalacia, even when given daily to chicks receiving a vitamin E deficient purified diet high in PUFA and free of antioxidants (Scott and Stoewsand, 1961). This diet developed complete oxidative rancidity during the course of the experiment; without the addition of DPPD to the diet, 75% of the chicks died of encephalomalacia.

3.4.3.2. *Nutritional Muscular Dystrophy in Chickens, Ducks, and Turkeys*

When lack of vitamin E is accompanied by a sulfur amino acid deficiency, chicks show signs of muscular dystrophy, particularly of the breast muscle, at about 4 weeks of age (Fig. 4). In vitamin E deficient ducks, a similar dystrophy occurs throughout all skeletal muscles.

Turkeys deficient in both vitamin E and selenium show extreme myopathy of the gizzard and heart muscles (Scott *et al.*, 1967). Nutritional muscular dystrophy in chicks is prevented by either vitamin E or the amino acid cystine, in ducks it is prevented only by vitamin E, while in turkeys selenium seems to be the primary factor required for prevention of dystrophy.

The initial histological change in the muscles is a hyaline degeneration. Mitochondria appear to undergo swelling, to coalesce, and to form intercytoplasmic globules. The muscle fibers later are disrupted transversely; extravasation separates groups of muscle fibers and individual fibers. The transmuted plasma usually contains erythrocytes and heterophilic leukocytes. In the chronic condition, reparative processes are observed simultaneously with evidence of degeneration. Proliferation of cell nuclei occurs together with fibroplasia, which leaves a scar in the degenerated muscle.

In turkeys, fibrosis is found in the degenerating gizzard, replacing the normal muscle cells. Little or no fibrosis appears to occur in the heart, where degeneration may destroy the heart muscle to such an extent that the vitamin E–selenium deficient poults die of pericardial hemorrhage.

3.4.3.3. *Exudative Diathesis*

An edema of subcutaneous tissues associated with abnormal permeability of the capillary walls was termed "exudative diathesis" by Dam and Glavind (1939)

Fig. 5. Exudative diathesis.

(Fig. 5). In severe cases, chicks stand with legs far apart as a result of the accumulation of fluid under the ventral skin. This greenish-blue viscous fluid is easily seen through the skin since it usually contains some blood components from slight hemorrhages throughout the breast and leg musculature and in the intestinal walls. Pericardial distension and sudden death have been noted, particularly in chicks receiving higher than normal levels of salt in the vitamin E–selenium deficient diet. A low ratio of albumin to globulins occurs (Goldstein and Scott, 1956). Exudative diathesis is prevented by either vitamin E or selenium. It is not affected by synthetic antioxidants or by the level of cystine in the diet.

3.4.3.4. Pancreatic Fibrosis

In chicks pancreatic fibrosis is not a vitamin E deficiency disease. It occurs with diets containing very high levels of vitamin E and is cured or prevented only by dietary selenium. Selenium deficiency causes poor growth, poor feathering, and fibrotic degeneration of the pancreas. Death usually occurs after markedly decreased absorption of lipids, including vitamin E. The pancreatic degeneration results in a decrease in pancreatic and intestinal lipase, causing failure of fat digestion. Bile flow is almost eliminated; the gallbladder is several times larger than normal. With bile and monoglycerides absent from the intestinal lumen, micelles fail to form, thus impairing the absorption of lipids. Addition of bile salts to the diet does not return fat digestion to normal levels, and only temporarily enhances the vitamin E absorption.

Free fatty acids and monoglycerides added to the basal diet with the bile salts improved absorption of vitamin E and survival during an experimental period of 4 weeks. This absorption prevented exudative diathesis, allowing the chicks to survive until lack of pancreatic enzymes caused death. The selenium requirement for prevention of pancreatic degeneration was found to depend on the vitamin E level in the diet. With very high dietary vitamin E levels of 100 IU/kg or more, as little as 0.02 mg Se (as sodium selenite) per kilogram of diet completely prevented pancreatic degeneration. However, when the vitamin E content of the diet was more nearly normal (10–15 IU/kg), 0.04–0.05 mg Se/kg of diet was required. Exudative diathesis did not develop so long as some vitamin E was being absorbed; it occurred only after absorption of vitamin E was decreased to very low levels by pancreatic degeneration. With the administration of fatty acids, monoolein, and bile salts, which markedly improved vitamin E absorption, exudative diathesis did not occur even though the pancreas under these dietary conditions continued to undergo severe atrophy and degeneration.

3.4.3.5. Vitamin E and Reproduction in Poultry

Although no gross symptoms of vitamin E deficiency have been reported in mature chickens, testicular degeneration occurs in males deprived of vitamin E over prolonged periods of time, and the hatchability of eggs is reduced markedly. Embryos from hens fed diets low in vitamin E may die as early as the fourth day of incubation.

Adamstone and Card (1934) reported that a vitamin E deficiency brought about temporary reproductive failure in females but permanent sterility in males involving actual destruction of the germinal elements of the testes.

While practical rations for chickens appear to contain vitamin E sufficient for normal fertility and hatchability, turkeys have been shown to require additional vitamin E for maximum reproduction. Breeding turkeys fed practical diets required, in addition to the vitamin E present in the basal diet, approximately 30 IU/kg of diet for maximum hatchability (Jensen *et al.*, 1956; Atkinson *et al.*, 1955). The vitamin E deficient embryos were smaller than normal, with a cloudy lens, an opaque spot between the lens and the cornea, and some areas of edema in the eye. In all of these studies, hatchability was increased by vitamin E supplementation from approximately 50–60% in eggs of turkey hens fed basal practical diets to over 80% in the eggs of turkey hens supplemented with vitamin E. Although the synthetic antioxidant BHT improved hatchability slightly, it was not a substitute for vitamin E for maximum hatchability (Jensen and McGinnis, 1957). Atkinson *et al.* (1955) showed that the vitamin E content of turkey egg yolks was roughly proportional to the vitamin E content of the diet. Synthetic antioxidants may not be transferred to the egg in sufficient quantities to prevent embryonic mortality.

3.4.4. Vitamin E Deficiency in Man

Vitamin E is an essential nutrient for humans. It is not synthesized in the body. Thus the level of vitamin E in the lipoproteins of plasma and in the phospholipids of vital mitochondrial, microsomal, and plasma membranes in humans, as in experimental animals, depends on (1) the amount of biologically active vitamin E being consumed, (2) the levels of prooxidants and antioxidants in the diet, (3) the adequacy of dietary selenium, and (4) the dietary intake of sulfur amino acids and other factors which may alter the vitamin E requirements of man.

In the United States, Canada, and other well-developed countries, the vitamin E intake of most populations has been considered adequate for maintenance, growth, activity, and reproduction in normal individuals. Exceptions are newborn infants, particularly premature or otherwise of low body weight, members of some low-income groups, or individuals practicing bizarre food faddism which entails consumption of low vitamin E foods.

Low levels of tocopherol have been found in plasma of many patients with absorptive defects. Thus patients with sprue, cystic fibrosis of the pancreas with accompanying steatorrhea, or any other disease which causes malabsorption of fat and steatorrhea also show a marked decrease in plasma tocopherol.

3.4.4.1. Vitamin E Deficiency in Infants and Children

Even when the vitamin E status of the mother is good, placental transfer of vitamin E to the developing fetus is poor. Plasma vitamin E concentrations of newborn infants are about one-third those of adults. In infants of low birth weight, particularly premature infants, the levels may be even lower (Dju *et al.*, 1958).

Premature infants usually have a plasma tocopherol level of about 0.2 mg/100 ml at birth, which decreases to about 0.15 mg within 30 days of receiving a formula consisting mainly of partially skimmed cow's milk. In Table III are shown the serum tocopherol levels in normal individuals, premature infants, and children with

Table III. Serum Tocopherol Levels in Individuals
of Various Ages

Age group	Mean serum tocopherol levels (mg/100 ml serum)	
	USA[a]	Canada[b]
Adults	0.85 ± 0.03	0.77
Postpartum mothers	1.33 ± 0.40	0
Children, 2–12 yr	0.72 ± 0.02	0.62
Infants, full term	0.22 ± 0.10	0.19
2 mo	0.33 ± 0.15	—
5 mo	0.42 ± 0.20	—
2 yr	0.58 ± 0.20	—
breast-fed, 2 mo	0.71 ± 0.25	—
premature	0.23 ± 0.10	0.20
premature, 1 mo	0.13 ± 0.05	—
Cystic fibrotics, 1–19 yr	0.15 ± 0.15	0.11
Biliary atresiacs, 3–15 mo	0.10 ± 0.10	—

[a]From Gordon *et al.* (1958).
[b]From Goldbloom (1960).

cystic fibrosis of the pancreas or biliary atresia who are receiving a formula diet consisting mainly of partially skimmed cow's milk.

In adult human males with plasma vitamin E levels of 1 mg/100 ml or higher, no hemolysis in red blood cells incubated with 2.5% H_2O_2 for 2 hr 50 min at 37°C occurred, while at plasma vitamin E levels of 0.5 mg/100 ml or less, hemolysis ranged from 40% to 70% (Horwitt et al., 1956).

In adults in the United States and Canada receiving normal diets, the vitamin E levels usually approximate those required to prevent erythrocyte hemolysis (Bieri et al., 1964). Vitamin E levels in both normal full-term and premature infants, however, are well below that required to protect the erythrocytes from hemolysis induced by hydrogen peroxide or dialuric acid. Premature infants with vitamin E deficiency have moderately shortened red cell survival times and, if given a diet containing polyunsaturated fats without added tocopherol, may develop a syndrome characterized by irritability, pitting edema, a variety of skin changes, and hemolytic anemia. This is corrected by supplementing the diet with an adequate level of vitamin E (Ritchie et al., 1968).

Desai et al. (1972) investigated the tocopherols and polyunsaturated fatty acid contents of various proprietary infant formulas and substitute products available in Canada. Wide variations were found from one product to another, and many were low in total vitamin E content. The formulas without milk fat contained large amounts of biologically less active γ- and δ-tocopherols. About 60% of the proprietary infant formulas had α-tocopherol/PUFA ratios of less than 0.4. These studies confirmed earlier findings (Oski and Barness, 1967) that some proprietary infant formulas contained widely varying amounts of tocopherol, and some contained insufficient vitamin E to prevent development of a deficiency state characterized by a low level of serum vitamin E and hemolytic anemia.

3.4.4.2. Vitamin E Deficiency in Adults

Horwitt (1960) found that a diet providing 3 mg α-tocopherol per day did not prevent erythrocyte hemolysis in adults, whereas one providing 18 mg vitamin E per day completely protected against hemolysis. Bunnell et al. (1965) assayed a variety of foods and menus for tocopherol content using modern analytical techniques; they found that the sample menus provided a daily intake ranging from 2.6 to 15.4 mg, with an average of 7.4 mg vitamin E per day. This value is approximately one-half of previous estimates of daily vitamin E intake, indicating that the relatively low vitamin E blood levels found in a portion of the population, even in well-developed countries such as the United States, depend on dietary habits and often may border on a deficiency state.

In some less well-developed parts of the world, dietary deficiency of vitamin E may be common (Rahman et al., 1964). Whereas Desai (1968) found only 5.9% of Canadian university students and 1.1% of other persons (of a total of 379 normal adults) to have plasma tocopherol concentrations below 0.5 mg/100 ml, Rahman et al. (1964) reported that 21% of the persons tested from a rural population of East Pakistan had values below 0.5 mg vitamin E/100 ml of plasma. Of the children and pregnant or lactating females, 29% and 26%, respectively, were in this low range. Deficient levels of vitamin E have been found among a considerable portion of the population in other developing countries.

It is of particular importance that the tocopherol level in postpartum mothers in the United States increased to a mean value of 1.33 mg/100 ml of serum, whereas in Pakistan lactating mothers showed a decrease in plasma vitamin E concentrations. Thus the oft-quoted statements that "Pregnant women have tocopherol levels considerably above the levels of nonpregnant adults" and that "The plasma tocopherol of full-term infants rises promptly on feeding of breast milk and more slowly with feeding of cow's milk" (Gordon and Nitowsky, 1970) apply only to mothers receiving adequate intakes of vitamin E.

Vitamin E deficiency diseases in adult humans have not been observed in the United States except in persons with steatorrhea. Under these conditions, stage I of tocopherol deficiency shows a gradual body depletion of vitamin E but no physiological abnormalities. During stage II, there are moderately severe tocopherol depletion, increased erythrocyte hemolysis, and creatinuria. In stage III, the tocopherol depletion becomes severe, ceroid is deposited in the smooth muscles, some patients develop "brown bowel," and the skeletal muscles may show lesions similar to those of nutritional muscular dystrophy in laboratory animals.

Thus it is evident that a simple deficiency of vitamin E produces no recognized severe deficiency syndrome, unless the deficiency is accompanied by some other pathological insult or stress.

Studies on the mode of action of vitamin E and selenium, discussed in the introduction to this chapter, now show that the major function of vitamin E appears to reside in the protection of phospholipids of the mitochondrial and microsomal membranes from free-radical attack. If free-radical formation is not checked by adequate vitamin E and selenium nutrition, it will lead to peroxidation and destruction of the organelles which are of vital importance in the production of antibodies and of other mechanisms required for normal recovery from pathological diseases and other environmental stresses.

Thus the major role of vitamin E is as the first line of defense against destruction of these defense mechanisms. Vitamin E acts together with selenium, which, via glutathione peroxidase, acts as the second line of defense by destroying any peroxides that do form.

3.4.5. Diseases Responsive to Vitamin E Therapy

In light of the role of vitamin E in protection of membranes of mitochondria, microsomes, and other organelles required for production of antibodies and similar body defense mechanisms, it is understandable that numerous reports have appeared in the medical literature indicating that vitamin E is helpful in preventing or curing a wide variety of diseases in humans. Diseases which have been observed repeatedly to be helped by supplemental dietary vitamin E under controlled conditions are presented in the following review.

3.4.5.1. Thrombophlebitis

In 1968 Haeger of Sweden reported studies with 227 men conducted over a 7-year period to determine if vitamin E (300–600 mg daily) would reduce peripheral occlusive arterial disease. The minimum time of observation was 2 years; 69% of the cases were observed for more than 3 years.

Of the 123 individuals not receiving tocopherol, 19 died. Of the 104 individuals receiving tocopherol, only nine died. Of the 104 "nontocopherol" patients living at the end of the experiment, it was necessary to amputate in 11 cases for pain or gangrene. In the 95 tocopherol-treated patients, only one amputation was performed. Sixteen percent of the individuals developed ischemic leg ulcers, toe gangrene, or other ischemic skin conditions. All patients were encouraged to exercise. In the tocopherol group, 30 patients increased the walking distance more than 100% (average 118%); many walked more than 1000 m per day. Treatment with other vasodilators or dicumarol promoted no better improvement of the peripheral occlusive arterial disease and no better walking distance than obtained with the placebo therapy.

Similar results were reported by Heyman and Stamm (1955), who administered vitamin E (300 mg dl-α-tocopherol) intramuscularly, daily for 10 days, to alternative operative patients during a 2-year period. Although no changes were observed in the thromboelastogram, the incidence of thromboembolism was 5% in 136 operations where α-tocopherol was not administered and only 1.7% in 59 α-tocopherol-treated cases. The authors suggested that α-tocopherol should be given for several days before the operation and for at least 15 days postoperatively.

Spencer (1961) studied the effect in groups of 25 patients with peripheral vascular disease of vitamin E and nylidrin hydrochloride (Arlidin) alone and combined. The patients presented with varicose ulcer, diabetic and ischemic ulcers, thrombophlebitis, cold feet, claudication, nocturnal cramps, and numb and burning feet. Treatment with vitamin E alone produced results which were considerably better than those obtained with the control group receiving no treatment, but the group receiving vitamin E alone responded more slowly than the group receiving either Arlidin alone or a combination of Arlidin and vitamin E.

The effects of vitamin E on thrombophlebitis are probably best summarized by Boyd and Marks (1967):

> Vitamin E depletion of humans produces no changes in blood vessels nor are vascular changes seen in malabsorption syndromes. There is no evidence that vitamin E prevents arteriosclerosis. But controlled studies of claudication show that it is valuable if at least 400 mg daily is given for at least three months. Similar dosage helps stasis ulcers. Vitamin E is not a vasodilator. It improves oxygen utilization, perhaps acting on the lipid membrane of the mitochondria. High dosage of vitamin E improves the survival of arteriosclerotics and should always be administered to such patients.

Numerous other studies indicate that vitamin E is beneficial in cases of thrombophlebitis, but often these experiments have dealt with only a few patients and are poorly controlled.

3.4.5.2. Intermittent Claudication

In studies on the value of α-tocopherol for treatment of intermittent claudication, Boyd and Marks (1963) divided patients into three grades: in grade I the calf pains spontaneously disappear and walking may be continued indefinitely; in grade II the calf pain stabilizes at some level of intensity and remains unchanged in spite of continued exercise; in grade III the calf pain becomes steadily worse until activity must be stopped. Only those patients were studied whose condition had remained stationary more than 6 months or who had been claudicating for more than 1 year and less than 3 years with no recent improvement. None of the patients had had arterial surgery, although some had had myoneurectomy and tenotomy of the tendoachilles. Forty-three patients took part in a blind, controlled trial in which they were given either 400 mg synthetic *dl*-*α*-tocopherol or a lactose placebo daily over a period of 13 weeks. In this controlled series the average walking distance before onset of pain was significantly improved in the *α*-tocopherol-treated group. The average walking distance of the placebo group (800 yards) was reduced at the end of the treatment because of deterioration, whereas the halting distance of the *α*-tocopherol-treated group had significantly increased above the 800 yards.

Haeger (1974) reported a study with 47 male subjects aged 43–82 years (mean age 67) with arteriographically proven occlusion and with claudication of at least grade II, where arterial "first flow" (previously shown to be the most reliable parameter) did not exceed 12 ml per 100 g per minute. None of the patients had any complicating disease. Thirty-three patients were treated with *α*-tocopherol and 14 were held as controls. Vitamin E was given at a daily dose of 300 mg. The controls had previously started treatment with anticoagulants or vasodilators. All patients were followed for 2–5 years at half-yearly intervals. Venous occlusion-plethysmography and controlled walking distance tests were made repeatedly with all patients. The first improvement appeared after 3–4 months. Maximum increase in circulation usually occurred after 12–18 months of therapy. The final results of the study are shown in Table IV.

In Canada, a double-blind study was conducted with 33 patients suffering from intermittent claudication and definite evidence of arterial disease in the leg (Clein *et al.*, 1962). Sixteen patients received *α*-tocopherol and 17 were given placebos. Definite evidence was obtained that *α*-tocopherol is useful in treatment of patients with distal bed occlusion claudication.

3.4.5.3. Cardiac Diseases

In a study with 171 dental practitioners and 128 wives of dentists, Charaskin and Ringsdorf (1970) reported that an increase in clinical findings paralleled age only in those subjects consuming less than the recommended dietary allowance of vitamin E; cardiovascular findings decreased in the group receiving the highest

Table IV. Effects of Vitamin E on Patients Suffering from Intermittent Claudication[a]

	Control	Vitamin E (300 mg/day)
Initial blood flow in lower legs, ml/100 g tissue/min	7.7 ± 5.6	7.6 ± 2.6
Blood flow after 20–25 months, ml/100 g/min	5.7 ± 2.2	10.2 ± 4.8
Improved blood flow, number of subjects	3	29
No change in blood flow, number of subjects	1	1
Diminished blood flow, number of subjects	10	2

[a]From Haeger (1974).

vitamin E treatment. Thus there was a statistically significant positive correlation between age and cardiovascular findings with suboptimal vitamin E intake but not with vitamin E therapy at various levels above 25 IU/day. In 1974, Anderson and Reid reported a double-blind experiment involving 50 patients, half of whom received 3200 IU vitamin E daily *per os* while an equal number received an indistinguishable placebo. Although no statistically convincing evidence was obtained that vitamin E is of value in the treatment of angina, a small beneficial effect could not be ruled out.

Ghabussi *et al.* (1974) reported studies on the effect of *dl*-α-tocopheryl acetate, 300 mg p.o. daily, in patients receiving strophanthin-K (SK). The subtoxic, full-effective SK dosage was 9.35 ± 0.39 mg with a daily subtoxic maintenance dose of 2.6 ± 0.16 mg. Pulse rate fell significantly in eight cases while the ECG PQ interval increased. Side effects included ventricular extrasystole (in 40% of the cases), vomiting (20%), and diarrhea (30%). Vitamin E increased the tolerance to SK; daily SK maintenance dose increased by 49.6% to 3.89 ± 0.23 mg, the subtoxic full-effect dose increasing by 45% to 13.56 ± 0.75 mg.

In a double-blind experiment with 22 males, aged 61–73 years, Toone (1973) reported that 11 subjects received placebos and 11 received α-tocopheryl succinate (400 IU) four times daily over a period of 2 years. All patients had severe angina pectoris and some had had a prior coronary occlusion. None was suitable for surgical bypass procedures. Over a 3-month period before the experiment began, all had given up smoking, walked 2–5 miles a day, lost weight, changed to a low-cholesterol, low-triglyceride diet, and avoided stress. In the tocopherol group, four patients with previous nitroglycerin intakes of 240, 200, 160, and 110 tablets per month, respectively, reduced their intake to one or two tablets per month; none of the placebo group did so. The difference was found to be significant. Thus in this study tocopherol treatment was reported to have a significant effect in helping the patients to cope with angina pectoris. There were no complications after the use of vitamin E therapy and no cases of toxicity.

After 24 years of studies on the use of α-tocopherol in degenerative cardiovascular disease, Vogelsang (1970) indicated that vitamin E appeared to help cardiac and arterial function, and that it should not be given to patients having hyperthyroid heart or be given along with iron medication or with laxatives containing mineral oil. He reported vitamin E to be synergistic with digitalis, which must be reduced to avoid digitalis intoxication when vitamin E is given. The author indicated that α-tocopherol could be used with nitrates, aminophylline, quinidine, procaine amide, diphenylhydantoin, lidocaine, digitalis (reduced dosage), reserpine, α-methyldopa, spironolactone, chlorothiazides and theophylline products, low molecular weight dextran, and insulin (reduced dosage, usually). In contrast, Helwig (1971) reported that vitamin E given in doses of 200–500 mg was effective in markedly *reducing* the toxicity of digitalis intoxication. Neither of these experiments was well controlled.

Olson (1973) reviewed the research on vitamin E and its relation to heart disease. He described the use of vitamin E and selenium in the treatment of animal disorders and pointed out that although the mode of action of vitamin E is not known with certainty, three hypotheses have been observed: (1) the antioxidant hypothesis suggests that α-tocopherol is a physiological antioxidant designed to protect polyunsaturated fatty acids and other easily oxidizable groups; (2) the respiratory chain hypothesis postulates that α-tocopherol plays a specific role in electron transport in mitochondria by serving as a catalyst for respiration; and (3) the genetic regulation hypothesis states that α-tocopherol controls in some way the transfer of genetic information from the chromosomes to the whole cell.

Although vitamin E deficiency is rare in humans, premature infants and children with malabsorption syndromes and steatorrhea, celiac disease, lipoprotein-nemia, fibrocystic disease of the pancreas, and biliary atresia have been shown to have one or more recognized signs of vitamin E deficiency.

In his review, Olson also recognizes that vitamin E has been recommended for the treatment of menstrual disorders, habitual abortion, burns, cyanotic congenital heart disease, angina pectoris, coronary thrombosis, hypertension, rheumatic fever and heart disease, indolent ulcers, diabetes mellitis complications, and glomerulonephritis. Overt toxicity has not been seen, but no good evidence has been obtained of a beneficial effect of vitamin E in these diseases since in most instances the studies included no proper controls. However, use of the vitamin against peripheral vascular disease with intermittent claudication, thrombophlebitis and thromboembolic states, and other cardiac vascular disorders was discussed in relation to its role in protecting vital membrane structures against peroxidative damage.

3.4.5.4. Eye Disorders

The effects of α-tocopherol in ocular diseases have been studied by a number of clinicians but few truly controlled studies have been reported. Rouher and Sole (1965) describe 8 years of treatment of evolving myopia. More than 1000 cases were treated with a combination of vitamin E and vitamin P (100–300 mg of each). In none of the 1000 cases did the myopia disappear. However, in 149 observations there were 80 stabilizations, 49 slight worsenings, and 20 failures. After 1 year of therapy there was a total increased myopia of 73 diopters in the 200 eyes of the treated patients, while a matched untreated group totaled 132 diopters for the 200 eyes. In 29 cases, stabilization of the myopia occurred with vitamin E treatment after other therapies had failed.

Pei-Fei Lee (1956) presented a clinical evaluation of α-tocopherol in the treatment of various ocular diseases. Long-term administration of high daily dosages of α-tocopherol was reported to be efficacious in hypertensive and diabetic retinopathies, senile macular degeneration, vitreous degeneration, and retrobulbar neuritis. Again, no controls were used in these studies, and not all patients improved even though all received vitamin E.

3.4.5.5. Skin Diseases

A number of reports in the literature indicate a beneficial effect of vitamin E in treatment of dystrophic epidermolysis bullosa (Smith and Michener, 1973) but not all such cases have been helped by supplemental vitamin E (Unger and Nethercott, 1973).

3.4.5.6. *Vitamin E Nutrition in Patients with Ulcers*

Parkhovnik (1972) of the USSR reported the results on vitamin E treatment of 122 patients with peptic ulcer. The vitamin E was given as a 30% solution in oil, administered intramuscularly by 15 injections on alternate days (approximately 300 IU vitamin E/day). Of 122 patients, 115 reported relief of pain; dyspepsia disappeared in all patients; tenderness in Boas's point disappeared in 118 patients (97%); acidity was normalized in 47%; craters disappeared in 58 of the 68 patients where they had been observed. The authors concluded that treatment with vitamin E proved to be more efficacious than any other therapy previously used, including procaine and other common treatments.

3.4.5.7. *Vitamin E and Diabetes*

Studies have been reported by Dalle Coste and Klinger (1955) in which *dl-α*-tocopherol was given at levels of 300–500 mg/day for periods of 2 months to 3 years to 41 diabetics. No change was seen in carbohydrate metabolism, but of nine cases of varicose ulcers seven which had resisted all other treatments began to improve after 30 days on vitamin E, eventually healing completely. Thirty-five of the patients had postphlebitic or varicose lesions and all appeared to be helped by the vitamin E therapy. After the first 20 days of treatment, the authors reported a decrease in edema and a lessening of the sense of pain.

The same year, DiNardo *et al.* (1955) studied total lipids, total and free cholesterol, phospholipids, and fasting blood sugar in eight diabetics given 400 mg *dl-α*-tocopherol for 20 days to 9 months. Two patients also received insulin. In four diabetics blood sugar was repeatedly tested up to $1\frac{1}{2}$ hr after glucose administration, then after glucose with insulin, some 5–6 hours after a single oral dose of 10 g of vitamin E had been given. No changes were demonstrated.

Vitamin E absorption and depletion were studied by McMasters *et al.* (1967) in 18 normal males and 20 obese diabetics. No clinical improvement in the diabetes was noted on giving 1000 mg mixed tocopherols daily for 14 days. In the normal males, the tocopherol level of the fat increased from 3.8 to 8.0 mg/100 ml and in plasma from 0.4 to 1.1 mg/100 ml. These levels fell rapidly when supplementation ceased. In the diabetics the tocopherol levels also rose after supplementation; the tocopherol content of the fatty tissues in the diabetics increased to a higher level than that of the normal subjects, but depletion also was faster. Yamakita (1968) reported on the serum level of total tocopherols and erythrocyte hemolysis using hydroperoxide in patients with diabetes mellitus. Some but not all patients with diabetes had serum tocopherol levels less than 0.6 mg/100 ml. In these patients the rate of hemolysis was higher than normal. Even in some diabetic patients with normal vitamin E levels there was an increased erythrocyte hemolysis when tested with hydroperoxide. This increased hemolysis was eliminated after treatment for 1 or 2 weeks with 300 mg *dl-α*-tocopheryl acetate daily.

3.4.5.8. *Vitamin E and Cancer*

Administration of *α*-tocopherol to patients with cervical cancer was reported by Graham *et al.* (1960) to cause an increase in sensitization response (SR levels).

In 99 consecutive patients with cervical cancer in Stockholm who received the usual radiotherapy, all with an initially poor SR were given either 100 mg of α-tocopherol orally per day or 25 mg of testosterone propionate intramuscularly three times a week. Of 59 patients who initially had shown a poor cytological response, 46 could be followed. Of these, 74% achieved a good cytological response on treatment. Tocopherol and testosterone appeared to be about equally effective. Of those with an initially good cytological response, 59% were cured for 4–5 years. Where an initially poor response was altered by therapy, the cure rate was 65%.

Fukuzumi (1970) has proposed that accumulation of oxidized lipids can be causative of cancer. He has shown that oxidized lipids are present in gastric cancer tissue. The incidence of gastric cancer is high in subjects who consume large amounts of smoked, baked, or dried fish. Since the fish contain high levels of unsaturated fats, these could readily initiate autoxidation to lipoperoxides in vitamin E deficient individuals. Fukuzumi suggested that gastric ulcer leads to gastric cancer by releasing hemoglobin, which is converted to heme, a substance known to catalyze biooxidation of unsaturated lipids. Incidence of gastric ulcers and gastric cancer is higher in males than in females. This is attributed by Fukuzumi to the fact that women possess higher levels of vitamin E, the natural antioxidant capable of suppressing lipid autoxidation.

3.4.5.9. Vitamin E in Gastrointestinal Diseases

Gontzea *et al.* (1970) in Rumania studied plasma tocopherol levels and erythrocyte hemolysis in 65 infants aged 2–8 months, of whom 34 had common digestive diseases and dystrophies of various degrees, and 31 were healthy controls. The vitamin E level of the blood was 45–50% lower and erythrocyte hemolysis 66–150% higher in the children with digestive diseases compared to the healthy children. Goransson *et al.* (1973) in Sweden recognized that subnormal plasma tocopherol levels may be caused by intake of low dietary tocopherol or high levels of polyunsaturated fatty acids, where gastrointestinal disorders also appeared to be associated with low plasma tocopherols. They reviewed work of others indicating that a plasma tocopherol of less than 0.225 mg/100 ml is always preceded by at least 9 months of gastrointestinal disease.

Kryukova *et al.* (1972) in Russia reported a marked drop in serum vitamin E levels as well as a decrease in catalase and peroxidase activities of the erythrocytes in patients with gastroduodenal ulcer, chronic gastritis, or gastric cancer. The low vitamin E levels were most marked in patients with cancer of the stomach, and in this instance there was a considerable degree of hemolysis associated with the disease.

These findings together with those of Fukuzumi (1970) raise the question of cause and effect. Are some gastric diseases the result of vitamin E deficiency or, under conditions of gastroduodenal ulcer, chronic gastritis, and gastric cancer, are the absorption and retention of vitamin E reduced? Both phenomena appear to be possible.

3.4.5.10. Vitamin E and Immunity

Heinzerling *et al.* (1974*a*) reported that supplementation of the diet of chicks with either 150 or 300 mg *d*-α-tocopheryl acetate caused a significant increase in antibody titer in chicks infected with *Escherichia coli*. The α_2-globulin fraction of serum also was increased. In other studies, Campbell *et al.* (1974) showed that α-tocopherol enhanced *in vitro* immune responses by populations of mouse spleen cells containing both adherent and nonadherent cells, and Heinzerling *et al.* (1974*b*) demonstrated that vitamin E supplementation of Purina mouse chow, at levels of 120–180 mg/kg (as *dl*-α-tocopheryl acetate), enhanced natural and immune protection of mice against *Diplococcus pneumoniae* and produced a marked increase in phagocytic index.

Whitaker *et al.* (1967) divided 60 protein–calorie malnourished children in Chiengmat, Thailand, into two groups. One group (41 children) received a hospital high-protein diet. The other group (19 children) received this diet supplemented with 250 mg vitamin E per day for 5 days. The children fed only the high-protein diet became progressively more anemic, while those fed supplemental vitamin E had a statistically significant rise in reticulocytes and hemoglobin levels. Although the anemia was of multiple etiology, it did not respond to a good diet unless that diet contained a high level of vitamin E.

3.4.5.11. Vitamin E and Liver Disease

Kater *et al.* (1969) determined serum tocopherol levels in 41 patients with alcoholic cirrhosis, 31 with fulminating hepatitis, 54 alcoholic patients without evidence of liver disease, 20 with acute alcoholic hepatitis but no cirrhosis, 9 with acute infectious hepatitis, 114 with severe debilitating disease not involving the liver, and 88 healthy blood bank donors who served as controls. Only marginal differences were found among controls and alcoholics with or without liver disease, patients with acute hepatitis, and those with debilitating nonhepatic diseases. However, the patients with cirrhosis and those with fulminant hepatitis had highly significant depressions in serum tocopherol levels as compared with the controls and all other groups.

3.4.5.12. Vitamin E and Leprosy

According to Bergel (1967), at the Leprosy Hospital in Rosario, Argentina, vitamin E is required for the prevention and cure of leprosy. Vitamin E acts by protecting the lipoprotein membrane of the lysosomes from free-radical peroxidation. Intact lysosomes are necessary for destruction of the lepra bacillus. No one receiving an adequate vitamin E intake has ever contracted leprosy, according to Bergel (1971), and adequate vitamin E is a first prerequisite for arrest and treatment of the disease.

3.4.5.13. Vitamin E and Selenium in Neuronal Ceroid Lipofuscinoses

Zeman (1974) and others have proposed that Batten's disease and other neuronal ceroid lipofuscinoses result from unrestrained, massive peroxidation of essential fatty acids *in situ*, culminating in an accumulation of large masses of lipid decomposition products polymerized with amino acid residues and other substances in an undefined complex.

The peroxidation etiology hypothesis for these diseases has led to treatments employing vitamin E, antioxidants, and selenium. Zeman (1974) treated seven patients suffering from Spielmeyer-Sjögren type of neuronal ceroid lipofuscinosis and reported encouraging results. Although damage already evident was not repaired, marked improvements in learning ability were demonstrated after treatment with daily doses of 2000 mg vitamin E plus BHT, ethoxyquin, vitamin C, and *dl*-methionine.

However, Siakotos *et al.* (1974) found blood and tissue levels of α-tocopherol to be higher than normal in patients suffering from Batten's disease. These workers suggested that some factor other than vitamin E must be concerned in prevention of this peroxidative disease.

In this regard, Armstrong *et al.* (1974) reported a decrease in peroxidase activity in white blood cells of patients with Batten's disease, whereupon Westermarck and Sandholm (1977) in Helsinki examined 12 patients suffering from neuronal ceroid lipofuscinosis and found the red blood cell level of the selenium enzyme, glutathione peroxidase, to be significantly lower in these patients than in normals. Supplementation of their diet with selenium (or intramuscular injection) produced a prompt rise in glutathione peroxidase activity to normal values. Five patients evidenced improvement in eating, mood, or speech after receiving selenium for one month.

3.4.5.14. Summary and Conclusions Regarding Vitamin E Deficiency Diseases in Man

Numerous reviews have dealt with vitamin E deficiencies in man (Anonymous, 1960, 1974; Horwitt, 1961; Hellstrom, 1961; Haubold and Heuer, 1962; Herting, 1966; Gerloczy, 1968; Draper, 1969; Losowsky and Kelleher, 1970; Shute, 1972; Goodhart and Shils, 1973; Bieri, 1975).

These reviewers present two extreme points of view: one group claims that vitamin E is the cure for almost every disease known to man; the other group takes the stand that vitamin E has not been proved scientifically to have any of the effects being claimed for it.

The true value of vitamin E, as indicated by the evidence presented in this chapter, lies between these two extremes.

It is regrettable that neither group of reviewers has evaluated fully all of the recent findings regarding this vitamin, such as (1) the vitamin E content of human foods and menus in the United States, Canada, and other countries throughout the world, (2) the fact that many areas of the world are deficient in both vitamin E and selenium, (3) the presence in foods of prooxidants which may destroy vitamin E, (4) the nutritional interrelationships which exist among vitamin E, selenium, and sulfur amino acids, and, *particularly*, (5) *the biochemical actions of vitamin E and selenium which are concerned simply with prevention of peroxidative damage to cells and subcellular elements, thereby aiding the body to maintain its normal defense mechanisms against disease and environmental insult.*

Thus, with the exception of hemolytic anemias of premature infants and various diseases associated with biliary obstruction, vitamin E does not appear to be a specific cure for any disease of humans. Yet, because it is necessary to maintain the health and integrity of all cells, vitamin E is needed for optimum functioning of the normal body mechanisms involved in defense against all diseases.

Analytical evidence indicates that many diets in the United States, Canada, and throughout the world may contain insufficient vitamin E to fully protect important lipid membranes in human beings.

3.5. *Metabolic Roles of Vitamin E*

Many functions of vitamin E have been suggested: (1) an enzymatic function in normal tissue respiration, possibly aiding in some unknown way the functioning of the cytochrome *c* reductase system (Nason and Lehman, 1956; Vasington *et al.*, 1960); (2) in microsomal drug hydroxylations (Carpenter, 1972; Diplock, 1974); (3) in normal phosphorylation reactions, especially of high-energy phosphate compounds such as creatine phosphate and adenosine triphosphate (Carpenter *et al.*, 1958; Fitch and Dinning, 1959; Nesheim *et al.*, 1959; Calvert *et al.*, 1961); (4) in metabolism of nucleic acids (Dinning and Day, 1958); (5) in synthesis of ascorbic acid (Caputto *et al.*, 1958; Kitabchi *et al.*, 1960); (6) in synthesis of ubiquinones (Edwin *et al.*, 1961; Green *et al.*, 1961); (7) in a structural role in the control of membrane permeability and stability (Lucy, 1972); (8) in regulation of succinate oxidation in rat liver mitochondria (Corwin and Schwarz, 1959); (9) in regulation of heme synthesis (Caasi *et al.*, 1972; Schwarz, 1972); and (10) for normal functioning of the thyroid gland (Weiser *et al.*, 1971). Most of these proposed actions of vitamin E may be secondary to its primary action, which resides in protection of the mitochondrial, microsomal, and other vital membranes. It is possible, of course, that future research will demonstrate some requirements for vitamin E in specific metabolic reactions, such as the reduction of iron for heme synthesis, but proof of a specific *in vivo* requirement for vitamin E for such functions is still lacking.

The early findings by Olcott and Mattill (1941) that α-tocopherol has antioxidant properties, of Davies and Moore (1941) that vitamin E reduces the rate of depletion of vitamin A in rat liver, and of Dam and Granados (1945), who found peroxides in adipose tissue of rats and chicks fed diets deficient in vitamin E and rich in unsaturated fatty acids, led to the general antioxidant theory for the function of vitamin E as described by Tappel in 1972. According to this theory, free radicals from lipid peroxidation, in the absence of vitamin E, react with other compounds through hydrogen removal, resulting in formation of peroxides aldehydes, and other decomposition products. Through these reactions a wide variety of tissue constituents are damaged, including structural and functional components of the cell and their component enzymes, particularly membranes of cells and subcellular particles such as mitochondria, microsomes, and lysosomes.

3.5.1. Action of Vitamin E Complementary to the Action of the Selenoenzyme Glutathione Peroxidase

As indicated earlier in this chapter, many deficiency diseases in experimental animals, such as exudative diathesis in chicks and liver necrosis in rats, may be completely prevented by supplementing the deficient diet with *either* vitamin E *or* selenium. It now has been amply demonstrated that the mechanism of action of selenium in prevention of exudative diathesis in chicks occurs through the glutathione peroxidase (GSHpx) of the chick plasma. A positive correlation was described by Scott *et al.* (1974) between glutathione peroxidase activities at 7 days of age and prevention of exudative diathesis measured at 13 days of age. Selenium, via glutathione peroxidase, appears to prevent exudative diathesis in chicks by destroying polyunsaturated fatty acid hydroperoxides as they are formed and before they have an opportunity to cause destruction of the plasma membranes of the capillary cells to an extent which would increase capillary permeability (Fig. 6). King *et al.* (1975) have shown that microsomal NADPH-cytochrome P_{450} reductase activity initiates lipid peroxidation in biological membranes because the enzyme generates superoxide anion radicals when catalyzing NADPH oxidation. The investigations indicated that the superoxide ion *per se* is not the radical species responsible for the lipid peroxidation in the microsomes but that the oxidative

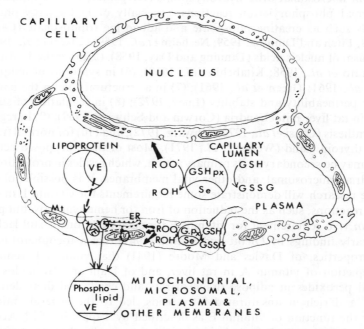

Fig. 6. Cross-section of a capillary, showing roles of vitamin E and selenium in protecting the membrane of the capillary cell wall. Vitamin E (VE) is intimately associated with the lipids of the plasma membrane, the mitochondria, and the endoplasmic reticulum, where it acts by preventing formation of free radicals. Selenium is an integral part of glutathione peroxidase (GSHpx), which destroys peroxides in the plasma and in the cytosol within the cell, thereby preventing any peroxides which may be formed from producing chain-reactive damage to the capillary cell wall.

attack on the membrane lipids was most likely due to hydroxyl free radicals produced by subsequent reactions of superoxide ion. Fridovich (1975) indicated that superoxide dismutase is present in all respiring cells and is essential for survival of aerobic cells, providing a defense against oxygen toxicity. Superoxide dismutase catalyzes the conversion $2O_2^- + 2H^+ \longrightarrow H_2O_2 + O_2$. King *et al.* observed that O_2^- itself is not reactive with lipids or with radical scavengers and hence cannot be the radical which initiates lipid peroxidation. Their results also indicate that xanthine oxidase, which forms superoxide anion during its activity under aerobic conditions, does not form singlet oxygen. However, if the hydrogen peroxide generated by the superoxide dismutase is not destroyed, it may react with superoxide ion in the presence of ferric ion to produce a hydroxyl free radical according to the following reaction:

$$H_2O_2 + O_2^- \xrightarrow{Fe^{3+}} \cdot OH + O_2 + OH^-$$

Removal of the H_2O_2 by glutathione peroxidase would markedly inhibit $\cdot OH$ radical formation. McCay *et al.* (1976) have provided evidence, using *in vitro* microsomal systems, that this destruction of hydrogen peroxide is responsible for the effect of glutathione peroxidase in protecting microsomal membranes.

Since it had been shown by Thompson and Scott (1970) that either selenium or vitamin E completely prevents exudative diathesis in chicks if fed at adequate levels and under conditions permitting their normal absorption and transport, the mechanism of action of vitamin E in preventing this disease was studied further. In chicks receiving graded levels of vitamin E, a direct relationship was shown to exist between hepatic mitochondrial lysis and lipid peroxidation, as measured by the thiobarbituric acid (TBA) method. These results demonstrated the applicability of the TBA determination as a suitable measure of lipid peroxidation of mitochondria and microsomal membranes under experimental conditions.

Lipid peroxidation of washed hepatic mitochondria and microsomes from newly hatched chicks of hens fed a diet low in vitamin E and selenium was very high. The mitochondria and microsomes had very little innate protection against lipid peroxidation. Baby chicks were fed a basal diet low in both vitamin E and selenium; some received vitamin E alone, others selenium alone; a fourth group received both vitamin E and selenium. After 14 days, the hepatic mitochondria and microsomes of the chicks fed the basal diet or diets with either vitamin E or selenium alone showed a high degree of *in vitro* peroxidation when incubated in the presence of ascorbic acid in tris-KCl buffer, pH 7.4, at 30° C for 10 min. Mitochondria or microsomes from chicks receiving *both vitamin E and selenium* showed no peroxidation under similar conditions, even when incubated for 2 hr or longer.

Using microsomes from chicks which had received only vitamin E, and adding supernatant from chicks which had received only selenium, evidence was obtained that the protective effect of vitamin E was because of the vitamin bound to the microsomal membranes, while the protective effect of selenium was because of glutathione peroxidase in the aqueous phase of either plasma or liver supernatant.

Supernatant from chicks receiving selenium caused a marked decrease in peroxidizability of the microsomes of chicks receiving only vitamin E. When the glutathione peroxidase in this supernatant was concentrated by use of Sephadex columns, the glutathione peroxidase fraction was inactive in protecting the microsomes of the chicks which had received only vitamin E, unless GSH was added to the incubation mixture. Microsomes from livers of the chicks fed vitamin E were not protected by GSH alone or by GSH plus the glutathione peroxidase fraction from the livers of the chicks which had received only vitamin E. These microsomes were almost completely protected, however, when incubated in the presence of GSH plus the glutathione peroxidase fraction from liver supernatant of chicks receiving both selenium and vitamin E.

In 1967, Moore *et al.* found that rats kept on vitamin E deficient diets for 8–10 months showed increased erythrocyte hemolysis and brown discoloration of the uterus, and as the deficiency became severe there was a rapid increase in rate of autolysis and lysosomal instability in the kidney tubules. In 1969 McCay suggested the presence of a microsomal system in liver (requiring NADPH and O_2) which actively oxidizes the polyunsaturated fatty acids (linoleic and arachidonic) in the β-position of microsomal phospholipids. As the level of dietary vitamin E increased, the rate of oxidation of the PUFA decreased. These findings may explain the positive correlation between dietary PUFA intake and the amount of vitamin E required to prevent symptoms of deficiency. McCay *et al.* (1971) then proposed that the function of vitamin E in this system involves the stabilization of microsomal membranes from free-radical attack during NADPH-dependent oxidations.

Fujita *et al.* (1973) found decreased linoleic acid and increased oleic acid in mitochondria of various organs of vitamin E deficient rats. Little or no change occurred in the brain. Intraperitoneal injection of α-tocopherol (50 mg/kg body weight/day) for 2 days caused fatty acid composition to return to normal in the vitamin E deficient rats.

Stache and Frimmer (1970) perfused livers of normal and tocopherol-deficient Holtzman rats with 200 ml of a buffered perfusion fluid and determined the release of β-glucuronidase and cathepsin in the perfusate. The amount of these enzymes released from the vitamin E deficient livers was significantly increased in relation to normal livers, thereby providing further evidence that vitamin E functions to maintain the integrity of subcellular membranes.

In 1970 Tinberg and Barber studied the peroxidation inhibition in structural protein–lipid micelle complexes derived from rat liver microsomal membranes by vitamin E. They compared these effects with those of the synthetic antioxidants butylated hydroxyanisole (BHA) and butylated hydroxytoluene (BHT). Twenty-five percent of the added vitamin E was bound to the microsomal structural protein in lipid membrane complexes; the addition of the vitamin completely protected the arachidonic and linoleic acid components of the membranes from peroxidation. Only the bound vitamin E was active in this respect. BHA and BHT were much less effectively bound to structural protein and lipid membranes than was vitamin E (less than 1% of the added BHA was bound), but both bound and free BHA did have an antioxidant activity against peroxidation of arachidonate and linoleate in the system.

These *in vitro* experiments may explain the much greater activity of vitamin E as compared with synthetic antioxidants for prevention of vitamin E deficiency diseases in man and animals. Vitamin E appears to be actively bound to the lipoprotein membranes of the cells and subcellular organelles, whereas the antioxidants are not. In view of the diversity of tissues affected by vitamin E deficiency, it is logical to assume that the vitamin possesses many different metabolic functions in the animal body, and is likely to affect the metabolism of other nutrients. No single primary metabolic role for vitamin E has yet been demonstrated other than the ability of adequate dietary levels to prevent *in vitro* peroxidation of erythrocytes and hepatic mitochondria, microsomes, and lysosomes.

Vitamin E deficiency symptoms for the most part are therefore secondary effects of the widespread nonspecific damage produced by peroxidative chain reactions. Administration in the diet of high levels of polyunsaturated fats such as cod liver oil or stripped vegetable oils, indicated in a previous section to have a marked effect on vitamin E deficiency diseases in various animals, also has a marked effect on autoxidation and formation of peroxides and yellow-brown pigments in animal tissues.

Although peroxides have been demonstrated to occur in adipose tissues, neither Green and Bunyan (1969) nor Glavind (1973) was able to demonstrate the presence of peroxides in parenchymatous organs of vitamin E deficient animals. Unfortunately, these workers were unaware of the action of selenium via glutathione peroxidase in destroying peroxides as quickly as they are formed. Their failure to observe peroxide formation in organs other than adipose tissue therefore may have been due to the presence of adequate selenium and thus a high level of glutathione peroxidase in these organs.

Scott *et al.* (1974) demonstrated that at least one important function of selenium is the role it plays as an integral part of glutathione peroxidase in protection of cellular and subcellular membranes from peroxidative damage. These findings, coupled with the evidence of the role of vitamin E within the membrane itself for the initial protection of the membrane against free radicals, confirms the hypothesis of Tappel (1972) that vitamin E reacts or functions as a chain-breaking antioxidant, thereby neutralizing free radicals and preventing peroxidation of lipids within the membranes.

The explanation for the rather specific effect of vitamin E in protection of lipid membranes appears to reside in the special capacity of these vital membranes to bind vitamin E. Witting (1965) presented a hypothetical reaction scheme of lipid peroxidation, postulating that vitamin E acted to protect lipid membranes. He suggested, however, that synthetic fat-soluble antioxidants having the same antioxidant properties as vitamin E should be equally as effective as the vitamin in protecting these membranes. Since α-tocopherol is needed for protection of mitochondrial and microsomal membranes even when the diet contains enough synthetic antioxidant, ethoxyquin, to completely protect chicks against encephalomalacia, it is clear that "d-α-tocopherol appears to have the advantage over other tocopherols and organic antioxidants in having the best access and longest retention in the tissues" (Century and Horwitt, 1965). Studies by Krishnamurthy and Bieri (1963) demonstrated that α-tocopherol is concentrated in phospholipids. As early as 1960, Csallany and Draper found that the synthetic organic antioxidant diphenyl-p-phenylenediamine (DPPD) is distributed throughout the body fat but that vitamin E is concentrated in the mitochondria and the microsomes.

Lucy (1972) suggested that "Vitamin E stabilizes membranes by virtue of specific physicochemical interactions between its phytyl side chain and the fatty acyl chains of polyunsaturated phospholipids, particularly those derived from arachidonic acid . . ." and "Interactions are proposed between the methyl groups of the phytyl chain and the *cis* double bonds of the fatty acid. Thus the methyl group at C'_4 of α-tocopherol can fit into a pocket provided by the *cis* double bond nearest the carboxyl group. The methyl group at C'_8 of tocopherol is then in register and can interact similarly with the third *cis* double bond."

Dinesen and Clausen (1971) and Molenaar *et al.* (1970) suggested that vitamin E is required for maintenance of the ultrastructure of the outer membranes of the mitochondria; that changes in both enzymatic action and peroxidation may result from changes in some common factor related to the function of the biological membranes and dependent on the presence of vitamin E. Vos *et al.* (1971) studied the ultrastructure of outer and inner mitochondrial membranes isolated from livers of normal and vitamin E deficient Peking ducklings. The purity of the fractions was assessed using the electron microscope and biochemical techniques. Gas chromatographic analyses of the composition of the inner mitochondrial membrane fractions revealed reduced levels of linoleic acid and arachidonic acid in vitamin E deficient as compared to normal ducks. Negative staining of the fractions showed specific structural changes in both outer and inner mitochondrial membranes of the vitamin E deficient ducks. The authors claimed that these results provided experimental support for an earlier theory that the reduced positive membrane contrast observed (after osmium tetroxide fixation) in tissues taken from vitamin E deficient animals is caused by a critical loss of specific polyunsaturated membrane-bound fatty acids.

Our results (Scott *et al.*, 1974), the results of McCay *et al.* (1971), and those of Tinberg and Barber (1970) appear to agree best with the postulations of Lucy, who suggests that the formation of a vitamin E–arachidonate complex in lipid membranes may have three functional consequences: (1) inhibition of peroxidative destruction of polyunsaturated fatty acids in cells and subcellular membranes, (2) prevention of permeability of biological membranes containing relatively high levels of polyunsaturated fatty acids, and (3) possible prevention of degradation of the membrane phospholipids by membrane-bound phospholipases. The results presented by Scott *et al.* (1974) lend support to the first two suggested functions of vitamin E in membranes.

3.5.2. Vitamin E and Xanthine Oxidase

More than 20 years ago, Dinning reported that vitamin E deficient rabbits and monkeys excreted increased amounts of allantoin and showed a marked increase in liver xanthine oxidase, while the uricase activity of the vitamin E deficient animals was normal. Catignani *et al.* (1974) postulated that vitamin E plays a role in the regulation of rabbit liver xanthine oxidase activity. The authors have suggested that in vitamin E deficiency there may be an alteration of the transcription of the messenger RNA for xanthine oxidase. However, Dinning (1953) suggested that the increase in xanthine oxidase may be secondary to a primary change in nucleic acid metabolism.

Prevention of lipid peroxidation may be important for the prevention of certain types of cancer. Carpenter and Howard (1974) reported that dietary vitamin

E is necessary for optimum activity of the microsomal, cytochrome P_{450}-dependent enzyme system of liver which carries out hydroxylations of a large number of drugs and *carcinogenic* hydrocarbons. Haber and Wissler (1962) found that α-tocopherol markedly reduced the carcinogenic activity of methylcholanthrene in mice, and Harman (1969) reported that the incidence of tumors in rats receiving dimethyl bromzanthracine plus vitamin E (20 mg daily) was reduced to 40% as compared to 74% with only 5 mg of vitamin E daily. The difference was statistically significant. Shamberger *et al.* (1974) reported that a single application of malonyl aldehyde to the shaved backs of mice caused an incidence of 52% tumors after daily treatment with 0.1% croton oil for 30 weeks. Analyses of skin of rats treated with DMBA, benzo[*a*]pyrine, or 3-methylcholanthrene, known carcinogens, showed increased levels of malonyl aldehyde in the tissues. Malonyl aldehyde has long been recognized as the thiobarbituric acid reactive product of peroxidation of linoleic or arachidonic acid. Carcinogenicity of DMBA, 3-methylcholanthrene, and benzo[*a*] pyrine may result in part from a catalysis by these compounds of lipid peroxidation and malonyl aldehyde formation.

A fatal liver necrosis was reported by Hove and Seibold (1955) in swine fed vitamin E deficient diets containing cod liver oil and low protein. The condition resembled a North European disease of swine, and one reported in Alabama and Georgia, where moldy corn had been fumigated regularly with a polychlorinated hydrocarbon compound. The damaging effects of carbon tetrachloride, other polychlorinated hydrocarbons, and some mycotoxin may result from *in vivo* catalysis of peroxidation.

Evidence has been obtained for a protective action of vitamin E against O_2 toxicity in patients receiving pure oxygen (Kimzey *et al.*, 1971), and the possible implications of this have been studied in astronauts (Loper *et al.*, 1969) and in underwater divers (Money and Sirone, 1972). Menzel *et al.* (1971) reported that trace amounts of air pollutants such as ozone and nitrous oxide rapidly oxidize polyunsaturated fatty acids and that vitamin E decreases the acute toxicity of both O_3 and NO_2. Fletcher and Tapel (1973) also found that α-tocopherol, alone or in combination with other antioxidants, exerted a protective effect in rats exposed to O_3 or NO_2. At levels of O_3 exposure of 0.7–0.8 ppm, α-tocopherol protection against lipid peroxidation in the lung was a reciprocal function of the logarithm of the dietary α-tocopherol as measured by endogenous lung thiobarbituric acid reactants. The reaction of NO_2 and α-tocopherol produced tocopheryl quinone together with small amounts of the dimer (Selander and Nilsson, 1972), indicating that α-tocopherol is converted to its normal metabolite when protecting lung tissue against NO_2. Thomas *et al.* (1968) studied the effects of NO_2 on lipids of rat lungs by exposing methyl linoleate to NO_2. Peroxidative changes appeared within 18 hr, reaching a maximum at 24 hr. Butylated hydroxytoluene (0.1%) added to the extracts of lung from animals killed 36–40 hr after exposure did not prevent peroxidative changes in the linoleate. Rats given 10 mg *dl*-α-tocopherol per day for 3 days prior to inhalation of NO_2 showed much less diene conjugation than those from the control group or receiving BHT. Even in these animals, however, α-tocopherol only partially prevented the lipid peroxidative changes induced by NO_2. Warshauer *et al.* (1974) determined the effects of vitamin E and O_3 on pulmonary antibacterial defense mechanisms.

Hemolysis due to O_3 treatment was greater than 95% in rats receiving the vitamin E deficient diets and less than 4% with vitamin E supplementation. Exposure to O_3 for 4 hr significantly decreased pulmonary bacteriocidal activity in both groups of rats.

Hope *et al.* (1975) showed that adequate vitamin E inhibits prostaglandin biosynthesis, apparently by preventing formation of endoperoxide intermediates in the conversion of arachidonic acid into PGE_2. Machlin *et al.* (1975) obtained evidence that such peroxidation triggers subsequent platelet aggregation and thrombocythemia.

These studies provide further evidence that the mechanism of action of vitamin E is involved with prevention of peroxidation of phospholipid membranes of the mitochondria, microsomes, and lysosomes, particularly those involved in defense mechanisms against environmental insult and pathological diseases.

3.5.3. Interrelationships with Other Factors

Interrelationships among vitamin E, prooxidants, and antioxidants are well established and generally recognized. The interrelationship of vitamin E and selenium also is now well established and has been discussed earlier. The relationship between vitamin E and cystine in prevention of nutritional muscular dystrophy in the chick appears to be unique, but may be found to apply also to other animals, once its nature has been elucidated.

Vitamin E also has been reported to be interrelated with the nutrition and metabolism of vitamin A, iron, and a large number of other nutrients, notably ascorbic acid (McCay *et al.*, 1959), vitamin D_2 (Selye *et al.*, 1964), vitamin B_{12} (Oski *et al.*, 1966), thiamin (D'Agostino, 1952), manganese (Lee *et al.*, 1962), and magnesium (Selisko, 1961).

3.5.3.1. Vitamin A

Hanna (1965) in Egypt and Fang *et al.* (1960) in Taiwan report that vitamin E at 200–500 mg per day was needed together with vitamin A (50,000–300,000 IU) for prevention of night blindness. McLaren *et al.* (1965) found that malnourished children with xerophthalmia had significantly lower serum vitamin E levels than malnourished children without ocular lesions.

It has long been known that vitamin E has a sparing effect on liver vitamin A. Søndergaard (1972) showed that the presence of vitamin E in the diet of rats resulted in the preservation of hepatic retinol during depletion of this vitamin, which occurred regardless of the presence or absence of polyunsaturated fatty acids (cod liver oil) in the diet. Conversely, the absence of vitamin E from the diet during retinol depletion resulted in a marked loss of hepatic retinol, particularly in the group receiving cod liver oil. In one study the effect of various antioxidants was compared with that of vitamin E. Only ethoxyquin had any retarding effect on vitamin A loss and 0.1% ethoxyquin had only about 60% of the activity of 0.01% *dl*-α-tocopheryl acetate. All antioxidants except ascorbic acid were somewhat effective when given with the diet containing cod liver oil. Thus the action of tocopherol in preservation of hepatic retinol appears to be independent of its antioxidant activity. This is in agreement with the conclusions of Cawthorne *et al.* (1968).

3.5.3.2. *Iron*

Goldberg and Smith (1958, 1960) gave rats high doses of iron (300–1650 mg/kg body weight) by intramuscular injecton with and without simultaneous administration of α-tocopherol and/or cod liver oil in various combinations. Increased *in vitro* erythrocyte hemolysis was produced by either the iron or the cod liver oil. Postmortem renal autolysis and increased deposition of brown polymers in the kidneys were observed in the rats given either iron or cod liver oil in the absence of vitamin E. This was largely prevented by administration of vitamin E. Tollerz and Lannek (1964) reported results which support the hypothesis that an overdose of iron acts as a prooxidant.

This is important in terms of the treatment of premature infants suffering from vitamin E dependent hemolytic anemias. Melhorne and Gross (1971) found significantly lower mean hemoglobin and serum tocopherol concentrations with higher reticulocyte counts and hydrogen peroxide erythrocyte fragility in the second month in infants with a gestational age of less than 36 weeks not receiving vitamin E supplements. Hemoglobin values were lowest and reticulocyte counts highest in the low gestational age group receiving iron supplements. Infants of the least gestational age were least able to achieve adequate tocopherol levels even with supplementation. In infants of less than 32 weeks gestational age, progressive improvement in intestinal tocopherol absorption was seen as full-term gestational age was approached. Thus administration of iron to vitamin E deficient premature infants actually caused a decrease in hemoglobin levels because of increased erythrocyte hemolysis.

3.6. *Vitamin E Requirements*

It is apparent from the nutritional and biochemical studies that the vitamin E requirement of animals, including man, depends on the dietary levels of prooxidants, antioxidants, selenium, sulfur amino acids, and other fat-soluble vitamins and on many other factors, such as the gestational age in premature infants and the efficiency of fat absorption. Many feel that the most important among these factors is the intake of polyunsaturated fatty acids. Thus Harris and Embree (1963) suggested that the daily tocopherol requirement is approximately 1 mg/0.06 g of PUFA consumed. This requirement indicates that the only role of vitamin E lies in prevention of peroxidation of dietary PUFA and that therefore no vitamin E is required when the diet is devoid of polyunsaturated fats. Of course, if an animal consumes over a long period of time a diet devoid of PUFA, death will eventually occur due to essential fatty acid deficiency. It is preferable, therefore, to set the minimum vitamin E requirement in a diet providing the level of linoleic acid which meets the minimum essential fatty acid requirements, and then one may assume that with each increase in PUFA above that basic essential fatty acid requirement the vitamin E requirement would be further increased by 1 mg/0.6 g of increased intake of PUFA.

Following the work of Noguchi *et al.* (1973), who demonstrated that both vitamin E and selenium are required for protection of mitochondrial and microsomal membranes of chick liver from peroxidation, Combs and Scott (1974*a*) undertook studies to determine the minimum dietary level of vitamin E required for protection of microsomal membranes. The basal diet contained all other nutrients required by the chick, including adequate sulfur amino acids, selenium, and sufficient corn oil to provide the essential fatty acid requirements of the chick. The basal diet also contained the synthetic antioxidant, ethoxyquin, at a level sufficient to prevent peroxidation of dietary PUFA and to prevent encephalomalacia. This diet was supplemented with vitamin E at levels of 0, 10, 25, 50, 100, or 150 IU *dl-α*-tocopheryl acetate per kilogram.

The results showed (1) that the level of tocopherol in the plasma increased significantly in a linear manner with increasing levels of dietary vitamin E until the level of 30 IU vitamin E per kilogram of diet was reached (Fig. 7) and (2) that with increasing dietary levels of vitamin E the hepatic microsomal peroxidation decreased significantly up to the level of 50 IU vitamin E/kg of diet, at which level peroxidation was almost completely prevented. By extrapolation of the linear regressions of both phases of the curve, the dietary requirement for vitamin E, in the presence of 0.15 ppm of selenium, for inhibition of *in vitro* ascorbic acid stimulated peroxidation in hepatic microsomes, was determined to be 30–50 IU/kg. Since no overt signs of vitamin E deficiency occur in chicks receiving 15 IU vitamin E/kg of diet with adequate selenium and sulfur amino acids (Scott, 1974), the level of vitamin E for optimum protection of microsomal membranes appears higher than that needed to prevent gross signs of vitamin E deficiency.

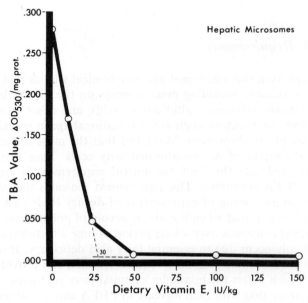

Fig. 7. Dietary vitamin E requirement of the chick determined at the subcellular level.

It is quite generally recognized that blood levels below 0.5 mg vitamin E per 100 ml plasma are deficient and that those above 1 mg/100 ml plasma are adequate. Whereas a dietary vitamin E level of 15 IU/kg produced a plasma tocopherol level of only 0.38 mg/100 ml, 30 IU vitamin E/kg of diet supported the plasma tocopherol level of 1.48 ± 0.45 mg/100 ml and at the same time reduced *in vitro* peroxidation of the microsomes to near zero levels (Combs and Scott, 1974*a*).

Horwitt (1960, 1962) found that erythrocyte hemolysis in experimental human subjects increased as the plasma tocopherol level dropped below approximately 1.0 mg/100 ml of plasma and reached a maximum (approximately 85% peroxidizability) as the plasma tocopherols dropped below 0.6 mg/100 ml. Horwitt studied the effect of graded levels of tocopherol supplementation in subjects who had been fed a vitamin E deficient diet for 54 months.In those receiving no tocopherol supplement, the plasma tocopherol dropped to 0.16 mg/100 ml after 138 days, and the *in vitro* peroxide hemolysis of the erythrocytes was 90%. Whereas 15 mg *d-α*-tocopheryl acetate or 20 mg *dl-α*-tocopheryl acetate per day for 138 days failed to increase the plasma tocopherol above 0.6–0.7 mg/100 ml of blood, administration of 60 mg *d-α*-tocopheryl acetate (70 IU) increased the plasma tocopherol level to 1.13 mg/100 ml plasma and reduced erythrocyte peroxide hemolysis to zero in 3 days. While these results indicate a somewhat higher vitamin E requirement for humans than that shown by Combs and Scott to be required in chicks, they do indicate that 20 IU per day in humans is not sufficient and that the requirement for normal blood level and prevention of erythrocyte hemolysis in these experiments lies between 20 and 80 IU vitamin E per day. Since man consumes approximately 0.5 kg of diet (dry basis) per day, this would indicate that the tocopherol requirement for establishment of a plasma tocopherol level of approximately 1 mg/100 ml of plasma lies between 40 and 160 IU/kg of diet. Combs *et al.* (1975) found that *in selenium-deficient chicks* a dietary vitamin E level of approximately 100 IU/kg was needed to produce a plasma tocopherol level of 1 mg/100 ml. Although the selenium status of the patients studied by Horwitt (1962) is unknown, it would appear from the studies with chicks (Combs *et al.*, 1975) that the dietary selenium intake of Horwitt's subjects may have been inadequate for maximum retention of vitamin E. A marginal selenium intake also increases the amount of vitamin E required for prevention of erythrocyte hemolysis.

The vitamin E requirement for human adults set forth by the National Research Council is 15 IU per day. Thus for a man consuming 0.5 kg of food per day the dietary requirement is 30 IU/kg of diet. This is in excellent agreement with the vitamin E requirement of chicks receiving adequate selenium and in the presence of normal dietary levels of essential fatty acids and sulfur amino acids. However, as indicated by Sauberlich *et al.* (1973), if the intake of linoleic acid is increased, e.g., from 16.5 g to 33 g daily, the supplement of *d-α*-tocopherol must be increased to 27.2 mg daily in order to prevent decrease in plasma tocopherol levels.

Although the National Research Council (USA) does not stipulate a vitamin E requirement for swine, Eggert *et al.* (1957) showed high mortality, yellow fat, and liver necrosis in pigs receiving a diet low in vitamin E and selenium. Supplementing this diet with either vitamin E (40 IU/kg) or selenium (1 ppm) completely prevented the deficiency syndrome.

Cordy (1954) prevented steatitis in kittens receiving a commercial food by giving 20 mg *dl*-α-tocopheryl acetate daily. Gershoff and Norkin (1962) found that prevention of *in vitro* erythrocyte hemolysis required a vitamin E supplementation of 136 IU/kg in diets containing 5% added tuna oil.

The minimum vitamin E requirements of various animals and man are shown in Table V. Some margins of safety over these requirements may be necessary in animals or human beings consuming larger than normal amounts of PUFA, in those exposed to higher than normal levels of oxygen, ozone, nitrites, or iron supplementation, or in persons suffering from steatorrhea.

3.7. Sources of Vitamin E

Vitamin E is present in the lipids of green leafy plants and the oils of seeds. Alfalfa meal, corn, and soybean represent the major sources of natural vitamin E for most animal feeds. In human nutrition the major sources of vitamin E are salad oils and dressings, shortenings, and margarines made from soybean, cottonseed, peanut, corn, and safflower oils. Unfortunately, much of the tocopherol in many of the best tocopherol sources is not α-tocopherol but is largely γ- or δ-tocopherol. Corn oil contains approximately 100 mg of total tocopherols per 100 g but only

Table V. Minimum Vitamin E Requirements

Animal	Requirements[a]
	IU/day
Man: Infants	5–10
Children	10–15
Adults	15
	IU/kg diet
Chickens: Starting and breeding	30
Growing and laying	10
Turkeys: Starting and growing	30
Breeding	40
Swine: Baby piglets	20
Growing pigs	10
Breeding gilts, sows, and boars	20
Cattle: Young and breeding	20
Horses: Young and breeding	20
Dogs	40
Cats	70
Minks and foxes	40
Fish: Salmon	30
Catfish	25[b]

[a]These levels are estimated to be adequate in complete diets (1) containing at least 0.1 ppm Se, (2) adequate in sulfur amino acids, and (3) containing no more than 1.5% linoleic acid. Vitamin E requirement increases in direct proportion to increased dietary intake of PUFA.

[b]In the presence of 125 mg ethoxyquin/kg diet; requirement in the absence of ethoxyquin was 100 IU/kg diet (Murai and Andrews, 1974).

*Table VI. Vitamin E Activities of Tocopherols, Tocotrienols,
and Tocopheramines*

		Gestation resorption, rats	Hemolysis, rats
Tocopherols:	α-	100	100
	β-	25–40	15–25
	γ-	8–19	3–18
	δ-	0.7	0.3
Tocotrienols:	α-	21	17
	β-	4	1–4
Tocopheramines:	*dl-α-*		86
	dl-β-		25
	dl-N-methyl-β-		132
	dl-γ-		20
	dl-N-methyl-γ-		125

10% is α-tocopherol. α-Tocopherol is the only tocopherol of nutritional significance, since as shown in Tables VI and VII the activities of β, γ- and δ-tocopherols
and the tocotrienols are generally less than 15–25% that of α-tocopherol. The
tocopheramines are not found in nature.

The tocopherol content of fresh oils is a poor indicator of the amount of
tocopherol in the diet, since much of the α-tocopherol may be destroyed in
cooking (Bunnell *et al.*, 1965). A particularly high loss of tocopherol appears
to occur during storage of foods which have been cooked in vegetable oils. Since
practically all commercially deep-fat-fried frozen foods such as french-fried
potatoes, scallops, shrimp, and chicken are prepared in vegetable oils or shortenings, these foods should supply a significant amount of α-tocopherol in the diet.
The studies of Bunnell and associates show, however, that this is not so. A great
loss of tocopherol was found in potato chips, french-fried potatoes, and other
foods even when stored at -12° C. Seventy-one percent of the tocopherol content
of oil extracted from potato chips was lost in storage for 1 month at room tempera-

*Table VII. Vitamin E Activity
of Tocopherols*

Compound	IU/mg
*dl-α-*Tocopheryl acetate	1[a]
*dl-α-*Tocopherol	1.1[a]
*dl-α-*Tocopheryl acid succinate	0.89[a]
*d-α-*Tocopheryl acetate	1.36[a]
*d-α-*Tocopherol	1.49[a]
*d-β-*Tocopherol	0.4
*dl-α-*Tocopherol	0.3
*d-γ-*Tocopherol	0.2
*dl-γ-*Tocopherol	0.15
*d-δ-*Tocopherol	0.016
*dl-δ-*Tocopherol	0.012

[a]National Formulary XIII.

ture and 63% was lost in 1 month at -12°C. Thus it was found that freezing does not protect α-tocopherol from destruction. Bunnell *et al.* (1965) pointed out that the formation of hydroperoxides is not prevented by low temperatures. Since hydroperoxides are produced as one of the first degradation steps in the oxidation of unsaturated fatty acids, and readily destroy tocopherol, this accounts for the large losses of tocopherols which occur during freezer storage, despite good commercial packaging practices.

The tocopherol content of animal feed ingredients and human foods is presented in Table VIII.

Bunnell *et al.* (1965), using the α-tocopherol content of the foods shown in Table VIII, estimated the α-tocopherol content of eight typical breakfast, luncheon, and dinner menus. These were found to range from 0.59 to 3.68, from 0.44 to 5.37, and from 1.61 to 6.38 mg α-tocopherol, respectively. Thus the daily intake of α-tocopherol could range between 2.6 mg and 15.4 mg, with an overall average of 7.4 mg, or approximately one-half the minimum NRC α-tocopherol requirement for an adult human. Bieri and Evarts (1973) calculated the average α-tocopherol intake of Americans to be approximately 9 mg per day, the average PUFA intake to be 21.2 g, and therefore the α-tocopherol to PUFA ratio to be 0.43 mg per gram of PUFA. Witting and Lee (1975) also found the average daily α-tocopherol intake of university students to be 7.5 mg and the linoleate intake to be 19.5 g per day (0.4 mg α-tocopherol/g PUFA). These values equilibrate to only 0.25 mg α-tocopherol per 0.6 g PUFA intake per day, a value far short of the 1 mg α-tocopherol per 0.6 g PUFA found by Harris and Embree (1963) for animals and confirmed by Horwitt (1960) in humans.

Although the vegetable oils represent our best natural source of vitamin E, they are also very high in polyunsaturated fatty acids, such that the ratio of α-tocopherol to PUFA in the vegetable oils is in the neighborhood of 0.2 mg α-tocopherol/0.6 g PUFA.

3.8. *Methods of Assay*

Determination of biologically active vitamin E is complicated by many interrelated factors. Chemical analyses must contend with the fact that eight different tocopherols occur in nature, and synthetic forms exist in many foods and feeds as *dl*-α-tocopherol or as *d*- and *dl*-α-tocopheryl acetates or succinates. Bioassays may be confounded by the levels of antioxidants, prooxidants, selenium, and other factors which may alter the biological activity of the vitamin.

However, enormous progress has been made during the past two decades in refinement of both the physicochemical assays, largely through use of thin-layer and gas chromatography for separation of the various isomers, through a better understanding of the factors which alter the bioassays for vitamin E. Bunnell (1967) has described in detail the various chemical and physicochemical assay methods for vitamin E. Bliss and György (1967) have presented details concerning currently accepted bioassays for the vitamin.

Table VIII. Vitamin E Content of Foods and Feedstuffs in Milligrams per 100 Grams

Food	α-Tocopherol	Total	Non-α-tocopherol + tocotrienols					
			β-Tᵃ	γ-T	δ-T	α-T-3ᵃ	β-T-3	γ-T-3
A. Major food sources of vitamin E								
1. Vegetable oils and products								
Cocoa butter, refined	2.6	12.9						
Coconut oil	1.0	3.1	0	0.6	—	0.5	0.1	1.9
Corn oil	12	53	—	52.5	0.5	—	—	—
Corn germ oil	23	105						
Cottonseed oil	32	31.3	—	31.3	—	—	—	—
Margarine	13	48						
Mayonnaise	6	3						
Olive oil	4	—						
Peanut butter	6.7	5						
Peanut oil	19	14	—	13.8	—	—	—	—
Rapeseed oil	18	39						
Rice bran oil	27	18						
Safflower oil	34	7		7				
Sesame oil	—	53	—	53	23	—	—	—
Shortening, processed	9.9	99	—	66	23	—	—	—
Soybean oil	10	85	10	63	22	—	—	—
Sunflower seed oil	25	6.5		6.5				
Wheat bran oil	20	75	10	—		11	54	
Wheat germ oil	115	77	66	—	—	2.6	8.1	—
2. Animal products								
Bacon	0.53	0.06		0.03		0.03		
Beef	0.3	0.3						
Beef liver	0.6	1.0						
Butter	1	0						
Cream, 20% fat	0.2	0						
Chicken	0.4	1.0						
Eggs	0.46	1.0	—	—	—	—	—	—

(Continued)

Table VIII. (Continued)

Food	α-Tocopherol	Non-α-tocopherol + tocotrienols						
		Total	β-T[a]	γ-T	δ-T	α-T-3[a]	β-T-3	γ-T-3
Ham	0.28	0.24	0.07					
Lard	1.5	0.15						
Liverwurst	0.35	0.34						
Milk	0.036	0.057		0.07				
Oil, anchovy	2.5	0						
cod liver	23	3						
herring	5	0						
menhaden	7.5	0						
Seafoods: haddock	0.6	0.6						
salmon	1.35	0.46						
scallops, cooked	0.6	5.6						
shrimp, cooked	0.6	6.0						
Turkey	0.8	0						
3. Plant sources								
Barley, grain	1.2	3.6	0.04	0.03	0.01	2.8	0.3	0.2
Corn, grain	2.0	3.7		1.5		0.7	—	1.5
Oats, grain	2.6	1.8	0.09			1.5	0.2	—
Rye, grain	1.6	2.0	0.4			1.5	0.8	—
Wheat, grain	1.0	3.9	0.7			0.4	2.8	
Bread, white	0.1	0.13	0.2			0.1	0.1	
whole wheat	0.45	1.75	0.25			0.15	1.3	
cornmeal	0.64	2.79	1.1			0.5		1.1
club crackers	0.8	0.37	0.04			0.03	0.3	
Hominy grits	0.31	0.86	0.10			0.06	0.7	
Oat cereal (dry)	0.6	0.93				0.9	0.03	
Oatmeal	2.27	1.7				1.6	0.1	
Onion rings, french fried	0.65	5.0	—	—		—	—	—
Peanuts	7.0	5.0	—	—	5.0			
Potatoes, french fried	0.3	1.0	—	—		—	—	—
Wheat and barley cereal	0.61	1.84	0.14			1.4	0.2	0.1

B. Vitamin E content of other foods[b]

	α-Tocopherol	Non-α-tocopherol + tocotrienols
1. Vegetables		
Cabbage	0.06	0.05
Carrots	0.11	0.1
Cauliflower	0.15	—
Celery	0.38	0.19
Corn, whole kernel, frozen	0.19	0.3
canned	0.05	0.04
Lettuce	0.06	0.11
Onion, fresh yellow	0.22	0.12
Peas, green, fresh	0.55	1.2
frozen	0.23	0.4
canned	0.02	0.02
Potatoes, raw	0.05	0.03
baked	0.03	0.02
Spinach, canned	0.02	0.04
Tomatoes, fresh	0.4	0.45
2. Meats		
Bologna	0.06	0.43
Lamb chop, broiled	0.16	0.16
Pork chops, fried	0.16	0.44
Pork sausage, fried	0.16	0.16
Salami	0.11	0.57
Veal cutlet, fried	0.05	0.19
3. Fruits and fruit juices		
Apple	0.31	0.2
Banana	0.22	0.2
Cantaloupe	0.14	0.17
Grapefruit juice, canned	0.04	0.14
Orange juice, fresh	0.04	0.16
Strawberries	0.20	0.20
Tomato juice, canned	0.22	0.49

(Continued)

Table VIII. (Continued)

Food	α-Tocopherol
C. Major vitamin E sources in feeds[c]	
Alfalfa, dehydrated, 17%	7.27
20%	8.90
sun cured, 13%	4.07
hay	5.27
Barley, whole (including T-3)	3.63
Brewer's grains, dried	2.68
Corn, whole yellow	1.99
Corn gluten feed	1.48
Corn gluten meal	2.59
Cottonseed meal	0.92
Distiller's dried corn solubles	5.58
Distiller's dried grain	3.05
Fat, animal	0.79
hydrolyzed animal and vegetable	5.68
poultry	1.89
Fish meals, herring	1.68
menhaden	0.57
Peruvian	0.21
others	0.56

Linseed meal	0.77
Malt sprouts	0.42
Meat and bone meal	0.08
Milo, whole	1.22
Molasses, cane	0.54
Oatmeal, feeding	2.36
Oats, whole	2.05
Peanut meal	0.29
Poultry by-products meal	0.22
Rice, brown	1.35
Rice bran	6.08
hulls	0.74
polishings	8.95
Safflower meal	0.08
Soybean hull flakes	0.66
Soybean meal, solvent process, 44%	0.30
50%	0.33
Soybean meal, unspecified	0.22
Wheat, whole	1.11
bran	1.71
middlings	2.68
mill run	3.17
red dog	3.55

[a] T, Tocopherols; T-3, tocotrienols.
[b] From Ames (1972), Bunnell et al. (1965), and Slover et al. (1969).
[c] From Bunnell et al. (1968).

3.8.1. Biological Assays

3.8.1.1. Resorption Gestation Assay

The need for vitamin E for prevention of fetal death has been recognized for many years as the most unequivocal assay for vitamin E. Briefly, it involves the rearing of female rats on vitamin E deficient diets from weaning or 12–14 days after birth until their body stores of vitamin E are depleted and the virgin females have reached 150 g body weight. They are then mated with normal males. After successful matings, the females are divided into groups and various levels of standard vitamin E preparations or unknown assay material are given by mouth. The vitamin supplements may be given as a single dose or divided doses on about the fourth day. If dosing is delayed beyond the tenth to twelfth day of pregnancy, vitamin E is ineffective. Supplements may be mixed in the diet but in this case care must be taken that the rats consume all of the diet. The females are killed after about 20–21 days. The uterus is opened and the numbers of living, dead, and resorbed fetuses are counted. Synthetic *dl-α*-tocopheryl acetate has been used as the International Standard, and 1 mg of this material, which is equivalent to 1 International Unit of vitamin E, is the amount needed to prevent resorption in approximately 50% of the rats receiving the vitamin E deficient diet.

3.8.1.2. Erythrocyte Hemolysis Test

In the erythrocyte hemolysis test, rats weighing about 100 g are fed a diet free of vitamin E for a depletion period of 3–4 weeks. A sample of blood is then taken from each rat and tested for hemolysis induction by addition of dialuric acid. Only animals showing a degree of hemolysis of 96–99% are retained in the test. In a preliminary trial with five animals per dose, dosages of vitamin E are determined which will reduce the hemolysis to a range of approximately 20–80% for both the standard and the test preparations. For the bioassay, two or three doses are selected from within this range for the standard and for each sample to be assayed. Each dose is dissolved in 0.2 ml olive oil and administered orally by stomach tube. After 40–44 hr the percent erythrocyte hemolysis is again measured. After a redepletion period of about 14 days, the rats can be used again for another bioassay, following the same procedures. The erythrocyte hemolysis test has been found to agree well with the results obtained by resorption gestation (Harris and Ludwig, 1949) and chick liver storage (Pudelkiewicz *et al.*, 1960).

3.8.1.3. Chick Liver Storage Assay for Vitamin E

Pudelkiewicz *et al.* (1960) used day-old White Plymouth Rock male chicks fed a vitamin E deficient diet for 13 days for assay of liver storage of vitamin E. The smallest and largest chicks were discarded, those remaining were distributed by weight into groups of eight chicks each; the tocopherol standards and the unknown materials to be assayed were then mixed with the basal diet and the chicks in all lots were pair-fed to the lot eating the least amount of feed over a period of 14 days. The tocopherol content of pooled livers from each group was then determined by chemical assay using the Emmerie-Engel reaction after careful extraction, molecular distillation, and chromatographic separation of the tocopherols.

3.8.2. Chemical Assays

Most chemical assays for tocopherols are based on the fact that when tocopherol reduces ferric to ferrous iron in the presence of α,α'-dipyridyl, a bathophenanthrolidone (Fabianek *et al.*, 1968), a red (or purple) complex is formed between the ferrous ions and the α,α'-dipyridyl. This red complex is easily measured in a spectrophotometer at 520 nm (Emmerie and Engel, 1938). The reaction is carried out in anhydrous, aldehyde-free ethanol, acetic acid, or a combination of xylene and *n*-propyl alcohol. In the absence of oxygen, the tocopherols are very readily distilled at fairly high temperatures under reduced pressure without destruction. This type of molecular distillation has been used in analytical procedures for separation of tocopherol from lipids (Quaife and Harris, 1948). Since light and alkaline pH also are catalysts in accelerating oxidation, care must be taken to protect the solutions from light during saponification.

The chemical assay methods involve saponification, solvent extraction of tocopherols from the sample, molecular distillation, and two-dimensional reverse-phase paper chromatography, or column chromatography with either Florex or secondary magnesium phosphate. Following specific elution techniques, the various fractions are assayed by the Emmerie-Engel method. Since separation of β- and γ-tocopherols is difficult with normal chromatography, nitroso derivatives of these compounds have been made which do separate on a zinc carbonate column.

3.8.2.1. Nitroso Reaction

All of the tocols except α-tocopherol and α-tocotrienol react with nitrous acid to form yellow nitroso derivatives. The difference between the total tocopherols as measured by the Emmerie-Engel method and the non-α forms as determined by the nitroso reaction gives the α-tocopherol plus α-tocotrienol.

3.8.2.2. Gas–Liquid Chromatography

Gas chromatographic methods originally developed by Bieri and Andrews (1963) have been further modified by Slover *et al.* (1967, 1969) and others.

In the gas chromatographic method, the individual tocopherols or tocotrienols are estimated as trimethylsilyl ethers using methyl silicon rubber SE-30 or Apiezon L as a liquid phase in concentrations from 1% to 10% and a temperature of 235°C. The α-tocopherols can be separated from the β- and γ-tocopherols, which come off together, and δ-tocopherol in a third fraction. It is possible to separate α-tocopherol from α-tocotrienol by gas chromatographic methods, and this method is also a convenient way of determining the amount of vitamin E present as free tocopherol and that esterified as its acetate. It is important that standards be run in each assay. Recoveries of added tocopherols should range between 95% and 100%.

3.8.2.3. Thin-Layer Chromatography

In thin-layer chromatography the degree of separation depends on the amount of foreign matter which is transferred to the plate, and complete separation therefore often is not accomplished. In using thin-layer chromatography for quantitative measurements, the spot or line is removed and the tocopherol is then eluted and measured by the Emmerie-Engel test. Thin-layer chromatographic methods generally show recoveries of only about 80% of added tocopherols.

3.9. Absorption, Transport, and Storage of Vitamin E

Studies on absorption, transport, and storage of vitamin E have shown (1) the essentiality of lipid–bile micelles in intestinal absorption, (1) the effect of various dietary lipids on absorption, (3) the digestion (hydrolysis) of *d*- and *dl*-α-tocopheryl acetates and succinates in the intestine, (4) the transport of vitamin E in the lipoproteins of plasma, and (5) the storage of the vitamin in adipose and parenchymatous tissue (Gallo-Torres, 1973).

3.9.1. Essential Role of Lipid–Bile Salt Micelles for Intestinal Absorption of Tocopherol

Thompson and Scott (1970) found that selenium is required to prevent atrophy of the pancreas. In selenium-deficient chicks, absorption of lipids including vitamin E was poor because of impairment, in the absence of fat hydrolysis, of normal lipid–bile micelle formation. Addition of bile acids and monoglycerides to the diet did not restore fat digestion to normal but markedly improved vitamin E absorption, thereby demonstrating that both bile and monoglyceride are necessary for the lipid–bile micelle formation needed for vitamin E absorption.

Gallo-Torres (1970) also reported the obligatory role of bile for the intestinal absorption of vitamin E into the lymph of rats. Only negligible amounts of radioactivity could be detected in the thoracic duct lymph when both bile and pancreatic juice were absent from the duodenum.

The lipid–bile micelle is required to transport the fat-soluble vitamin E across the "unstirred water layer" representing the aqueous phase in the intestinal lumen immediately adjacent to the brush border of the microvilli. Vitamin E is absorbed, together with free fatty acids, monoglycerides, and other fat-soluble vitamins, by penetrating the epithelial cell through the apical plasma membrane of the absorptive cells in the brush border. Studies on absorption and transport of α-tocopherol by Davies *et al.* (1971) show that, for normal utilization of absorbed α-tocopherol, normal lipoprotein transport mechanisms are involved. Transfer of vitamin E from the cell thus requires several stages. It must, in mammals, pass through the lateral or basal plasma membrane of the cell and through the basal lamina and then enter the fluid of the lamina propria. From this the vitamin enters the capillaries of the lymph and is transported in the chylomicrons (Dobbins, 1975). Only small amounts of tocopherol apparently are transported from the intestine via the portal vein in mammals, whereas all of the tocopherol absorption in birds occurs via the portal vein directly to the liver. Absorption across the intestinal membrane in birds apparently occurs by the same mechanism as that in mammals, but in the case of birds the vitamin E is transported directly to the liver by "portomicrons" (Bensadoun and Rothfeld, 1972) in the portal blood.

Wiss and Weber (1965) and Gallo-Torres *et al.* (1973) have shown a relationship between the level of fat in the diet and absorption of vitamin E. In confirmation of the studies of Thompson and Scott (1970), Gallo-Torres *et al.* (1973) found that monoolein or triolein in normal rats increased the absorption of vitamin E. However, dietary linoleic acid at high levels caused a marked decrease in the percentage of the administered dose of vitamin E that appeared in the lymphatics. Akerib and Sterner (1971) have confirmed this inhibition of vitamin E absorption by polyunsaturated fatty acids. Although under normal dietary conditions vitamin E present in oils or in the natural state requires lipid–bile micelles for absorption, Schmandke and Schmidt (1965) found that *dl*-α-tocopherol is absorbed twice as well from an aqueous-miscible emulsion as from an oil solution on oral administration. Harries and Muller (1971) found water-miscible α-tocopheryl acetate to be superior to fat-soluble preparations in oral treatment of children with cystic fibrosis. A daily dose of 1 mg α-tocopheryl acetate/kg body weight corrected preexisting vitamin E deficiencies within 2 months and was adequate for subsequent maintenance of adequate vitamin E blood levels.

Schmandke *et al.* (1969) studied the intestinal absorption of α-tocopherol in 30 male subjects receiving graded levels of the vitamin ranging from 10 to 2000 mg, *per os*. The results, shown in Table IX, indicate good absorption in normal humans of levels up to 1500 mg. The normal excretion rate of α-tocopherol plus α-tocoquinone was 3.5 ± 2.6 mg/day, of which $50.6 \pm 23.6\%$ was present in feces as α-tocoquinone.

It is possible that certain types of lipids such as the polyunsaturated fatty acids may compete with vitamin E for sites of absorption and transport, either in the lipid–bile micelle or during ingress and egress from the intestinal cell. Thus it was shown by Combs and Scott (1974*b*) that vitamin A at 10^6 IU/kg of diet reduced plasma tocopherol levels to approximately one-tenth those obtained with the same diet in the absence of added vitamin A.

Relative absorption of α-, β-, γ-, and δ-tocopherols was studied by Pearson and Barnes (1970) by administration of 500–800 μg of each tocopherol in 0.2 ml

Table IX. Intestinal Absorption of
α-Tocopherol[a]

Dose (mg)	Amount absorbed	
	mg	Percent of dose
10	9.7 ± 1.3	96.9 ± 13.0
30	26.1 ± 3.9	87.3 ± 13.5
50	36.2 ± 5.8	72.6 ± 11.6
100	81.5 ± 6.1	81.5 ± 6.1
500	337.3 ± 96.8	67.6 ± 19.4
1500	1053.8 ± 301.3	70.3 ± 27.9
2000	1104.0 ± 576.1	55.2 ± 28.8

[a]From Schmandke *et al.* (1969).

triolein injected into small intestinal loops of hooded female rats and determination of the amount that disappeared from the intestinal loop. It was found that 32% of the α-tocopherol, 18% β-tocopherol, 30% γ-tocopherol, and 1.8% δ-tocopherol disappeared over the 6-hr experimental period. Bieri and Poukka (1970) found that the red cells of patients with α,β-lipoproteinemia were capable of maintaining a normal α-tocopherol content even though the plasma level of vitamin E was greatly depressed. The ratio of micromoles α-tocopherol to millimoles PUFA in red cells was found to be about 1:850.

3.9.2. Studies on Absorption and Retention of d- and l-Epimers of α-Tocopherol

Scott and Desai (1964) showed that the relative anti-muscular dystrophy activities of the d- and l-epimers of α-tocopheryl acetate determined in chick bioassay were 1.46 and 0.36, respectively, as compared with the reference standard dl-α-tocopheryl acetate (unofficial Animal Nutrition Research Council nutrition standard), which was assigned an antimuscular dystrophy value of 1 unit per milligram. Plasma tocopherol analyses showed that the prevention of muscular dystrophy was correlated directly with the total plasma tocopherol levels and that when 0.9–1.0 mg tocopherol per 100 ml of blood was present the muscular dystrophy was completely prevented regardless of the form of tocopherol included in the diet. Desai *et al.* (1965) showed, in confirmation of studies by Weber *et al.* (1964) with rats, that l-α-tocopherol is absorbed as well as or better than the d-form of the vitamin. It appeared therefore that the differences in biopotency must be due to differences in retention whereby the d-epimer is retained much better than the l-epimer in the blood and perhaps in our tissues of the body. The results indicated the existence of an active carrier of d-α-tocopherol in the blood and tissues which has a greater affinity for the d-epimer than for l-α-tocopherol. Experiments by Desai and Scott (1965) and Scott (1965) comparing the plasma tocopherol levels obtained from oral administration of d- and l-α-tocopheryl acetates in the presence of graded levels of dietary selenium indicated that selenium is involved in some unknown way in the retention of d-α-tocopherol in plasma. Whether the differences in plasma levels of d- and l-epimers of α-tocopherol are the result of differences (1) in rate of excretion, (2) in rate of destruction, (3) in the affinity of the epimers for specific carriers, or (4) in chemical activity influenced by the structural configurations remains to be determined. It is clear, however, that selenium bears a synergistic relationship to vitamin E in two different ways: first by backing up the effect of vitamin E in preventing chain-reactive peroxidation of vital phospholipids and second by promoting in some unknown way a higher blood level of d-α-tocopherol. Combs and Scott (1974b) found that antioxidants, including vitamin E, increase the utilization of dietary selenium in the chick.

3.9.3. Storage

The evidence on storage of α-tocopherol indicates that while tocopherol is present in adipose tissue its presence there may be due simply to its lipid solubility. However, Century and Horwitt (1965), Krishnamurthy and Bieri (1963), Csallany and Draper (1960), and Lucy (1972) have shown that d-α-tocopherol appears to be concentrated in the phospholipids of mitochondria and microsomes, where it appears to be held by some definite attachment which may be intimately concerned with its activity in prevention of peroxidation of these organelles.

3.10. High Dietary Intakes of Vitamin E

Numerous experiments have been conducted in an attempt to determine if very high dietary intakes of vitamin E are toxic. The clinical studies are largely in agreement with the very recent investigations of Farrell and Bieri (1975) on "Megavitamin E Supplementation in Man." In a group of 28 adults ingesting 100–800 IU α-tocopherol per day for an average of 3 years, laboratory screening for toxic side effects by performance of 20 standard clinical blood tests failed to reveal any disturbances in liver, kidney, muscle, thyroid gland, erythrocytes, leukocytes, coagulation parameters, or blood glucose. It was concluded that megavitamin E supplements produce no toxic side effects. Approximately one-half of the subjects claimed a general feeling of improved health or well-being, although no specific beneficial effects were noted. The other half indicated no change in health status throughout the period of high vitamin E intake.

It is possible that high vitamin E intakes may induce deficiencies of the other fat-soluble vitamins if these are present at borderline levels in the diet. Thus March *et al.* (1973) reported that in chicks receiving a vitamin D deficient diet and the *minimum* NRC requirements of vitamin K and vitamin A, the inclusion of 1000–2200 IU vitamin E kg diet depressed growth, reduced hematocrit values, increased the prothrombin times, and reduced bone calcification. These effects of massive doses of vitamin E in relation to minimum levels of the other fat-soluble vitamins provide further evidence of competition for absorption among the fat-soluble vitamins at the lipid–bile micelle level in the intestinal tract.

Large doses of vitamin E not only are nontoxic but also apparently counter act toxicity due to excessive doses of vitamin A. McCuaig and Motzok (1970) reported that in animals receiving about 1000 times the normal nutritional levels of these vitamins, growth was depressed and chicks died on the high vitamin A intake unless also given vitamin E. Massive doses of vitamin E up to 10,000 IU/kg diet were not toxic in these experiments.

Jenkins and Mitchell (1975) found that dl-α-tocopheryl acetate (6000 IU/kg diet) counteracted most toxic effects of hypervitaminosis A (2.9×10^6 IU vitamin A/kg diet) in male Holtzman rats. Liver vitamin A increased with increasing levels of dietary vitamin E. Blood urea nitrogen and plasma cholesterol levels were unchanged, even though excess vitamin A significantly increased adrenal weight and vitamin E reduced it.

A highly significant correlation was found between tocopherol content and stability of milk against oxidized flavors (Krukovsky *et al.*, 1949, 1952). Milk containing less than 2.0 mg tocopherol/100 g fat showed poor keeping qualities. The tocopherol content of fresh pasteurized milk could be increased by feeding the cow a diet high in vitamin E. The shelf life of cold-storage chickens and turkeys was shown to be markedly increased by high dietary levels of vitamin E (Mecchi *et al.*, 1956). The required dietary vitamin E level for optimum reduction of rancidity in chickens was found by Marusich *et al.* (1975) to be 40 IU/kg diet for 8 weeks or 160 IU/kg diet for the last 5 days before marketing. Turkeys require higher levels: 200 IU/kg diet for 4 weeks or 400 IU/kg diet during the last 3 weeks was needed for optimum delay in onset of rancidity. It is evident that body storage of vitamin E: is vital, both for protection of lipid membranes *in vivo* and for the keeping qualities of meat and milk.

3.11. Conclusions

Vitamin E, discovered more than 50 years ago, has remained for half a century one of the great enigmas of nutritional science. While severe deficiencies of the vitamin affect the health and integrity of almost every tissue and produce a wide variety of deficiency diseases in experimental animals, no specific overt vitamin E deficiency disease has ever been described in humans.

Vitamin E nutrition is intimately interrelated with that of selenium and depends also on the levels of prooxidants and antioxidants present in the diet.

Evidence is available showing the necessity of adding vitamin E to many animal diets, particularly diets for breeding animals and those fed to very young animals.

At the cellular level, the requirement for vitamin E has been demonstrated in both experimental animals and man. Studies to determine the optimum vitamin E level for maximum protection of mitochondrial and microsomal membranes showed this requirement to be approximately 30 Internationl Units of vitamin E per kilogram of diet. This is in excellent agreement with the NRC (1974) recommended dietary allowance (RDA) of 15 IU/day in man, as a minimum requirement. However, in view of increasing exposures of man to atmospheric pollutants, the 1969 RDA of 30 IU vitamin E/day may be more realistic of actual needs.

Good experimental evidence indicates a need for supplemental vitamin E in diets of pregnant and lactating mothers, newborn infants, particularly premature infants, and older people suffering from circulatory disturbances and intermittent claudication. Analyses of typical menus indicate that the average American diet falls far short of providing the 15 International Units of vitamin E per day specified as the recommended dietary allowance of adult man by the Food and Nutrition Board of the National Research Council (1974).

While megavitamin E supplementation of normal human diets has been shown to produce no detectable ill effects, the evidence presented in this chapter supports the view that the minimum vitamin E requirement of normal individuals is no higher than 30 IU/kg diet. Higher levels are indicated for persons exposed to oxygen or to environmental pollutants such as O_3 and NO_2. Menzel *et al.* (1975) reported that addition of low concentrations of methyl oleate ozonide to erythrocytes of vitamin E deficient volunteers resulted in production of Heinz bodies, which were significantly reduced in the erythrocytes after the volunteers had ingested 100 mg *d*-α-tocopheryl acetate daily for 1 week. Therapeutic levels of vitamin E are indicated for patients having intermittent claudication or other evidence of prolonged deficiency. In patients with biliary insufficiency or cystic fibrosis of the pancreas, the vitamin may be administered parenterally (Newmark *et al.*, 1975) or orally together with bile salts and monoglycerides to promote normal absorption.

Vitamin E is nature's best lipid-soluble antioxidant. Its vitamin function stems, however, from its specificity of action in the animal body. The phospholipids of mitochondria, endoplasmic reticulum, and plasma membranes possess special affinities for α-tocopherol. The vitamin appears to locate at specific sites within these membranes and to prevent or inhibit initiation of lipid peroxidation. In vitamin E deficient animals, these organelles may undergo varying degrees of loss of activity, depending on the extent of the deficiency. Because the organelles most readily affected by vitamin E deficiency are those which function to ensure normal metabolism and to provide defense mechanisms against environmental insults, supplementation of deficient diets with vitamin E often produces beneficial effects by aiding the animal or man to better cope with various stresses and diseases.

3.12. References

Adamstone, F. B., 1947, Histologic comparison of brains of vitamin-A-deficient, vitamin E-deficient chicks, *Arch. Pathol.* **43**:301–312.

Adamstone, F. B., and Card, L. E., 1934, The effects of vitamin E deficiency on the testes of the male fowl (*Gallus domesticus*), *J. Morphol.* **56**:339–359.

Akerib, M., and Sterner, W., 1971, Inhibition of vitamin E absorption by a lipid fraction, *Int. J. Vit. Nutr. Res.* **41**:42–43.

Ames, S. R., 1972, Tocopherols: Occurrence in foods, in: *The Vitamins*, Vol. 5 (W. H. Sebrell, Jr., and R. S. Harris, eds.), pp. 233–248, Academic Press, New York.

Anderson, T. W., and Reid, D. B. W., 1974, A double-blind trial of vitamin E in angina pectoris, *Am. J. Clin. Nutr.* **27**:1174–1178.

Andrews, E. D., Hartley, W. J., and Grant, A. B., 1968, Selenium-responsive diseases of animals in New Zealand, *New Zealand Vet. J.* **16**:3–17.

Anonymous, 1960, "Therapy" with vitamin E, *Nutr. Rev.* **18**:227–229.

Anonymous, 1974, Supplementation of human diets with vitamin E, *Nutr. Rev. Suppl.* **32**:35–38.

Armstrong, D., Dimmitt, S., Van Wormer, D. E., and Okla, T., 1974, Studies in Batten disease, *Arch. Neurol.* **30**:144–152.

Aterman, K., 1959, Selenium, liver necrosis and depigmentation of incisors in the rat, *Br. J. Nutr.* **13**:38–41.

Atkinson, R. L., Ferguson, T. M., Quisenberry, J. H., and Couch, J. R., 1955, Vitamin E and reproduction in turkeys, *J. Nutr.* **55**:337–397.

Barker, M. O., Brin, M., and Hainsselin, L., 1973, A rapid micromethod for the functional evaluation of vitamin E status in rats with the fragiligraph, *Biochem. Med.* **8**:1–10.

Baustad, B., and Nafstad, I., 1972, Haematological response to vitamin E in piglets. *Br. J. Nutr.* **28**:183–190.

Bensadoun, A., and Rothfeld, A., 1972, The form of absorption of lipids in the chicken, *Gallus domesticus*, *Proc. Soc. Exp. Biol. Med.* **141**:814–817.

Bergel, M., 1967, Lysosomes and their relation to vitamin E and leprosy, *Leprosy Rev.* **38**:189–198.

Bergel, M., 1971, Personal communication to the author, Buenos Aires, Argentina.

Bergel, F., Jacob, A., Todd, A. R., and Work, T. S., 1938, Synthetic experiments in the coumaran and chroman series: The structure of tocopherols, *J. Chem. Soc.* **1938**:1375–1382.

Berneske, G. M., Butson, A. R. C., Gauld, E. N., and Levy, D., 1960, Clinical trial of high dosage vitamin E in human muscular dystrophy, *Can. Med. Assoc. J.* **82**:418–421.

Bieri, J. G., 1969, Biological activity and metabolism of *N*-substituted tocopheramines: Implications on vitamin E function, in: *The Fat-Soluble Vitamins* (H. F. DeLuca and J. W. Suttie, eds.), pp. 307–316, University of Wisconsin Press, Madison.

Bieri, J. G., 1975, Vitamin E, *Nutr. Rev.* **33**:161–167.

Bieri, J. G., and Andrews, E. L., 1963, The determination of α-tocopherol in animal tissues by gas liquid chromatography, *Iowa State J. Sci.* **38**:3–12.

Bieri, J. G., and Evarts, R. P., 1973, Tocopherol and fatty acids in American diets, *J. Am. Dietet. Assoc.* **62**:147–151.

Bieri, J. G., and Poukka, R. K. H., 1970, Red cell content of vitamin E and fatty acids in normal subjects and patients with abnormal lipid metabolism, *Int. J. Vit. Res.* **40**:344–350.

Bieri, J. G., Teets, L., Belavady, B., and Andrews, E. L., 1964, Serum vitamin E levels in a normal adult population in the Washington, D. C. area, *Proc. Soc. Exp. Biol. Med.* **117**:131–133.

Blaxter, K. L., 1955, Vitamin E and experimental muscular degeneration in calves, in: *Proceedings of the Third International Congress on Vitamin E.*, p. 622.

Blaxter, K. L., Brown, F., Wood, W. A., and MacDonald, A. M., 1953, The nutrition of the young Ayrshire calf. 14. Some effects of natural and synthetic anti-oxidants on the incidence of muscular dystrophy induced by cod-liver oil, *Br. J. Nutr.* **7**:337–349.

Bliss, C. K., and György, P., 1967, Bioassays of vitamin E, in: *The Vitamins* (P. György and W. N. Pearson, eds.), pp. 304–316, Academic Press, New York.

Bonetti, E., and Stirpe, F., 1963, Effect of selenium on muscular dystrophy in vitamin E-deficient rats and guinea pigs, *Proc. Soc. Exp. Biol. Med.* **114**:109–115.

Bonetti, E., Abbondanza, A., Della Corte, E., Novello, F., and Stirpe, F., 1975, Studies on the formation of lipid peroxides and on some enzymic activities in the liver of vitamin E-deficient rats, *J. Nutr.* **105**:364–371.

Bowman, O. M., and Dyer, I. A., 1960, Experimentally induced muscular dystrophy: Blood creatine levels and histopathological changes in dystrophic rabbits, *J. Nutr.* **72**:289–292.

Boyd, A. M., and Marks, J., 1963, Treatment of intermittent claudication—Reappraisal of the value of alpha-tocopherol, *Angiology* **14**:198–208.

Boyd, A. M., and Marks, J., 1967, Vitamin E and circulation, in: *Symposium on Vitamins A, E and K*, West Berlin.

Brin, M., Horn, L., and Barker, M. O., 1974, Relationship between fatty acid composition of erythrocytes and susceptibility to vitamin E deficiency, *Am. J. Clin. Nutr.* **27**:945–951.

Bunnell, R. H., 1967, Vitamin E assay by chemical methods, in: *The Vitamins* (P. György and W. N. Pearson, eds.), pp. 261–304, Academic Press, New York.

Bunnell, R. H., Keating, J., Quaresimo, A., and Parman, G. K., 1965, Alpha-tocopherol content of foods, *Am. J. Clin. Nutr.* **17**:1–10.

Bunnell, R. H., Keating, J. P. and Quaresimo, A. J., 1968, Alpha-tocopherol content of feedstuffs, *J. Agr. Food Chem.* **16**:659–664.

Bunyan, J., Green, J., Edwin, E. E., and Diplock, A. T., 1960a, Studies on vitamin E. 3. The relative activities of tocopherols and some other substances *in vivo* and *in vitro* against dialuric acid-induced haemolysis of erythrocytes, *Biochem. J.* **75**:460–467.

Bunyan, J., Green, J., Edwin, E. E., and Diplock, A. T., 1960b, Studies on vitamin E. 5. Lipid peroxidation in dialuric acid-induced haemolysis of vitamin E-deficient erythrocytes, *Biochem. J.* **77**: 47–51.

Caasi, P. I., Hauswirth, J. W., and Nair, P. P., 1972, Biosynthesis of heme in vitamin E deficiency, *Ann. N. Y. Acad. Sci.* **203**:93–102.

Cahn, R. S., Ingold, C. K., and Prelog, V., 1956, Specification of asymmetric configuration in organic chemistry, *Experientia* **12**:81–94.

Calvert, C. C., Monroe, R. A., and Scott, M. L., 1961, Studies on phosphorus metabolism in dystrophic chicks, *J. Nutr.* **73**:355–362.

Campbell, P. A., Cooper, H. R., Heinzerling, R. H., and Tengerdy, R. P., 1974, Vitamin E enhances *in vitro* immune response by normal and nonadherent spleen cells, *Proc. Soc. Exp. Biol. Med.* **146**:465–469.

Caputto, R., McCay, P. B., and Carpenter, M. P., 1958, Requirement of Mn^{++} and Co^{++} for the synthesis of ascorbic acid by liver extracts of animals deprived of tocopherol, *J. Biol. Chem.* **223**: 1025–1029.

Carpenter, M. P., 1972, Vitamin E and microsomal drug hydroxylations, *Ann. N.Y. Acad. Sci.* **203**:81–92.

Carpenter, M. P., and Howard, C. N., 1974, Vitamin E, steroids, and liver microsomal hydroxylations, *Am. J. Clin. Nutr.* **27:**966–979.

Carpenter, M. P., McCay, P. B., and Caputto, R., 1958, Creatine-phosphate utilization by muscle extracts of rabbits on vitamin E-deficient diets, *Proc. Soc. Exp. Biol. Med.* **97:**205–209.

Catignani, G. L., Chytil, F., and Darby, W. J., 1974, Vitamin E deficiency: Immunochemical evidence for increased accumulation of liver xanthine oxidase, *Proc. Natl. Acad. Sci. USA* **71:**1966–1968.

Cawthorne, M. A., Bunyan, J., Diplock, A. T., Murrell, E. A., and Green, J., 1968, On the relationship between vitamin A and vitamin E in the rat, *Br. J. Nutr.* **22:**133–143.

Century, B., and Horwitt, M. K., 1960, Role of diet lipids in the appearance of dystrophy and creatinuria in the vitamin E deficient rat, *J. Nutr.* **72:**357–367.

Century, B., and Horwitt, M. K., 1965, Biological availability of various forms of vitamin E with respect to different indices of deficiency, *Fed. Proc.* **24:**906–911.

Cheraskin, E., and Ringsdorf, W. M., Jr., 1970, Daily vitamin E consumption and reported cardio-vascular findings, *Nutr. Rep. Intern.* **2:**107.

Christensen, F., Dam, H., Gortner, R. A., Jr., and Søndergaard, E., 1956, *In vitro* hemolysis of erythrocytes from vitamin E deficient rats and chicks, *Acta Physiol. Scand.* **35:**215–224.

Christensen, F., Dam, H., Prange, I., and Søndergaard, E., 1958, The effect of selenium on vitamin E-deficient rats, *Acta Pharmacol. Toxicol.* **15:**181–188.

Clein, L. J., Williams, H. T. G., and Macbeth, R. A., 1962, The use of tocopherol in the treatment of intermittent claudication, *Can. Med. Assoc. J.* **86:**215–216; *Can. Med. Assoc. J.* **87:**538–541.

Coffin, D. L., and Holzworth, J., 1954, "Yellow fat" in two laboratory cats: Acid-fast pigmentation associated with a fish-base ration, *Cornell Vet.* **44:**63–71.

Cohen, G., and Hochstein, P., 1962, Hydrogen peroxide detoxification in acatalasic and normal erythrocytes, *Fed. Proc.* **21:**69 (abstr.).

Combs, G. F., Jr., and Scott, M. L., 1974*a*, Dietary requirements for vitamin E and selenium measured at the cellular level in the chick, *J. Nutr.* **104:**1292–1296.

Combs, G. F., Jr., and Scott, M. L., 1974*b*, Antioxidant effects on selenium and vitamin E function in the chick, *J. Nutr.* **104:**1297–1303.

Combs, G. F., Jr., Noguchi, T., and Scott, M. L., 1975, Mechanisms of action of selenium and vitamin E in protection of biological membranes, *Fed. Proc.* **34:**2090–2095.

Cordy, D. R., 1954, Experimental production of steatitis (yellow fat disease) in kittens fed a commercial canned cat food and prevention of the condition by vitamin E, *Cornell Vet.* **44:**310–318.

Corwin, L. M., and Schwarz, K., 1959, An effect of vitamin E on the regulation of succinate oxidation in rat liver mitochondria, *J. Biol. Chem.* **234:**191–197.

Csallany, A. S., 1971, A reappraisal of the structure of a dimeric metabolite of α-tocopherol, *Int. J. Vit. Nutr. Res.* **41:**376–384.

Csallany, A. S., and Draper, H. H., 1960, Determination of *N,N'*-diphenyl-*p*-phenylenediamine in animal tissues, *Proc. Soc. Exp. Biol. Med.* **104:**739–742.

Csallany, A. S., and Draper, H. H., 1963, The structure of a dimeric metabolite of *d-α*-tocopherol isolated from mammalian liver, *J. Biol. Chem.* **238:**2912–2918.

Csallany, A. S., Draper, H. H., and Shah, S. N., 1962, Conversion of *d-α*-tocopherol-C_{14} to tocopheryl-*p*-quinone *in vivo*, *Arch. Biochem. Biophys.* **98:**142–145.

D'Agostino, L., 1952, Influence of alpha-tocopherol on urinary elimination of thiamine, *Bull. Soc. Ital. Biol. Sper.* **28:**157–159.

Dalle Coste, P., and Klinger R., 1955, α-Tocopherol in diabetic diseases of the blood vessels, *Riforma Med.* **69:**853–856.

Dam, H., 1944, Studies on vitamin E-deficiency in chicks, *J. Nutr.* **27:**193–211.

Dam, H., 1957, Influence of antioxidants and redox substances on signs of vitamin E deficiency, *Pharmacol. Rev.* **9:**1–16.

Dam, H., 1962, Interrelations between vitamin E and polyunsaturated fatty acids in animals, *Vitamins Hormones* **20:**527–549.

Dam, H., and Glavind, J., 1939, Alimentary exudative diathesis, a consequence of vitamin E-avitaminosis, *Nature (London)* **143:**810–818.

Dam, H., and Granados, H., 1945, Peroxidation of body fat in vitamin E deficiency, *Acta Physiol. Scand.* **10:**162–171.

Dam, H., Prange, I., and Søndergaard, E., 1952, Muscular degeneration (white striations of muscles) in chicks reared on vitamin E deficient, low fat diets, *Acta Pathol. Microbiol. Scand.* **31:**172.

Davies, A. W., and Moore, T., 1941, Interaction of vitamins A and E, *Nature (London)* **147**:794–796.

Davies, T., Kelleher, J., Smith, C. L., and Losowsky, M. S., 1971, The effect of orotic acid on the absorption, transport and tissue distribution of α-tocopherol in the rat, *Int. J. Vit. Nutr. Res.* **41**: 360–367.

Davis, G. L., Maynard, L. A., and McCay, C. M., 1938, *Cornell University Agricultural Experiment Station Memoir*, No. 217.

Desai, I. D., 1968, Plasma tocopherol levels in normal adults, *Can. J. Physiol. Pharmacol.* **46**:819–22.

Desai, I. D., and Scott, M. L., 1965, Mode of action of selenium in relation to biological activity of tocopherols, *Arch. Biochem. Biophys.* **110**:309–315.

Desai, I. D., Parekh, C. K., and Scott, M. L., 1965, Absorption of *d*- and *l*-α-tocopheryl acetates in normal and dystrophic chicks, *Biochim. Biophys. Acta* **100**:280–272.

Desai, I. D., O'Leary, L. P., and Schwartz, N., 1972, Vitamin E status of proprietary infant formulas available in Canada, *Nutr. Rep. Intern.* **6**:83–92.

DeWitt, W. B., and Schwarz, K., 1958, Multiple dietary necrotic degeneration in the mouse, *Experientia* **14**:1–7.

Dicks, M. W., 1965, Vitamin E content of foods and feeds for human and animal consumption, *Univ. Wyoming Agr. Exp. Sta. Bull.* **435**:1–194.

DiNardo, A., Gatti, E, and Pace, G, 1955, Vitamin E and diabetes mellitus, *Acta Vitaminol.* **9**:172–180.

Dinesen, B., and Clausen, J., 1971, Changes in the activity of nuclear RNA-polymerase in the liver and brain of the rat as consequences of a deficiency of vitamin E and of linoleic acid, *Acta Agr. Scand. Suppl.* **19**:142–151.

Dinning, J. S., 1953, An elevated xanthine oxidase in livers of vitamin E-deficient rabbits, *J. Biol. Chem.* **202**:213–215.

Dinning, J. S., and Day, P. L., 1957, Vitamin E deficiency in the monkey. I. Muscular dystrophy, hematologic changes, and the excretion of urinary nitrogenous constituents, *J. Exp. Med.* **105**: 395–401.

Dinning, J. S., and Day, P. L., 1958, Vitamin E deficiency in the monkey. III. The metabolism of sodium formate C^{14}, *J. Biol. Chem.* **233**:240–242.

Diplock, A. T., 1974, Vitamin E, selenium, and the membrane-associated drug-metabolizing enzyme system of rat liver, *Vitamins Hormones* **32**:445–461.

Dju, M. H., Mason, K. E., and Filer, L. G., 1958, Vitamin E in human tissues from birth to old age, *Am. J. Clin. Nutr.* **6**:50–60.

Dobbins, W. O., III, 1975, Human intestinal epithelium as a biological membrane, in: *Pathobiology of Cell Membranes* (B. F. Trump and A. V. Arstila, eds.), pp. 429–453, Academic Press, New York.

Draper, H. H., 1957, Ineffectiveness of selenium in the treatment of nutritional muscular dystrophy in the rabbit, *Nature (London)* **180**:1419.

Draper, H. H., 1969, Vitamin E in human nutrition, in: *Symposium on the Biochemistry, Assay and Nutritional Value of Vitamin E*, p. 69, Association of Vitamin Chemists, Chicago.

Draper, H. H., Bergan, J. G., Chiu, Mei, Csallany, A. S., and Boaro, A. V., 1964, A further study of the specificity of the vitamin E requirement for reproduction, *J. Nutr.* **84**:395–400.

Edwin, E. E., Dilock, A. T., Bunyan, J., and Green, J., 1961, Studies on vitamin E. 6. The distribution of vitamin E in the rat and the effect of α-tocopherol and dietary selenium on ubiquinone and ubichromenol in tissues, *Biochem. J.* **79**:91–105.

Eggert, R. G., Patterson, E., Akers, W. T., and Stokstad, E. L. R., 1957, The role of vitamin E and selenium in the nutrition of the pig, *J. Anim. Sci.* **16**:1037.

Eijkman, C., 1897, Eine beriberi-ähnliche Krankheit der Hühner, *Virchows Arch. Pathol. Anat. Physiol.* **148**:523–532.

Emmel. V. M., 1957, Studies on the kidney in vitamin E deficiency, *J. Nutr.* **61**:51–65.

Emmerie, A., and Engel, C., 1938, Colorimetric determination of *d,l*-α-tocopherol (vitamin E), *Nature (London)* **142**:873.

Evans, H. M., 1962, The pioneer history of vitamin E, *Vitamins Hormones* **20**:379–387.

Evans, H. M., and Bishop, K. S., 1922, On the existence of a hitherto unrecognized dietary factor essential for reproduction, *Science* **56**:650–651.

Evans, H. M., Burr, G. O., and Althausen, T. L., 1927, Antisterility vitamin, fat-soluble E, *Mem. Univ. Calif.* **8**:1–176.

Evans, H. M., Emerson, O. H., and Emerson, G. A., 1936, The isolation from wheat germ oil of an alcohol, α-tocopherol, having the properties of vitamin E, *J. Biol. Chem.* **113**:319–332.

Evans, H. M., Emerson, O. H., Emerson, G. A., Smith, L. I., Ungnade, H. E., Prichard, W. W., Austin, F. L., Hoehn, H. H., Opie, J. W., and Wawzonek, S., 1939, The chemistry of vitamin E. XIII. Specificity and relationship between chemical structure and vitamin E activity, *J. Org. Chem.* **4**:376.

Fabianek, J., DeFilippi, J., Richards, T., and Herp, A., 1968, Micromethod for tocopherol determination in blood serum, *Clin. Chem.* **14**:456–462.

Fang, H. H., Hall, A. L., and Hwang, F. T. F., 1960, Combined effects of vitamins A, D, and E on dark adaptation in man, *Am. J. Optm.* **37**:93.

Farrell, P. M., and Bieri, J. G., 1975, Megavitamin E supplementation in man, *Am. J. Clin. Nutr.* **28**:1381–1386.

Fernholz, E., 1938, Constitution of alpha-tocopherol, *J. Am. Chem. Soc.* **60**:700–705.

Fieser, L. F., Tishler, M., and Wendler, N. L., 1940, The method used for vitamin K synthesis also applicable for the synthesis of α-tocopherol, *J. Am. Chem. Soc.* **62**:996.

Fitch, C. D., 1968, Experimental anemia in primates due to vitamin E deficiency, *Vitamins Hormones* **26**:501–514.

Fitch, C. D., and Dinning, J. S., 1959, Phosphate metabolism in nutritional muscular dystrophy and hyperthyroidism, *Proc. Soc. Exp. Biol. Med.* **100**:201–203.

Fitch, C. D., and Dinning, J. S., 1963, Vitamin E deficiency in the monkey. V. Estimated requirements and the influence of fat deficiency and antioxidants on the syndrome, *J. Nutr.* **79**:69–78.

Fitch, C. D., and Dinning, J. S., 1964, Vitamin E deficiency in the monkey. VI. Metabolism of creatine-1-C^{14}, *Proc. Soc. Exp. Biol. Med.* **115**:986–989.

Fite, G. L., 1954, The pathology of dietary liver necrosis—A preliminary report, *Ann. N.Y. Acad. Sci.* **57**:831.

Fletcher, B. L., and Tappel, A. L., 1973, Protective effects of dietary α-tocopherol in rats exposed to toxic levels of ozone and nitrogen oxide, *Environ. Res.* **6**:165–175.

Fodor, G., and Kemény, G. L., 1965, On the hepato-protective effect of selenium in carbon tetrachloride poisoning in albino rats. *Experientia* **21**:666–667.

Frampton, V. L., Skinner, W. A., Jr., and Bailey, P. S., 1954, The production of tocored upon the oxidation of *dl*-α-tocopherol with ferric chloride, *J. Am. Chem. Soc.* **76**:282–284.

Fridovich, I., 1975, Superoxide dismutases, *Ann. Rev. Biochem.* **44**:147.

Friedman, L., Weiss, W., Wherry, F., and Kline, O. L., 1958, Bioassay of vitamin E by the dialuric acid hemolysis method, *J. Nutr.* **65**:143–160.

Fujita, T., Yasuda, M., Kitamura, Y., and Shimamura, S., 1973, Fatty acid composition of mitochondria of various organs from rats fed diets deficient in vitamin E (in Japanese), *J. Pharm. Soc. Jpn.* **92**:670–676.

Fukuzumi, K., 1970, Significance of lipoperoxides in gastric cancer, *Fette Seifen Anstrich.* **72**:853–855.

Funk, C., 1912, The etiology of the deficiency diseases, *J. State Med.* **20**:341–368.

Gallagher, C. H., 1961, Protection by antioxidants against lethal doses of carbon tetrachloride, *Nature (London)* **192**:881–882.

Gallo-Torres, H. E., 1970, Obligatory role for bile for the intestinal absorption of vitamin E, *Lipids* **5**:379–384.

Gallo-Torres, H. E., 1973, Studies on the intestinal lymphatic absorption, tissue distribution and storage of vitamin E, *Acta Agr. Scand. Suppl.* **19**:97–104.

Gedigk, P., 1959, Lipogenous pigments, *Verhandl. Deutsch. Ges. Pathol.* **42**:430.

Gerber, G. B., Koszalka, T. R., Gerber, G., and Miller, L. L., 1964, Creatine synthesis in perfused liver of vitamin E-deficient rats, *Proc. Soc. Exp. Biol. Med.* **116**:884–887.

Gerloczy, F., 1968, Vitamin E deficiency in men and its mechanism, *Orv. Hetilap* **109**:897.

Gershoff, S. N., and Norkin, S. A., 1962, Vitamin E deficiency in cats, *J. Nutr.* **77**:303–308.

Ghabussi, P., Kraemer, K. D., and Hochrein, H., 1974, Toleranzsteigerung von k-Strophanthin durch *dl*-α-Tocopherol-acetat: Klinische Untersuchungen an dekompensierten Herzkranken. *Arzneimittel-Forsch.* **24**:202–205.

Gitler, C., Sunde, M. L., and Baumann, C. A., 1958, Effects of certain necrosis-preventing factors on hemolysis in vitamin E-deficient rats and chicks, *J. Nutr.* **65**:397–407.

Glavind, J., 1973, The biological antioxidant theory and the function of vitamin E, *Acta Agr. Scand. Suppl.* **19**:105–112.

Gloor, U., Würsch, J., Schwieter, U., and Wiss, O., 1966, Resorption, retention, distribution and metabolism of *dl*-α-tocopheramine, *dl*-*N*-methyl-α-tocopheramin and α-tocopherol in comparison to *dl*-α-tocopherol in the rat, *Helv. Chim. Acta* **49**:2303–2312.

Goettsch, M., and Pappenheimer, A. M., 1931, Nutritional muscular dystrophy in the guinea pig and rabbit, *J. Exp. Med.* **54**:145–165.

Goldberg, L., and Smith, J. P., 1958, Changes associated with excess amounts of iron in rat viscera, *Br. J. Exp. Pathol.* **39**:59.

Goldberg, L., and Smith, J. P., 1960, Vitamin E and A deficiencies in relation to iron overloading in the rat, *J. Pathol. Bacteriol.* **80**:173.

Goldbloom, R. B., 1960, Investigations of tocopherol deficiency in infancy and childhood: Studies of serum tocopherol levels and of erythrocyte survival. *Can. Med. Assoc. J.* **82**:1114–1117.

Goldstein, J., and Scott, M. L., 1956, An electrophoretic study of exudative diathesis in chicks, *J. Nutr.* **60**:349–359.

Gontzea, I., Rujinski, A., and Busca, G., 1970, Tocopherol in infants with gastrointestinal disease, *Pediatria* **19**:215–219.

Goodhart, R. S., and Shils, M. E., 1973, Vitamin E in modern nutrition in health and disease, in: *Dietotherapy*, pp. 1694–1699, Lea and Febiger, Philadelphia.

Goransson, G., Norden, A., and Akesson, B., 1973, Low plasma tocopherol levels in patients with gastro-intestinal disorders, *Scand. J. Gastroenterol.* **8**:21.

Gordon, H. H., and Nitowsky, H. M., 1970, Vitamin E, in: *The Pharmacological Basis of Therapeutics*, 4th ed., Macmillan, New York.

Gordon, H. H., Nitowsky, H. M., Tildon, J. T., and Levin, S., 1958, Studies of tocopherol deficiency in infants and children. V. An interim summary, *Pediatrics* **21**:673–681.

Graham, J. B., Graham, R.-M., and Kottmaier, H. L., 1960, Potentiation of radiotherapy by supplemental agents in cancer of the uterine cervix—4 year results, *Acta Intern. Union Against Cancer* **16**:1291–1293.

Granados, H., and Dam, H., 1945, Inhibition of pigment deposition in incisor teeth of rats deficient in vitamin E from birth, *Proc. Soc. Exp. Biol. Med.* **59**:295–296.

Green, J., and Bunyan, J., 1969, Vitamin E and the biological antioxidant theory, *Nutr. Abstr. Rev.* **39**:321–345.

Green, J., Diplock, A. T., Bunyan, J., and Edwin, E. E., 1961, Studies on vitamin E. 8. Vitamin E, ubiquinone and ubichromenol in the rabbit, *Biochem. J.* **79**:108–111.

Green, J., Bunyan, J., Cawthorne, M. A., and Diplock, A. T., 1969, Vitamin E and hepatotoxic agents. 1. Carbon tetrachloride and lipid peroxidation in the rat, *Br. J. Nutr.* **23**:297–307.

Grijns, G., 1901, *Geneesk. Tijdschr. Ned. Ind.* 1, as cited in McCollum, E. V., 1957, *A History of Nutrition* (H. B. Glass, ed.), p. 228, Riverside Press, Cambridge, Mass.

György, P., and Rose, C. S., 1949, Tocopherol and hemolysis *in vivo* and *in vitro*, *Ann. N.Y. Acad. Sci.* **52**:231–239.

György, P., Cogan, G., and Rose, C. S., 1952, Availability of vitamin E in the newborn infant, *Proc. Soc. Exp. Biol. Med.* **81**:536–540.

Haber, S. L., and Wissler, R. W., 1962, Effect of vitamin E on carcinogenicity of methylcholanthrene, *Proc. Soc. Exp. Biol. Med.* **111**:774–775.

Haeger, K., 1968, The treatment of peripheral occlusive arterial disease with α-tocopherol as compared with vasodilator agents and antiprothrombin (dicumarol), *Vasc. Dis.* **5**:199.

Haeger, K., 1974, Long-time treatment of intermittent claudication with vitamin E, *Am. J. Clin. Nutr.* **27**:1179–1181.

Hanna, M. B., 1965, Management of night-blindness with vitamins A and E, *Bull. Ophthalmol. Soc. Egypt* **58**:219.

Harman, D., 1969, Vitamin E inhibits dimethyl-benzanthracene cancer in rats, *Clin. Res.* **17**:125.

Harries, J. T., and Muller, D. P. R., 1971, Absorption of different doses of fat soluble and water miscible preparations of vitamin E in children with cystic fibrosis, *Arch. Dis. Childh.* **46**:341–344.

Harris, P. L., and Embree, N. D., 1963, Quantitative consideration of the effect of polyunsaturated fatty acid content (P.U.F.A.) of the diet upon the requirement of vitamin E, *Am. J. Clin. Nutr.* **13**:385–392.

Harris, P. L., and Ludwig, M. I., 1949, Relative vitamin E potency of natural and synthetic alpha-tocopherol (and esters), *J. Biol. Chem.* **179**:1111–1115; *J. Biol. Chem.* **180**:611–614.

Harris, P. L., Ludwig, M. I., and Schwarz, K., 1958, Ineffectiveness of factor 3-active selenium compounds in resorption-gestation bioassay for vitamin E, *Proc. Soc. Exp. Biol. Med.* **97**:686–688.

Haubold, D. H., and Heuer, E., 1962, On the therapeutic possibilities of vitamin E, *Die Kapsel* **11**:37.

Hayes, K. C., Nielsen, S. W., and Rousseau, J. E., Jr., 1969, Vitamin E deficiency and fat stress in the dog, *J. Nutr.* **99**:196–209.

Heinzerling, R. H., Nockels, C. F., Quarles, C. L., and Tengerdy, R. P., 1974a, Protection of chicks against *E. coli* infection by dietary supplementation with vitamin E, *Proc. Soc. Exp. Biol. Med.* **146**:279–283.

Heinzerling, R. H., Tengerdy, R. P., and Wick, L. L., 1974b, Vitamin E increases protection of mice against *Diplococcus pheumoniae* type I (DPI) infection, *Fed. Proc.* **33(3):** Part I, 763 (abstr.).

Hellstrom, J. G., 1961, Vitamin E—A review, *Med. Serv. J.* **17**:238.

Helwig, H. P., Hochrein, H., and Helwig, B., 1971, Vitamin E in der Behandlung der Digitalis-Intoxikation, *Arzneimittel-Forsch.* **21**:335–342.

Herting, D. C., 1966, Perspective on vitamin E, *Am. J. Clin. Nutr,* **19**:210.

Heyman, A., and Stamm, O., 1955, Influence of alpha tocopherol on post-operative thromboelasto-gram, *Gynaecologia* **140**:224–227.

Holst, A., and Frølich, T., 1907, Experimental studies relating to ship beriberi and scurvy, *J. Hyg.* **7**:634–671.

Hope, W. C., Dalton, C., Machlin, L. J., Filipski, R. J., and Vane, F. M., 1975, Influence of dietary vitamin E on prostaglandin biosynthesis in rat blood, *Prostaglandins* **10**:557–567.

Hopkins, F. G., 1906, The analyst and the medical man, *Analyst (London)* **31**:385.

Horwitt, M. K., 1960, Vitamin E and lipid metabolism in man, *Am. J. Clin. Nutr.* **8**:451–461.

Horwitt, M. K., 1961, Vitamin E in human nutrition, *Borden's Rev. Nutr. Res.* **22**:1–2.

Horwitt, M. K., 1962, Interrelations between vitamin E and polyunsaturated fatty acids in man, *Vitamins Hormones* **20**:541–558.

Horwitt, M. K., Harvey, C. C., Duncan, G. D., and Wilson, W. C., 1956, Effects of limited tocopherol intake in man with relationships to erythrocyte hemolysis and lipid oxidations, *Am. J. Clin. Nutr.* **4**:408–418.

Hove, E. L., 1948, Interrelationship between α-tocopherol and protein metabolism. III. The protective effect of vitamin E and certain nitrogenous compounds against CCl_4 poisoning in rats, *Arch. Biochem. Biophys.* **17**:467–474.

Hove, E. L., and Siebold, H. R., 1955, Liver necrosis and altered fat composition in vitamin E-deficient swine, *J. Nutr.* **56**:173–186.

Hove, E. L., Fry, G. S., and Schwarz, K., 1958, Ineffectiveness of factor 3-active selenium compounds in muscular dystrophy of rabbits on vitamin E-free diets, *Proc. Soc. Exp. Biol. Med.* **98**:27–29.

Inglett, G. E., and Mattill, H. A., 1955, Oxidation of hindered 6-hydroxychromans, *J. Am. Chem. Soc.* **77**:6552–6554.

Irving, J. T., 1959, Curative effects of selenium upon the incisor teeth of rats deficient in vitamin E, *Nature (London)* **184**:645–646.

Isler, O., Schudel, P., Mayer, H., Würsch, J., and Ruëgg, R., 1962, Chemistry of vitamin E, *Vitamins Hormones* **20**:389–405.

Jenkins, M. Y., and Mitchell, G. V., 1975, Influence of excess vitamin E on vitamin A toxicity in rats, *J. Nutr.* **105**:1600–1606.

Jensen, L. S., and McGinnis, J., 1957, Studies on the vitamin E requirement of turkeys for reproduction, *Poultry Sci.* **36**:1344–1350.

Jensen, L. S., Scott, M. L., Heuser, G. F., Norris, L. C., and Nelson, T. S., 1956, Studies on the nutrition of breeding turkeys. 1. Evidence indicating a need to supplement practical turkey rations with vitamin E, *Poultry Sci.* **35**:810–816.

John, W., 1937, Cumotocopherol, a new factor of the vitamin-E group, *Z. Physiol. Chem.* **250**:11–24.

John, W., and Emte, W., 1941, Antisterility factors (vitamin E). VIII. Some new oxidation products of tocopherol, *Z. Physiol. Chem.* **268**:85–103.

Jungherr, E., and Pappenheimer, A. M., 1937, Nutritional myopathy of the gizzard in turkeys, *Proc. Soc. Exp. Biol. Med.* **37**:520–526.

Karrer, P., and Isler, O., 1938, U.S. Patent 2,411,967.

Karrer, P., and Isler, O., 1941, U.S. Patent 2,411,969.

Karrer, P., Fritzsche, H., Ringier, B. H., and Salomon, H., 1938, Synthesis of α-tocopherol (vitamin E), *Nature (London)* **141**:1057; *Nutr. Abstr. Rev.* **8**:358.

Kater, R.-M. H., Unterecker, W. J., and Davidson, C. S., 1969, Serum tocopherol concentrations in liver diseases, *Gastroenterology* **56**:1216.

Kimzey, S. L., Fischer, C. L., and Mengel, C. A., 1971, The protective action of vitamin E against O_2 toxicity, *Abstr. Papers Am. Chem. Soc.*, No. 161, AGFD29.

King, M. M., Lai, E. K., and McCay, P. B., 1975, Singlet oxygen production associated with enzyme-catalyzed lipid peroxidation in liver microsomes, *J. Biol. Chem.* **250**:6496.

Kitabchi, A. E., McCay, P. B., Carpenter, M. P., Trucco, R. E., and Caputto, R., 1960, Formation of malonaldehyde in vitamin E deficiency and its relation to the inhibition of gulonolactone oxidase, *J. Biol. Chem.* **235**:1591–1598.

Krishnamurthy, S., and Bieri, J. G., 1961, An effect of dietary selenium on hemolysis and lipid autoxidation of erythrocytes from vitamin E deficient rats, *Biochem. Biophys. Res. Commun.* **4**:384–387.

Krishnamurthy, S., and Bieri, J. G., 1963, The absorption, storage and metabolism of α-tocopherol-C^{14} in the rat and chicken, *J. Lipid Res.* **4**:330–336.

Krukovsky, V. N., and Loosli, J. K., 1952, Further studies on the influence of tocopherol supplementation on the vitamin content of the milk fat, stability of milk and milk and fat production, *J. Dairy Sci.* **35**:834.

Krukovsky, V. N., Loosli, J. K., and Whiting, F., 1949, The influence of tocopherols and cod liver oil on the stability of milk, *J. Dairy Sci.* **32**:196.

Kryukova, L. V., Federov, V. V., and Kalashnikova, V. A., 1972, Tocopherol concentration and some properties of erythrocytes in gastric disease, *Vopr. Pit.* **31**(2):24–26.

Lee, Pei-Fei, 1956, Alpha-tocopherol in ocular diseases, *Summary* **8**:85.

Lee, Y. C. P., Kuha, K. T., Visscher, M. B., and King, J. T., 1962, Role of manganese and vitamin E deficiency in mouse paralysis, *Am. J. Physiol.* **203**:1103.

Levander, O. A., and Argrett, L. C., 1969, Effects of arsenic, mercury, thallium and lead on selenium metabolism in rats, *Toxicol. Appl. Pharmacol.* **14**:308–314.

Lillie, R. D., Ashburn, L. L., Sebrell, W. H., Jr., Daft, F. S., and Lowry, J. V., 1942, Histogenesis and repair of the hepatic cirrhosis in rats produced on low protein diets and preventable with choline, *Public Health Rep.* **57**:502–508.

Loper, D. C., Chapin, R. E., and Vanderveen, J. E., 1969, Serum alpha tocopherol levels in spacemen, *Fed. Proc.* **28**:2793 (abstr.).

Losowsky, M. S., and Kelleher, J., 1970, Vitamin E research on man in the United Kingdom, *Int. Z. Vitaminforsch.* **40**:107.

Lucy, J. A., 1972, Functional and structural aspects of biological membranes: A suggested structural role for vitamin E in the control of membrane permeability and stability. Part I. Cellular biochemistry of vitamin E, *Ann. N.Y. Acad. Sci.* **203**:4–11.

Machlin, L. J., Filipski, R., Willis, A. L., Kuhn, D. C., and Brin, M., 1975, Influence of vitamin E on platelet aggregation and thrombocythemia in the rat, *Proc. Soc. Exp. Biol. Med.* **149**:275–277.

Madsen, L. L., McCay, C. M., and Maynard, L. A., 1935, Synthetic diets for Herbivora, with special reference to the toxicity of cod-liver oil, *Cornell Univ. Agr. Exp. Sta. Mem.* **178**:3–53.

March, B. E., Wong, E., Seier, L., Sim, J., and Biely, J., 1973, Hypervitaminosis E in the chick, *J. Nutr.* **103**:371–377.

Martin, A. J. P., and Moore, T., 1939, Some effects of prolonged vitamin E deficiency in the rat, *J. Hyg.* **39**:643–650.

Marusich, W. L., DeRitter, E., Ogrinz, E. F., Keating, J., Mitrovic, M., and Bunnell, R. H., 1975, Effect of supplemental vitamin E in control of rancidity in poultry meat, *Poultry Sci.* **54**:831–844.

Mason, K. E., 1933, Differences in testis injury and repair after vitamin A deficiency, vitamin E deficiency and inanition, *Am. J. Anat.* **52**:153.

Mason, K. E., 1944, Physiological action of vitamin E and its homologues, *Vitamins Hormones* **2**:107–153.

Mason, K. E., 1954, The tocopherols. Effects of deficiency. A. In animals, in: *The Vitamins*, Vol. III, pp. 514–541 (W. H. Sebrell, Jr., and R. S. Harris, eds.), Academic Press, New York.

Mason, K. E., and Hartsough, G. R., 1951, "Steatitis" or "yellow fat" in mink and its relation to dietary fats and inadequacy of vitamin E, *J. Am. Vet. Med. Assoc.* **119**:72.

Mason, K. E., and Young, J. O., 1967, Effectiveness of selenium and zinc in protecting against cadmium-induced injury of the rat testis, in: *Selenium in Biomedicine* (O. H. Muth and J. E. Oldfield, eds.), Avi, Westport, Conn.

Mayer, H., and Isler, O., 1971, Synthesis of vitamin E, in: *Methods in Enzymology*, Vol. 18 (Part C) (S. P. Colowick, ed.), p. 241, Academic Press, New York.

Mayer, N. P., Schudel, P., Rüegg, R., and Isler, O., 1963, Synthesis of vitamin E, *Helv. Chim. Acta* **46:**650.

McCay, P. B., 1969, Enzymic oxidation of polyunsaturated fatty acids of membrane phospholipids: Regulatory effect of α-tocopherol, in: *Proceedings of the Eighth International Congress of Nutrition*, pp. 106–110, Excerpta Medica Foundation, Amsterdam.

McCay, P. B., Carpenter, M. P., Kitabchi, A. E., and Caputto, R., 1959, Activation by tocopherol of synthesis of ascorbic acid by liver extracts, *Arch. Biochem. Biophys.* **82:**472.

McCay, P. B., Poyer, J. L., Pfeifer, P. M., May, H. E., and Gilliam, J. M., 1971, A function for α-tocopherol: Stabilization of the microsomal membrane from radical attack during TPNH-dependent oxidations, *Lipids* **6:**297–306.

McCay, P. B., Gibson, D. O., Fong, K. L., and Hornbrook, K. R., 1976, The effect of glutathione peroxidase activity on lipid peroxidation in biological membranes, *Biochim. Biophys. Acta* **431:**459–468.

McCollum, E. V., 1925, Studies on experimental rickets, *J. Biol. Chem.* **65:**97.

McCollum, E. V., 1957, *A History of Nutrition*, pp. 359–372, Riverside Press, Cambridge, Mass.

McCollum, E. V., and Davis, M., 1913, Necessity of certain lipins in the diet during growth, *J. Biol. Chem.* **15:**167.

McCollum, E. V., and Davis, M., 1916, The essential factors in the diet during growth, *J. Biol. Chem.* **23:**231–246.

McCuaig, L. W., and Motzok, I., 1970, Excessive dietary vitamin E: Its alleviation of hypervitaminosis A and lack of toxicity, *Poultry Sci.* **49:**1050–1052.

McLaren, D. S., Shirajian, E., Tchalian, M., and Khoury, G., 1965, Xerophthalmia in Jordan, *Am. J. Clin. Nutr.* **17:**117–130.

McMasters, V., Howard, T., Kinsell, L. W., Van der Veen, J., and Olcott, H. S., 1967, Tocopherol storage and depletion in normal and diabetic humans, *Am. J. Clin. Nutr.* **20:**622–626.

Mecchi, E. P., Pool, M. F., Behman, G. A., Hamachi, M., and Klose, A. A., 1956, The role of tocopherol content in the comparative stability of chicken and turkey fat, *Poultry Sci.* **1956:**1238–1246.

Melhorn, D. K., and Gross, S., 1971, Vitamin E-dependent anemia in the premature infant. I. Effects of large doses of medicinal iron. II. Relationship between gestational age and absorption of vitamin E, *J. Pediatr.* **79:**569–588.

Menzel, D. B., Roehm, J. N., and Lee, S. D., 1971, Vitamin E: The biological and environmental antioxidant, *J. Agr. Food Chem.* **20:**481–486.

Menzel, D. B., Slaughter, R. J., Bryant, A. M., and Jauregui, H. O., 1975, Prevention of ozonide-induced Heinz bodies in human erythrocytes by vitamin E, *Arch. Environ. Health* **30:**234–236.

Metzger, H. J., and Hagen, W. A., 1927, The so-called stiff lambs, *Cornell Vet.* **17:**35–44.

Milhorat, A. T., 1954, Therapy in muscular dystrophy, *Med. Ann. Dist. Columbia* **23:**15–22.

Molenaar, I., Vos, J., Jager, F., and Hommes, F. A., 1970, The influence of vitamin E deficiency on biological membranes, *Nutr. Metab.* **12:**358–370.

Money, D. F. L., and Sirone, P. J., 1972, Under-water diving, oxygen poisoning and vitamin E, *New Zealand Med. J.* **476:**34–35.

Moore, T., Sharman, I. M., and Symonds, K. R., 1958, Kidney changes in vitamin-E deficient rats, *J. Nutr.* **65:**183–198.

Moore, T., Sharman, I. M., Stanton, M. G., and Dingle, J. T., 1967, Influences of vitamin E-deficiency and its duration on stability of lysosomes in rat kidneys, *Biochem. J.* **103:**923.

Morton, R. A., 1968, The history of vitamin research, *Int. Z. Vit. Res.* **38:**5–44.

Murai, T., and Andrews, J. W., 1974, Interactions of dietary α-tocopherol, oxidized menhaden oil and ethoxyquin on channel catfish (*Ictalurus punctatus*), *J. Nutr.* **104:**1416–1431.

Murty, H. S., Caasi, P. I., Brooks, S. K., and Nair, P. P., 1970, Biosynthesis of heme in the vitamin E-deficient rat, *J. Biol. Chem.* **245:**5498–5504.

Muth, O. H., 1960, Carbon tetrachloride poisoning of ewes on a low selenium ration, *Am. J. Vet. Res:* **21:**86–87.

Muth, O. H., Oldfield, J. E., Remmert, L. F., and Schubert, J. R., 1958, Effects of selenium and vitamin E on white muscle disease, *Science* **128:**1090.

Nafstad, I., and Tollersrud, S., 1970, The vitamin E-deficiency syndrome in pigs. I. Pathological changes, *Acta Vet. Scand.* **11**:452–480.

Nason, A., and Lehman, I. R., 1956, The role of lipides in electron transport. II. Lipide cofactor replaceable by tocopherol for enzymatic reduction of cytochrome *c*, *J. Biol. Chem.* **222**:511–530.

National Research Council, 1974, *Recommended Dietary Allowances*, 8th ed., National Academy of Sciences, Washington, D.C.

Nelan, D. R., and Robeson, C. D., 1962, Oxidation of α-tocopherol with potassium ferricyanide and its application to distinguishing *d*-α-tocopherol from synthetic *dl*-α-tocopherol, *Nature (London)* **193**:477.

Nesheim, M. C., Leonard, S. L., and Scott, M. L., 1959, Alterations in some biochemical constituents of skeletal muscle of vitamin E-deficient chicks, *J. Nutr.* **68**:359–369.

Newmark, H. L., Pool, W., Bauernfeind, J. C., and De Ritter, E., 1975, Biopharmaceutic factors in parenteral administration of vitamin E, *J. Pharm. Sci.* **64**:655–657.

Noguchi, T., Cantor, A. H., and Scott, M. L., 1973, Mode of action of selenium and vitamin E in prevention of exudative diathesis in chicks, *J. Nutr.* **103**:1502–1511.

Obel, A. L., 1953, Studies on the morphology and etiology of so-called toxic liver dystrophy (hepatosis dietetica) in swine, *Acta Pathol. Microbiol. Scand.*, Suppl. 94.

Olcott, H. S., 1938, The paralysis in the young of vitamin E deficient female rats, *J. Nutr.* **15**:221–225.

Olcott, H. S., and Mattill, H. A., 1931, The unsaponifiable lipids of lettuce. III. Antioxidants, *J. Biol. Chem.* **93**:65–70.

Olcott, H. S., and Mattill, H. A., 1941, Constituents of fats and oils affecting the development of rancidity, *Chem. Rev.* **29**:257–268.

Oldfield, J. E., Muth, O. H., and Schubert, J. R., 1960, Selenium and vitamin E as related to growth and white muscle disease in lambs, *Proc. Soc. Exp. Biol. Med.* **103**:799–800.

Olson, R. E., 1973, Vitamin E and its relation to heart disease, *Circulation* **48**:179–184.

Osborne, T. B., and Mendel, L. B., 1913, The relation of growth to the chemical constitution of the diet, *J. Biol. Chem.* **15**:311–326.

Osborne, T. B., and Mendel, L. B., 1914, Amino acids in nutrition and growth, *J. Biol. Chem.* **17**:325–349.

Oski, F. A., and Barness, L. A., 1967, Vitamin E deficiency: A previously unrecognized cause of hemolytic anemia in the premature infant, *J. Pediatr.* **70**:211.

Oski, F. A., Myerson, R., Barness, L. A., and Williams, W. J., 1966, A possible interrelationship between vitamin E and vitamin B_{12}, *Am. J. Clin. Nutr.* **18**:307.

Pappenheimer, A. M., 1940, Certain nutritional disorders of laboratory animals due to vitamin E deficiency, *J. Mt. Sinai Hosp. N.Y.* **7**:65–76.

Pappenheimer, A. M., and Goettsch, M., 1931, A cerebellar disorder in chicks, apparently of nutritional origin, *J. Exp. Med.* **53**:11–16.

Pappenheimer, A. M., and Goettsch, M., 1933, Nutritional myopathy in ducklings, *J. Exp. Med.* **59**:35–42.

Parizek, J., 1957, The destructive effect of cadmium ion on testicular tissue and its prevention by zinc, *J. Endocrinol.* **15**:56–63.

Parizek, J., and Zahor, Z., 1956, Effect of cadmium salts on testicular tissue, *Nature (London)* **177**:1036–1038.

Parkhovnik, M. B., 1972, Therapeutic efficacy of vitamin E in peptic ulcer, *Vrach. Delo.* **8**:68.

Pearson, C. K., and Barnes, M. C., 1970, Absorption of tocopherols by small intestinal loops of the rat *in vivo*, *Int. J. Vit. Res.* **40**:19–22.

Pennock, J. F., Hemming, F. W., and Kerr, J. D., 1964, A reassessment of tocopherol chemistry, *Biochem. Biophys. Res. Commun.* **17**:542–548.

Proctor, J. F., Hogue, D. E., and Warner, R. G., 1958, Selenium, vitamin E and linseed oil meal as preventatives of muscular dystrophy in lambs, *J. Anim. Sci.* **17**:1183–1184.

Prout, W., 1827/1828, On the ultimate composition of simple alimentary substances, *Philos. Mag.* **2**:(2) 144; **3**:33, 107 (1828).

Pryor, W. A., 1971, Free radical pathology, *Chem. Eng. News* **49(23)**:34–36, 41–51.

Pudelkiewicz, W. J., Matterson, L. D., Potter, L. M., Webster, L., and Singsen, E. P., 1960, Chick tissue-storage bioassay of alpha-tocopherol: Chemical analytical techniques, and relative biopotencies of natural and synthetic alpha-tocopherol, *J. Nutr.* **71**:115–121.

Quaife, M. L., and Harris, P. L., 1948, Chemical assay of foods for vitamin E content, *Anal. Chem.* **20**:1221–1224.

Rahman, M. M., Hossain, S., Talukdar, S. A., Ahmad, K., and Bieri, J. E., 1964, Serum vitamin E levels in the rural population of East Pakistan, *Proc. Soc. Exp. Biol. Med.* **117**:133–135.

Recknagel, R. O., and Ghoshal, A. K., 1966, Quantitative estimation of peroxidative degeneration of rat liver microsomal and mitochondrial lipids after carbon tetrachloride poisoning, *Exp. Mol. Pathol.* **5**:413–426.

Ritchie, J. H., Fish, M. B., McMasters, V., and Grossman, M., 1968, Edema and hemolytic anemia in premature infants: A vitamin E deficiency syndrome, *New Engl. J. Med.* **279**:1185–1190.

Robeson, C. D., and Nelan, D. R., 1962, Isolation of an *l*-epimer of natural *d*-α-tocopherol, *J. Am. Chem. Soc.* **84**:3196.

Robinson, K. L., and Coey, W. E., 1951, A brown discoloration of pig fat and vitamin E deficiency, *Nature (London)* **168**:997–998.

Rose, C. S., and György, P., 1950, Tocopherol requirements of rats by means of the hemolysis test, *Proc. Soc. Exp. Biol. Med.* **74**:411–415.

Rose, C. S., and György, P., 1952, Specificity of hemolytic reaction in vitamin E-deficient erythrocytes, *Am. J. Physiol.* **168**:414.

Rotruck, J. T., Pope, A. L., Ganther, H. E., and Hoekstra, W. G., 1972, Prevention of oxidative damage to rat erythrocytes by dietary selenium, *J. Nutr.* **102**:689–696.

Rotruck, J. T., Pope, A. L., Ganther, H. E., Swanson, A. B., Hafernan, D. G., and Hoekstra, W. G., 1973, Selenium: Biochemical role as a component of glutathione peroxidase, *Science* **179**:588–590.

Rouher, F., and Sole, P., 1965, Eight years treatment of evolving myopia, *Bull. Soc. Ophthalmol. France* **65**:1086.

Rubel, T., 1969, Vitamin E manufacture, *Chemical Process Rev. No. 39*, Noyes Data Corp., Park Ridge, N.J.

Sauberlich, H. E., Dowdy, R. P., and Skala, J. H., 1973, Laboratory tests for the assessment of nutritional status: Vitamin E, *CRC Crit. Rev. Clin. Lab. Sci.*, September, pp. 288–294.

Schmandke, H., and Schmidt, G., 1965, Absorption of α-tocopherol from oily and aqueous solutions, *Int. Z. Vitaminforsch.* **35**:128–130.

Schmandke, H., Sima, C., and Manne, R., 1969, Die Resorption von α-Tokopherol beim Menschen, *Int. Z. Vitaminforsch.* **39**:796–798.

Schwarz, K., 1944, Tocopherol as a liver-protecting substance, *Z. Physiol. Chem.* **281**:106–108.

Schwarz, K., 1951, A hitherto unrecognized factor against dietary necrotic liver degeneration in American yeast (factor 3), *Proc. Soc. Exp. Biol. Med.* **78**:852–856.

Schwartz, K., 1958, Effect of antioxidants on dietary necrotic liver degeneration, *Proc. Soc. Exp. Biol. Med.* **99**:20–24.

Schwarz, K., 1972, The cellular mechanisms of vitamin E action: Direct and indirect effects of α-tocopherol on mitochondrial respiration, *Ann. N.Y. Acad. Sci.* **203**:45–52.

Schwarz, K., and Foltz, C. M., 1957, Selenium as an integral part of factor 3 against dietary necrotic liver degeneration, *J. Am. Chem. Soc.* **79**:3292–3293.

Schwarz, K., Mertz, W., and Simon, E. J., 1958, *In vitro* effect of tocopherol metabolites on respiratory decline in dietary necrotic liver degeneration, *Biochim. Biophys. Acta* **32**:484–491.

Schwarz, K., Stesney, J. A., and Foltz, C. M., 1959, Relation between selenium traces in L-cystine and protection against dietary liver necrosis, *Met.* **8**:88.

Scott, M. L., 1953, Prevention of the enlarged hock disorder in turkeys with niacin and vitamin E, *Poultry Sci.* **32**:670–677.

Scott, M. L., 1965, Comparative biological effectiveness of *d*-, *dl*- and *l*-forms of α-tocopherol for prevention of muscular dystrophy in chicks, *Fed. Proc.* **24**:901–905.

Scott, M. L., 1974, Lesions of vitamin E and selenium deficiencies in poultry and their pathogenesis, *Folia Veter. Latina* **4**:113–134.

Scott, M. L., and Desai, I. D., 1964, The relative anti-muscular dystrophy activity of the *d*- and *l*-epimers of α-tocopherol and of other tocopherols in the chick, *J. Nutr.* **83**:39–43.

Scott, M. L., and Stoewsand, G. S., 1961, A study of the ataxias of vitamin A and vitamin E deficiencies in the chick, *Poultry Sci.* **40**:1517–1523.

Scott, M. L., Hill, F. W., Norris, L. C., Dobson, D. C., and Nelson, T. S., 1955, Studies on vitamin E in poultry nutrition, *J. Nutr.* **56**:387–402.

Scott, M. L., Bieri, J. G., Briggs, G. M., and Schwarz, K, 1957, Prevention of exudative diathesis by factor 3 in chicks on vitamin E-deficient torula yeast diets, *Poultry Sci.* **36:**1155 (abstr.).

Scott, M. L., Olson, G., Krook, L, and Brown, W. R., 1967, Selenium-responsive myopathies of myocardium and of smooth muscle in the young poult, *J. Nutr.* **91:**573-583.

Scott, M. L., Noguchi, T., and Combs, G. F., Jr., 1974, New evidence concerning mechanisms of action of vitamin E and selenium, *Vitamins Hormones* **32:**429–444.

Seidel, J. C., and Harper, A. E., 1960, Some observations on vitamin E deficiency in the guinea pig, *J. Nutr.* **70:**147–155.

Selander, H., and Nilsson, J. L. G., 1972, Vitamin E and air pollution: Nitrogen dioxide oxidation of α-tocopherol model compound, *Acta Pharm. Soc.* **9:**125–128.

Selisko, O., 1961, Is magnesium a vitamin E synergist? *Naturwissenschaften* **48:**556.

Selye, H., Tuchweber, B., and Gabbiana, G., 1964, Further studies on anacalciphylaxis, *J. Am. Geriat. Soc.* **12:**207–214.

Seward, C. R., Gaughan, G., and Hove, E. L., 1966, Effect of selenium on incisor depigmentation and carbon tetrachloride poisoning in vitamin E-deficient rats, *Proc. Soc. Exp. Biol. Med.* **121:** 850–852.

Shamberger, R. J., Andreone, T. L., and Willis, C. E., 1974, Antioxidants and cancer. IV. Initiating activity of malonaldehyde as a carcinogen, *J. Natl. Cancer Inst.* **53:**1771–1773.

Shute, E. V., 1972, Proposed study of vitamin E therapy, *Can. Med. Assoc. J.* **106:**1057.

Siakotos, A. N., Koppang, N., Youmans, S., and Bucana, C., 1974, Blood levels of α-tocopherol in a disorder of lipid peroxidation: Batten's disease, *Am. J. Clin. Nutr.* **27:**1152–1157.

Simon, E. J., Eisengart, A., and Milhorat, A. T., 1955, A metabolite of vitamin E from human urine, *Fed. Proc.* **14:**281 (abstr.).

Singsen, E. P., Bunnell, R. H., Kozeff, A., Matterson, L. D., and Jungherr, E. L., 1953, Studies on encephalomalacia in chicks: The protective action of diphenyl-*p*-phenylene diamine against encephalomalacia, *Poultry Sci.* **32:**924–925.

Skinner, W. A., 1964, Vitamin E oxidation with free radical initiators: Azobisisobutyronitrile, *Biochem. Biophys. Res. Commun.* **15:**469–472.

Skinner, W. A., and Parkhurst, R. M., 1971, Reaction products of tocopherols, *Lipids* **6:**240–244.

Skinner, W. A., Johnson, H. L., Ellis, M., and Parkhurst, R. M., 1971, Relationship between antioxidant and antihemolytic activities of vitamin E derivatives *in vitro*, *J. Pharm. Sci.* **60:**643–645.

Slover, H. T., Shelley, L. M., and Burks, T. L., 1967, Identification and estimation of tocopherols by gas liquid chromatography, *J. Am. Oil Chem. Soc.* **44:**161–166.

Slover, H. T., Lehmann, J., and Valis, R. J., 1969, Vitamin E in foods: Determination of tocols and tocotrienols, *J. Am. Oil Chem. Soc.* **46:**417–420.

Smith, E. B., and Michener, W. M., 1973, Vitamin E in treatment of dermolytic bulbous dermatosis, *Arch. Dermatol.* **108:**254–256.

Smith, L. I., Ungnade, H. E., and Prichard, W. W., 1938, The structure and synthesis of alpha-tocopherol, *Science* **88:**37–38.

Smith, L. I., Irwin, W. B., and Ungnade, H. E., 1939*a*, Chemistry of vitamin E. XVII. The oxidation products of α-tocopherol and of related 6-hydroxychromans, *J. Am. Chem. Soc.* **61:**2424–2429.

Smith, L. I., Ungnade, H. E., Stevens, J. R., and Christman, C. C., 1939*b*, Chemistry of vitamin E. XVIII. Condensation of phenols and hydroquinones with allylic alcohols, allylic aldehydes and conjugated dienes, *J. Am. Chem. Soc.* **61:**2615–2618.

Søndergaard, E., 1967, Selenium and vitamin E interrelationships, in: *Selenium in Biomedicine* (O. H. Muth and J. E. Oldfield, eds.), pp. 365–381, Avi, Westport, Conn.

Søndergaard, E., 1972, The influence of vitamin E on the expenditure of vitamin A from the liver, *Experientia* **28:**773–774.

Spencer, A. M., 1961, Report of a study of nylidrin HCl and vitamin E in the treatment of peripheral vascular disease, *J. Am. Podiatr. Assoc.* **51:**341–344.

Stache, S., and Frimmer, M., 1970, Freisetzung lysosomaler Enzyme aus der isoliert perfundierten Lebel von vitamin E-mangel Ratten, *Int. J. Vit. Res.* **40:**145–152.

Stokstad, E. L. R., Patterson, E. L., and Milstrey, R., 1957, Factors which prevent exudative diathesis in chicks on torula yeast diets, *Poultry Sci.* **36:**1160 (abstr.).

Stowe, H. D., and Whitehair, C. K., 1963, Gross and microscopic pathology of tocopherol-deficient mink, *J. Nutr.* **81:**287–300.

Sure, B., 1924, Dietary requirements for reproduction. II. The existence of a specific vitamin for reproduction, *J. Biol. Chem.* **58**:693–709.

Tappel, A. L., 1968, Will antioxidant nutrients slow aging processes? *Geriatrics* **23**:97–105.

Tappel, A. L., 1970, Polyunsaturated lipid peroxidation in aging processes, *J. Am. Oil Chem. Soc.* **47(7)**:123.

Tappel, A. L., 1972, Vitamin E and free radical peroxidation of lipids, *Ann. N.Y. Acad. Sci.* **203**:12–28.

Tappel, A. L., 1975, Lipid peroxidation and fluorescent molecular damage to membranes, in: *Pathology of Cell Membranes* (B. F. Trump and A. U. Urstila, eds.), pp. 145–170, Academic Press, New York.

Thomas, B. H., and Cannon, C. Y., 1937, Reproduction on rations free from vitamin E, *Proc. Am. Soc. Anim. Prod.* **30**:59.

Thomas, H. V., Mueller, P. K., and Lyman, R. L., 1968, Lipoperoxidation of lung lipids in rats exposed to nitrogen dioxide, *Science* **159**:532–534.

Thompson, J. N., and Scott, M. L., 1970, Impaired lipid and vitamin E absorption related to atrophy of the pancreas in selenium deficient chicks, *J. Nutr.* **100**:797–809.

Tinberg, H. M., and Barber, A. A., 1970, Studies on vitamin E action: Peroxidation inhibition in structural protein–lipid micelle complexes derived from rat liver microsomal membranes, *J. Nutr.* **100**:413–418.

Tollerz, G., and Lannek, N., 1964, Protection against iron toxicity in vitamin E-deficient piglets and mice by vitamin E and synthetic antioxidants, *Nature (London)* **201**:846–847.

Toone, W. M., 1973, The use of vitamin E in ischemic heart disease, *New Engl. J. Med.* **289**: 979.

Unger, W. P., and Nethercott, J. R., 1973, Epidermolysis bullosa dystrophica unsuccessfully treated with vitamin E and oral corticosteroids, *Can. Med. Assoc. J.* **108**:1136.

Vasington, F. D., Reichard, S. M., and Nason, A., 1960, Biochemistry of vitamin E, *Vitamins Hormones* **18**:43–87.

Vogelsang, A., 1970, Twenty-four years using α-tocopherol in degenerative cardiovascular disease, *Angiology* **21**:275–279.

Vos, J., Molenaar, I., Searle-Van Leeuwen, M., and Hommes, F. A., 1971, Cellular membranes in vitamin E-deficiency: An ultrastructural and bio-chemical study of isolated outer and inner mitochondrial membranes, *Acta Agr. Scand. Suppl.* **19**:192–203.

Warshauer, D., Goldstein, E., Hoeprich, P. D., and Lippert, W., 1974, Effect of vitamin E and ozone on the pulmonary antibacterial defense mechanisms, *J. Lab. Clin. Med.* **83**:228–240.

Weber, F., and Wiss, O., 1963, Metabolism of vitamin E in the rat, *Helv. Physiol. Pharmacol. Acta* **21**: 131–134.

Weber, F., Gloor, U., Würsch, J., and Wiss, O., 1964, Synergism of d-(2R,4'R,8'R-) and 1-(2S,4'RS, 8'RS-)-α-tocopherol during absorption, *Biochem. Biophys. Res. Commun.* **14**:186–188.

Weiser, H., Achterrath, U., and Boghtu, W., 1971, Vitamin E and the thyroid system, *Acta Agr. Scand. Suppl.* **19**:208–218.

Westermarck, T., and Sandholm, M., 1977, Decreased erythrocyte glutathione peroxidase activity in neuronal ceroid lipofuscinosis (NCL)—corrected with selenium supplementation, *Acta Pharmacol. Toxicol.* **40**:70–74.

Whitaker, J. A., Fort, E. G., Vimokesant, S., and Dinning, J. S., 1967, Hematologic response to vitamin E in the anemia associated with protein-calorie malnutrition, *Am. J. Clin. Nutr.* **20**: 783–789.

Whiting, F., Willman, J. P., and Loosli, J. K., 1949, Tocopherol (vitamin E) deficiency among sheep fed natural feeds, *J. Anim. Sci.* **8**:234–242.

Whittle, K. J., Dunphy, P. J., and Pennock, J. F., 1966, The isolation and properties of delta-tocotrienol from Hevea latex, *Biochem. J.* **100**:138–145.

Wiss, O., and Weber, F., 1965, Ernährungsphysiologische Beziehungen zwischen dem Vitamin E und den ungesaltigten Fettsauren, *Z. Ernahrungswiss.* **4**:152–159.

Witting, L. A., 1965, Biological availability of tocopherol and other antioxidants at the cellular level, *Fed. Proc.* **24**:912–916.

Witting, L. A., and Lee, L, 1975, Dietary levels of vitamin E and polyunsaturated fatty acids and plasma vitamin E, *Am. J. Clin. Nutr.* **28**:571–576.

Wright, P. L., and Bell, M. C., 1964, Selenium-75 metabolism in the gestating ewe and fetal lamb: Effects of dietary α-tocopherol and selenium, *J. Nutr.* **84**:49–57.

Yamakita, Y., 1968, Serum level of total tocopherol and hemolysis test using hydroperoxide in patients with diabetes mellitus, *J. Osaka City Med. Center* **17**:17–27.

Yoshida, M., and Hoshii, H., 1972, Preventive effect of *dl*-α-tocopheryl acetate against nutritional encephalomalacia in chicks induced by dilauryl succinate, *Agr. Biol. Chem.* **36**:755.

Yoshida, M., Hoshii, H., and Morimoto, H., 1970, Interrelationships between dilauryl succinate feeding and nutritional encephalomalacia in starting chicks, *Agr. Biol. Chem.* **34**:1301.

Zeman, W., 1974, Studies in the neuronal ceroid lipofuscinoses, *J. Neuropathol. Exp. Neurol.* **33**:1–12.

Chapter 4

Vitamin K

J. W. Suttie

4.1. Introduction

4.1.1. Historical Background

The discovery of vitamin K resulted from a series of experiments by Henrik Dam on the possible essentiality of cholesterol in the diet of the chick. Dam (1929*a,b*, 1930) was investigating reports that chicks did not thrive on diets which had been extracted with nonpolar solvents to remove the sterols, and noted that such animals developed subdural or muscular hemorrhages and that blood taken from these animals clotted slowly. It was shown in these initial experiments that addition of cholesterol, lemon juice, yeast, or cod liver oil to the diet did not decrease the incidence of this lesion. The disease was subsequently observed by McFarlane *et al.* (1931), who described a clotting defect seen when chicks were fed ether-extracted fish or meat meal in an attempt to establish the vitamin A and D requirement of the chick. Progress in the elucidation of the nature of this disorder was confused to some extent when Holst and Halbrook (1933) reported that fresh cabbage would prevent the disease and concluded that the missing factor was vitamin C. Later Dam (1934) found that synthetic ascorbic acid would not cure the defect and concluded that the disease was due to a lack of an unknown growth factor found in seeds and cereals. Reports by Cook and Scott (1935*a,b*) claiming that the disease was caused by a toxic factor in fish meal complicated the interpretation of available observations, but Dam continued to study the distribution and lipid solubility of the active component in vegetable and animal sources, and in 1935 proposed (Dam, 1935*a,b*) that the antihemorrhagic vitamin of the chick was a new fat-soluble vitamin, which he called "vitamin K." Not only was *K* the first letter of the alphabet which was not used to describe an existing or a postulated vitamin activity at that time, but it was also the first letter of the German word Koagulation. Dam's reported discovery of a new vitamin was followed by an independent report by Almquist and Stokstad (1935*a,b*) describing their success in curing the hemorrhagic disease with ether extracts of alfalfa and clearly

J. W. Suttie • Department of Biochemistry, College of Agricultural and Life Sciences, University of Wisconsin—Madison, Madison, Wisconsin 53706.

pointing out that microbial action in fish meal and bran preparations could lead to the development of antihemorrhagic activity.

Indirect evidence indicated that the clotting defect might be related to a lowered concentration of plasma prothrombin. Dam *et al.* (1936) succeeded in preparing a crude plasma prothrombin fraction and demonstrating that its activity was decreased when it was obtained from vitamin K deficient chick plasma. Quick (1937) also recognized that the clotting defect in chicks fed the hemorrhagic diet resulted from a lack of prothrombin. About this time it was recognized that a clinically observed hemorrhagic condition resulting from obstructive jaundice, or biliary problems, was due to poor utilization of vitamin K by these patients; and the bleeding episodes were attributed to a lack of plasma prothrombin. It was not recognized at this time that the synthesis of plasma proteins other than prothrombin might be vitamin K dependent, and it was widely believed that the defect in the plasma of animals fed vitamin K deficient diets was solely due to a lack of prothrombin. It should be realized that whole blood clotting times and the one-stage Quick "prothrombin time" assays used at that time were not specific for prothrombin and that any crude preparation of prothrombin also contained the other K-dependent clotting factors. It is not surprising, therefore, that it was not established for some time that an increased concentration of prothrombin alone would not cure this defect. A real understanding of the various factors involved in regulating the generation of thrombin from prothrombin did not begin until the mid 1950s; and during the next 10 years Factor VII (Owen *et al.*, 1951; Koller *et al.*, 1951), Factor X (Hougie *et al.*, 1957), and Factor IX (Aggeler *et al.*, 1952; Biggs *et al.*, 1952) were discovered and subsequently shown to be dependent on vitamin K for their synthesis.

Following Dam's clear demonstration of the need for a dietary antihemorrhagic factor, a number of groups attempted to isolate and characterize this new vitamin. Dam collaborated with Karrer of the University of Zurich in the isolation of the vitamin, and by 1939 they succeeded (Dam *et al.*, 1939) in isolating the vitamin as a yellow oil from alfalfa. Although they described some of its properties, they did not recognize it as a quinone derivative. A group at the University of California led by Almquist was also isolating the vitamin from alfalfa, and the Washington University group of Doisy was purifying it from both alfalfa and putrefied fish meal. It was soon recognized (Karrer and Geiger, 1939; McKee *et al.*, 1939) that the active preparations were quinones, and Almquist and Klose (1939a) demonstrated that phthicol (2-methyl-3-hydroxy-1,4-naphthoquinone), which had previously been isolated from *Mycobacterium tuberculosis*, had biological activity. Vitamin K_1 was characterized as 2-methyl-3-phytyl-1,4-naphthoquinone and synthesized by Doisy's group (MacCorquodale *et al.*, 1939), and their identification was confirmed by independent syntheses of this compound by Karrer *et al.* (1939), Almquist and Klose (1939b), and Fieser (1939a,b). The Doisy group also isolated a form of the vitamin from putrefied fish meal which in contrast to the oil isolated from alfalfa was a crystalline product. Subsequent studies demonstrated that this compound, called "vitamin K_2," contained an unsaturated side chain, and it was characterized (Binkley *et al.*, 1940) as 2-methyl-3-farnesylfarnesyl-1,4-naphthoquinone. This was the structure assumed to be correct for many years, but Isler *et al.* (1958) later demonstrated that a crystalline form of the vitamin isolated by Doisy's method and shown by mixed melting point determination to be identical to Doisy's compound contained seven, not six, isoprenyl units and was in fact 2-methyl-3-(all *trans*-farnesylgeranylgeranyl)-1,4-naphthoquinone or vitamin $K_{2(35)}$, not vitamin $K_{2(30)}$. Isler *et al.* (1958) also demonstrated that the crude product obtained from putrefied fish meal did contain some vitamin $K_{2(30)}$. The elucidation of the structure of vitamin K was an extremely competitive research area with a number of large groups involved, and it is difficult at this time to fairly assign priority and credit. Some indication of the effort involved can be gained from an inspection of the large number of papers published in the 1939 issues of the *Journal of the American Chemical Society*, the *Journal of Biological Chemistry*, and *Helvetica Chimica Acta*, as well as numerous notes and full papers in various other journals during that year. The result of this concentration of effort in characterization and synthesis was an extremely rapid solution to what was a very difficult problem in organic chemistry.

It was recognized rather early that sources of the vitamin such as putrefied fish meal contained a number of different vitamins in the K_2 series with differing-chain-length polyprenyl groups at the 3-position. Although it was suggested that animal tissues were able to form the alkylated forms of vitamin K from the parent menadione, this was not definitely shown until Martius and Esser (1958) demonstrated that they could isolate a radioactive polyprenylated form of the vitamin from tissues of rats fed radioactive menadione.

4.1.2. Available Review Articles

Information regarding the then current state of knowledge of vitamin K chemistry and action is available in a number of extensive journal reviews and edited collections of articles dealing with the subject. Almquist (1941) and Dam (1942) have reviewed the literature in this field shortly after the discovery of the vitamin; and, at a somewhat later period, Dam (1948) and Isler and Wiss (1959) surveyed the vitamin K literature in the *Vitamins and Hormones* series. Both the first and second editions of *The Vitamins* (Sebrell and Harris, 1954, 1971) contain multiauthored sections dealing with various aspects of chemistry and biological activity of vitamin K, and a 1966 symposium held in honor of Professor Dam has been published (Harris *et al.*, 1966) in *Vitamins and Hormones*. None of these extensive sources of information contains any literature pertinent to the current understanding of the mechanism of action of vitamin K, a field which had developed almost entirely in the last 10 years. The published proceedings of more recent symposia held in honor of Harry Steenbock (DeLuca and Suttie, 1970) and R. A. Morton (Harris *et al.*, 1974) are, however, valuable sources of information dealing with the early phases of our current knowledge of the molecular action of vitamin K. A review of recent advances in an understanding of the mechanism of action of the vitamin, as well as an extensive review of the chemistry and activation of prothrombin, is also available (Suttie and Jackson, 1977), as is a more historical perspective of this field (Almquist, 1975).

4.1.3. Nomenclature

Nomenclature of those compounds possessing vitamin K activity has been sub-ject to a number of modifications since the discovery of the vitamin. The nomen-clature in general use at the present time is that adopted by an IUPAC-IUB (1966) subcommittee on nomenclature of quinones. The term "vitamin K" is used as a ge-neric descriptor for 2-methyl-1,4-naphthoquinone (I) and all derivatives of this compound which exhibit an antihemorrhagic activity in animals fed a vitamin K deficient diet. The compound 2-methyl-3-phytyl-1,4-naphthoquinone (II) is gen-

erally called "vitamin K_1" or "phylloquinone." The compound first isolated from putrefied fish meal and called "vitamin K_2" is one of a series of vitamin K's with un-saturated side chains called "multiprenylmenaquinones" which are found in animal tissues and bacteria. This particular form of the vitamin has seven iso-prenoid units or 35 carbons in the side chain and has been called "vitamin $K_{2(35)}$" or "menaquinone-7" (MK-7) (III). Vitamins of the menaquinone series with up to 13 prenyl groups have been identified, as well as several partially saturated mem-bers of this series. The parent compound of the vitamin K series, 2-methyl-1,4-naphthoquinone, has often been called "vitamin K_3," but is more commonly designated as "menadione." This nomenclature is summarized in Table I. Nomen-clature in this field has recently been complicated by an attempt of the Interna-tional Union of Nutritional Sciences (IUNS) to alter the IUPAC nomenclature. This revision suggests that menadione be called "menaquinone," phylloquinone be called "phytylmenaquinone" (PMQ), and a vitamin of the K_2 series, such as $K_{2(20)}$, be called "prenylmenaquinone-4" (MQ-4) rather than "menaquinone-4" (MK-4). These differences have not been resolved, but at the present time most workers in the field see little advantage in a change from the IUPAC nomenclature.

4.1.4. Isolation and Chemical Characterization

Vitamin K can be isolated from biological material by standard methods used to obtain physiologically active lipids. Initial extractions are usually made

Table I. Comparison of Vitamin K Nomenclature

Chemical name	Old	IUPAC (abbr.)	IUNS (abbr.)
2-Methyl-1,4-naphthoquinone (I)	K_3	Menadione	Menaquinone
2-Methyl-3-phytyl-1,4-naphthoquinone (II)	K_1	Phylloquinone (K)	Phytylmenaquinone (PMQ)
2-Methyl-3-multiprenyl-1,4-naphthoquinone (class)	$K_{2(n)}$	Menaquinone-n (MK-n)	Prenylmenaquinone-n (MQ)
2-Methyl-3-farnesylgeranylgeranyl-1,4-naphthoquinone (III)	$K_{2(35)}$	Menaquinone-7 (MK-7)	Prenylmenaquinone-7 (MQ-7)

with the use of some type of dehydrating conditions such as the chloroform–methanol technique of Bligh and Dyer (1959) or by first grinding the wet tissue with anhydrous sodium sulfate and then extracting it with hexane or ether. Crude nonpolar solvent extracts of this type contain large amounts of contaminating lipid, and further purification and identification of vitamin K are facilitated by a preliminary fractionation of the crude lipid extract on hydrated silicic acid (Matschiner *et al.*, 1967). A number of the forms of the vitamin can be separated by reverse-phase partition chromatography as described by Matschiner and Taggart (1967), and identification of different forms of the vitamin can be accomplished on thin-layer silica gel plates impregnated with either silver nitrate or paraffin (Matschiner and Amelotti, 1968). A large number of thin-layer and paper chromatographic systems have been used for the identification of various members of the vitamin K series, and these have been reviewed by Mayer and Isler (1971) and Sommer and Kofler (1966). The latter reference contains an extensive tabulation of the R_f values of the various vitamin K isoprenalogs in different paper and thin-layer chromatographic systems.

All separations involving concentrated extracts of vitamin K should be carried out in subdued light to minimize ultraviolet decomposition of the vitamin. Compounds with vitamin K activity are also sensitive to alkali but are relatively stable to oxidizing atmosphere and heat and can be vacuum-distilled with little decomposition.

Phylloquinone is an oil at room temperature, while the various menaquinones can easily be crystallized from organic solvents and have melting points from 35°C to 60°C depending on the length of the isoprenoid chain. The various forms of the vitamin can readily be characterized by a number of physical methods. They exhibit an ultraviolet spectrum which is characteristic of the naphthoquinone nucleus with four distinct peaks between 240 and 280 nm, and a less sharp absorption at around 320–330 nm. The extinction coefficient ($E_{1\,cm}^{1\%}$) decreases with chain length and has been used as a means of determining the length of the side chain. The molar extinction value, ϵ, for both phylloquinone and the various menaquinones is about 19,000. These compounds also exhibit characteristic infrared and nuclear magnetic resonance absorption spectra. Again, the characteristic features of the spectra are largely those of the naphthoquinone ring. Nuclear magnetic resonance analysis of phylloquinone has been used to firmly establish that natural phylloquinone is the *trans* isomer and can be used to establish the *cis/trans* ratio in synthetic mixtures of the vitamin. Mass spectroscopy has been

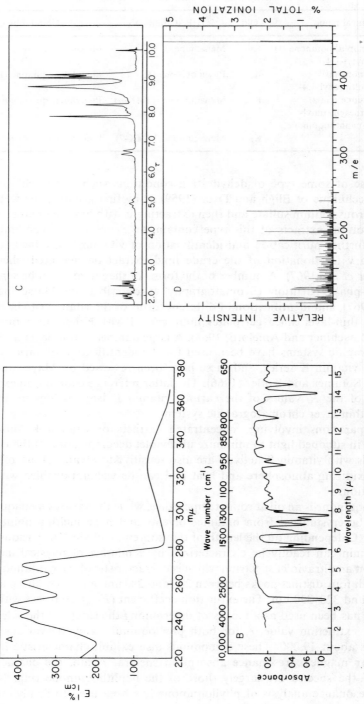

Fig. 1. A: Ultraviolet absorption spectrum of phylloquinone in petroleum ether. B: Infrared absorption spectrum of phylloquinone in CDCl₃ at 60 Mc. D: Mass fragmentation spectrum of phylloquinone, the parent molecular ion, at m/e 450.

useful in determining the length of side chain and the degree of saturation of vitamins of the menaquinone series isolated from natural sources. The ultraviolet, infrared, nuclear magnetic resonance, and mass fragmentation spectra of phylloquinone are shown in Fig. 1. The spectra of vitamins of the menaquinone series are similar, and many of the available spectral data have been collected and summarized by Sommer and Kofler (1966).

4.1.5. Blood Coagulation

Hemostasis, the cessation of bleeding, is a complex phenomenon, which is accomplished by a combination of vasorestriction of the injured vessel at the site of injury, platelet aggregation to form a platelet plug, and triggering of the coagulation process which ultimately results in the formation of a fibrin clot. This last response encompasses the complex series of reactions often referred to as the "blood-clotting mechanism." As four of the proteins involved in the reactions which have been identified as components in this complex are dependent on vitamin K for their synthesis, an understanding of their action has been intimately involved in an understanding of the metabolic role of vitamin K. A basis for an understanding of the role of the multitude of factors involved in coagulation came with the realization (Davie and Ratnoff, 1964; MacFarlane, 1964) that many of the proteins involved could be looked at as zymogens which could be activated by a specific protease and that these modified proteins could in turn activate still other zymogens. This basic cascade or waterfall theory of blood coagulation has now been altered to provide a model (Fig. 2) which has been extremely useful in focus-

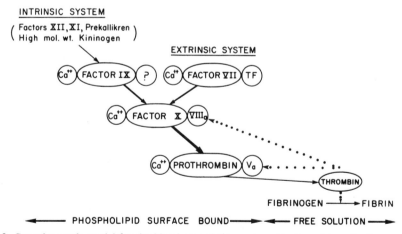

Fig. 2. Staged cascade model for the blood coagulation system. Prothrombin activation can be triggered by events initiated either by factors present in blood (intrinsic system) or by the release of a tissue factor (TF) from an injury (extrinsic system). At each stage, except for the two initial stages of the intrinsic and extrinsic systems, a proteinase, in conjunction with the accessory components of the activation complex, converts the substrate (enclosed by the ellipse) from an inactive to a highly active proteolytic enzyme which then becomes the proteinase in the subsequent stage. The four components enclosed in ellipses are the "vitamin K dependent" clotting proteins, Factor VII, Factor IX, Factor X, and prothrombin. From Suttie and Jackson (1977).

ing attention and directing efforts toward a clearer understanding of this complex phenomenon. The basic biochemistry involved in this process has been reviewed (Davie and Fujikawa, 1975; Esnouf, 1976). A key step in this process is, of course, the generation of thrombin from prothrombin (Factor II) by the action of activated Factor X (Factor X_a). These two proteins, as well as Factors VII and IX, depend on vitamin K for their synthesis.

Prothrombin activation has most often been studied by utilizing the protein prepared from bovine plasma. Bovine prothrombin is a single-chain protein with a molecular weight of 72,000–74,000. The protein has been sequenced (Magnusson *et al.*, 1975), and although only isolated portions of this sequence have been confirmed in other laboratories, there is reason to believe that the published sequence is in all important aspects correct. Only about half of the mass of the prothrombin molecule is required to produce thrombin, and the rest is lost during the activation process. Intensive investigations by a number of laboratories during the past few years have now resulted in a general agreement on the number and order of proteolytic events associated with thrombin generation. As this subject has been comprehensively dealt with in a recent review (Suttie and Jackson, 1977), these findings will be summarized but not extensively referenced here. The complete activation of prothrombin by the so-called prothrombinase complex [prothrombin, Factor X_a, Factor V, phospholipid, and calcium ions] results in the formation of two-chain thrombin from the carboxyl-terminal end of the molecule and two activation peptides called "fragment 1" and "fragment 2" from the amino-terminal end of the molecule. If the action of [X_a, V, PL, Ca^{2+}] on prothrombin is stopped at an intermediate point, three other peptides can be identified: fragment 1·2, the complete nonthrombin portion of the molecule; prethrombin-1, a peptide containing thrombin and fragment 2; and prethrombin-2, a peptide which forms two-chain thrombin upon cleavage of an internal disulfide loop. Overlapping

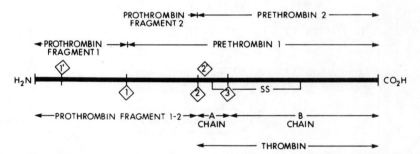

Fig. 3. Schematic structure of prothrombin. The three peptide bonds that are cleaved in bovine prothrombin during or as a result of prothrombin activation are designated by ①, ②, and ③. Another bond, ①', can be cleaved in both bovine and human prothrombins; however, neither the precise location nor the agent responsible has been determined. Cleavages ② and ③ are catalyzed by Factor X_a to yield active two-chain thrombin, and cleavage at ① is catalyzed by thrombin. By convention, "prethrombin" implies that the amino acid sequence ultimately giving rise to thrombin is contained in the polypeptide, and "fragment" implies that the amino acid sequence of the polypeptide is not finally associated with thrombin. In human prothrombin an additional bond, ②', is cleaved, which results in the formation of an A chain of bovine thrombin that is 13 residues shorter than the A chain of bovine thrombin. From Suttie and Jackson (1977).

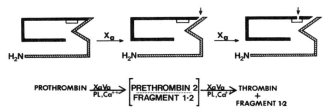

Fig. 4. Mechanism of prothrombin conversion to thrombin. Factor X_a, in conjunction with Factor V_a, phospholipid, and calcium ions, cleaves two peptide bonds in the prothrombin molecule to form thrombin. The first cleavage generates a precursor of thrombin with the same amino acid sequence (prethrombin-2) but which is inactive. Concomitant with formation of prethrombin-2 is the formation of prothrombin fragment $1 \cdot 2$—the "pro" half of prothrombin. These two peptides are held together in a tight noncovalent association until cleavage of the second bond results in the generation of enzyme activity and a two-chain structure for thrombin. The dark, stippled, and hatched regions of the molecule represent the prethrombin-2, fragment-2, and fragment-1 peptides, respectively, as shown in Fig. 3. From Suttie and Jackson (1977).

peptides are formed because both Factor X_a and thrombin can cleave prothrombin at the specific points shown in Fig. 3.

As there are three peptide bonds cleaved and two proteases involved in the conversion of prothrombin to its ultimate products, there are a number of potential pathways the reaction could take. At the present time there appears to be a general consensus among various investigators that the pathway shown in Fig. 4 is the important one. Factor X_a first cleaves prothrombin to form prethrombin-2 and fragment 1·2. These two peptides do not dissociate, but are held together by a noncovalent interaction, and Factor X_a can cleave prethrombin-2 to form two-chain thrombin. As the reaction proceeds, the thrombin generated can cleave fragment 1·2 to fragment 1 and fragment 2, and can also cleave prothrombin to form prethrombin-1 and fragment 1. These degradations of the prothrombin molecule account for the complete mixtures of peptides which are seen in partial activation mixtures. Although the thrombin portion of the molecule has the important physiological role of cleaving fibrinogen to form fibrin, the nonthrombin portion of the molecule also has important functions. The amino-terminal region of prothrombin (fragment 1) and the analogous region on the light chain of Factor X_a contain the vitamin K dependent acidic groups which are responsible for the calcium-dependent binding of these proteins to the phospholipid surface. The fragment 2 region contains the site of the prothrombin–Factor V_a interaction. These interactions are of critical importance in achieving a rapid rate of thrombin generation.

The activation of prothrombin from species other than the bovine has not been as extensively studied. It appears (Downing *et al.*, 1975) that human prothrombin is activated in an analogous fashion with the added complexity that a small peptide can be cleaved by thrombin from prethrombin-2 to form what could be called "prethrombin-2'," and ultimately a slightly altered thrombin. The activation of rat prothrombin has been studied by Grant and Suttie (1976*b*), and it appears to be subjected to the same cleavages as have been determined for bovine prothrombin. The generation of the prethrombin-2' species observed in human prothrombin activation would not have been detected in their study.

4.2. Biological Activity and Physiology

4.2.1. Structure–Activity Correlations

Following the discovery of vitamin K in the early 1930s, a large number of related compounds were synthesized and their biological activities compared. Much of the information available on the biological activity of a large number of compounds synthesized by L. F. Fieser has been brought together in a single paper (Fieser *et al.*, 1941), and these data and others have been reviewed and summarized elsewhere (Weber and Wiss, 1971; Griminger, 1966). The data from various studies are somewhat difficult to compare because of variations in methods of assay, but a number of generalities were apparent rather early.

The first simple compound discovered (Almquist and Klose, 1939*a*) with antihemorrhagic activity was 2-methyl-3-hydroxy-1,4-naphthoquinone (IV), commonly called "phthicol." This compound has only weak activity, and it was soon recognized that menadione (I), 2-methyl-1,4-naphthoquinone, was much more active, and most investigators of the biological activities of synthetic naphthoquinones related the activity of other compounds to that of menadione. The 2-methyl group is usually considered essential for activity, and alterations at this position, such as the 2-ethyl derivative (V), result in inactive compounds. This is not due to the inability of the 2-ethyl derivative to be alkylated, as 2-ethyl-3-phytyl-1,4-naphthoquinone is also inactive. There is some evidence that the methyl

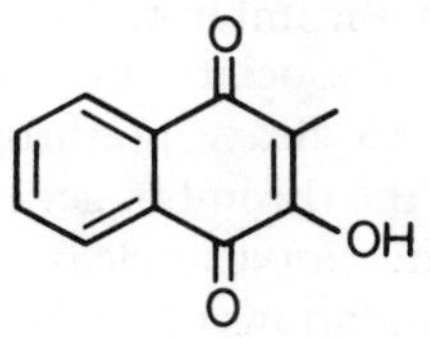

IV

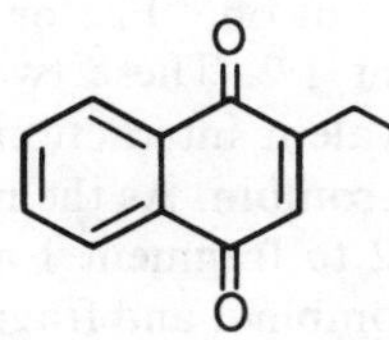

V

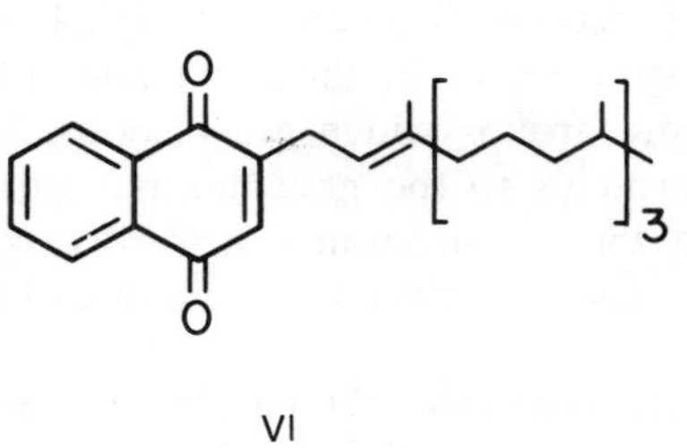

VI

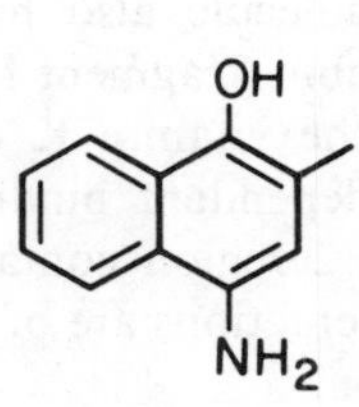

VII

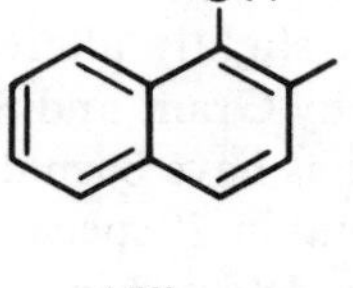

VIII

group may not be absolutely essential, as 2-phytyl-1,4-naphthoquinone (VI) has considerable activity. The available data suggest, but do not conclusively establish, that VI functions as a demethyl phylloquinone with some biological activity rather than being methylated to form phylloquinone. Although there were early suggestions that menadione might be functioning as a vitamin, it is now usually assumed that it is alkylated to a biologically active menaquinone either by intestinal microorganisms or by tissue alkylating enzymes. The range of compounds which can be utilized by animals (or intestinal bacteria) is wide, and compounds such as 2-methyl-4-amino-1-naphthol (VII) and even 2-methyl-1-naphthol (VIII) have biological activity similar to that of menadione. The reduced form of the substituted 1,4-naphthoquinones, the corresponding hydroquinones, will spontaneously oxidize to the quinone form, but they can be stablized by esterification. Compounds such as the diphosphate, the disulfate, the diacetate, and the dibenzoate of reduced vitamin K's have been prepared and they are active forms.

Studies with substituted 2-methyl-1,4-naphthoquinones have revealed that polyisoprenoid side chains are the most effective substituents at the 3-position. The biological activity of phylloquinone, 2-methyl-3-phytyl-1,4-naphthoquinone (II), is reduced by saturation of the double bond to form 2-methyl-3-(β,γ-dihydrophytyl)-1,4-naphthoquinone (IX). This compound is, however, considerably more active than 2-methyl-3-octadecyl-1,4-naphthoquinone (X), which has an unbranched alkyl side chain of similar size. Natural phylloquinone is the *trans* isomer,

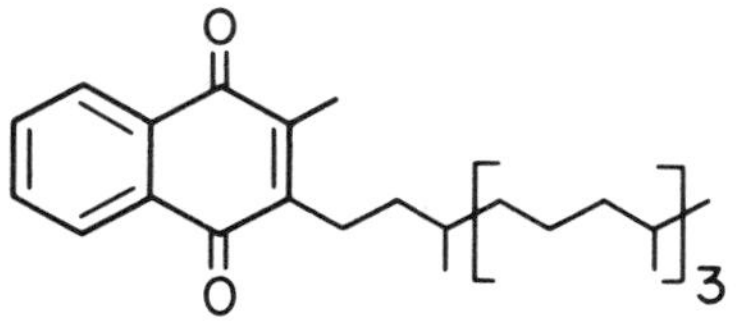

IX

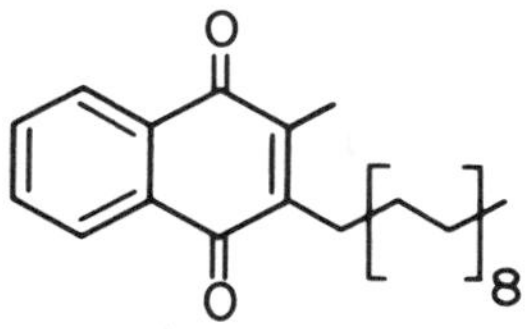

X

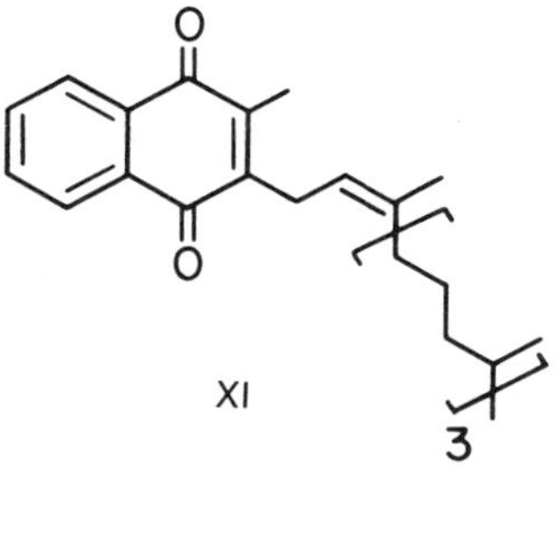

XI

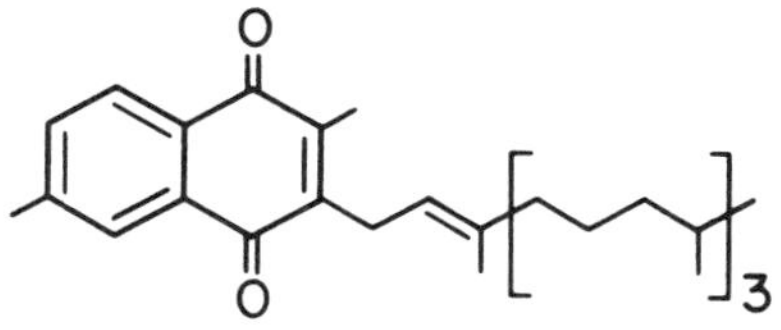

XII

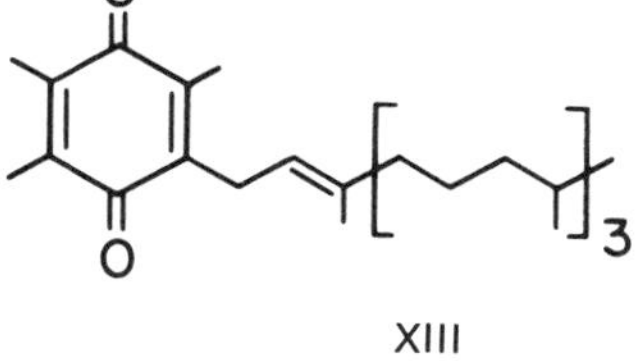

XIII

Table II. Relationship between Vitamin K Activity and Structure[a]

Compound	Effective dose (μg)
2-Methyl-1,4-naphthoquinone (I)	0.3
2-Methyl-3-phytyl-1,4-naphthoquinone (II)	1
2-Methyl-1-naphthol (VIII)	1
2-Methyl-3-(β-γ-dihydrophytyl)-1,4-naphthoquinone (IX)	8
2-Methyl-3-octadecyl-1,4-naphthoquinone (X)	25
2-Phytyl-1,4-naphthoquinone (VI)	50
2-Methyl-3-hydroxy-1,4-naphthoquinone (IV)	500
2-Ethyl-1,4-naphthoquinone (V)	Inactive
2,6-Dimethyl-3-phytyl-1,4-naphthoquinone (XII)	Inactive
2,3,5-Trimethyl-1,4-benzoquinone (XIII)	Inactive

[a]The data represent a curative assay of vitamin K deficient chicks and are based on an 18 hr response to the administration of the different compounds into the crop in 0.1 ml of peanut oil. An effective dose was that which reduced the whole blood-clotting time from in excess of 60 min to less than 10 min. The largest dose tested was 1000 μg. From Fieser *et al.* (1941).

and, although there has been some confusion in the past, it now seems (Matschiner and Bell, 1972) that the *cis* isomer of phylloquinone (XI) is essentially inactive. It has also been reported that 2,3-dimethyl-1,4-naphthoquinone has considerable biological activity, but there is no indication whether this compound has activity on its own, or whether it is subjected to demethylation and subsequent alkylation before exerting a biological effect. The naphthoquinone nucleus cannot be altered appreciably, as methylation to form 2,6-dimethyl-3-phytyl-1,4-naphthoquinone (XII) results in loss of activity, and the benzoquinone most closely corresponding to phylloquinone, 2,3,5-trimethyl-6-phytyl-1,4-benzoquinone (XIII), is inactive.

Many of the data available on these compounds were obtained by the use of an 18-hr curative test on vitamin K deficient chickens following oral administration, and the activities of a number of compounds based on this assay are summarized in Table II. This type of assay allows sufficient time for metabolic alterations of the administered form to an active form of the vitamin, and the observation

Table III. Relative Activity of Vitamin K Isoprenalogs[a]

Number of C atoms in side chain	Relative biological activity (molar) phylloquinone = 100		
	Phylloquinone series	Menaquinone series	
	Oral (chick)	Oral (chick)	Intracardial (rat)
10	10	15	<2
15	30	40	—
20	100	100	13
25	80	>120	15
30	50	100	170
35	—	70	1700
40	—	68	—
45	—	60	2500
50	—	25	1700

[a]Data from Weber and Wiss (1971) and Matschiner and Taggart (1968).

of an antihemorrhagic activity of a compound by this assay does not mean that it was in fact substituting for the vitamin at its active site.

There has been considerable interest in assessing the relative biological activity of forms of the vitamin with different numbers of isoprene units in the side chain. The relative effectiveness of various compounds appears to be dependent on the route of administration. Weber and Wiss (1971) have reviewed data obtained by the oral administration of the vitamin to vitamin K deficient chicks which shows that isoprenalogs with three to five isoprenoid groups in either a menaquinone- or a phylloquinone-type compound have maximum activity. The lack of effectiveness of higher isoprenalogs in this type of assay may be due to the relatively poor absorptions of these compounds. Matschiner and Taggart (1968) have shown that when the intracardial injection of vitamin K to deficient rats is used as a criterion, the very high molecular weight isoprenalogs of the menaquinone series are the most active, and maximum activity was observed with MK-9 (Table III).

4.2.1.1. Activity in Practical Diets

The major use of vitamin K in the animal industry is in poultry diets. As chicks are very sensitive to vitamin K restriction, and antibiotics which decrease intestinal vitamin synthesis are often added to poultry diets, this is a practical necessity. The addition of phylloquinone has usually been considered too expensive for this purpose and a form of menadione is used. Menadione itself possesses high biological activity in a deficient chick, but its effectiveness depends on the presence of lipids in the diet to promote adsorption, and it is not particularly stable in feed products. Because of this, water-soluble forms are used. Menadione forms a water-soluble sodium bisulfite addition product, menadione sodium bisulfite (MSB) (XIV), which has been used commercially but which is also somewhat unstable in mixed feeds. In the presence of excess sodium bisulfite, MSB crystallizes with an additional mole of sodium bisulfite as a complex called "menadione sodium

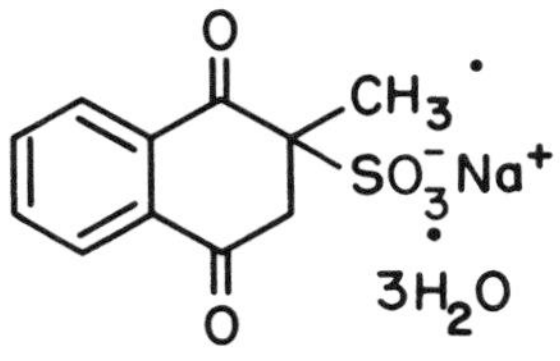

XIV 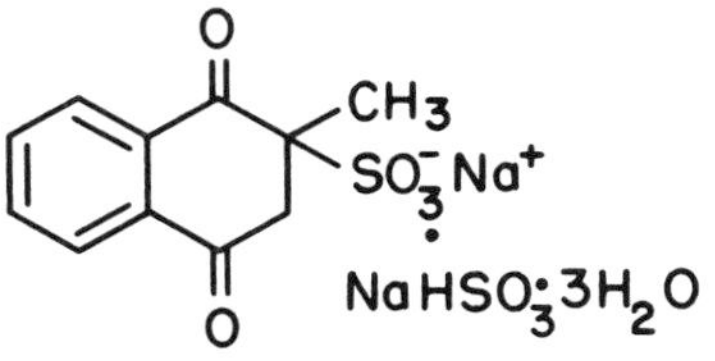   XV

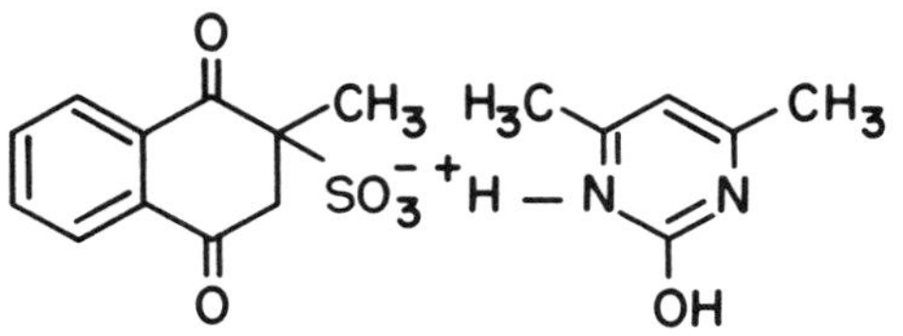

XVI

bisulfite complex" (MSBC) (XV), which has increased stability and is widely used in the poultry industry. A third water-soluble compound is a salt formed by the addition of dimethylpyrimidinol to MSB, which is called "menadione pyrimidinol bisulfite" (MPB) (XVI). There has been some disagreement in the feed industry regarding the relative biopotency of these compounds. Much of this has been the result of comparisons made on the basis of the weight of the salts rather than on the basis of menadione content. There are, however, some indications (Griminger, 1965; Dua and Day, 1966) that, on a molar basis, MPB is somewhat more effective. This difference was not observed by Macy and Winkler (1971), and the current data would suggest that these water-soluble compounds have nearly the same activity on a molar basis. MPB has also been shown (Seerley *et al.*, 1976) to be an effective form of the vitamin in swine rations.

4.2.1.2. *In Vitro Systems Activity*

In vitro systems are now available (Shah and Suttie, 1974; Esmon *et al.*, 1975*b*) which should eventually enable an elucidation of true structural requirements of the vitamin. The current data obtained using these systems, which are discussed in detail in Section 4.6.7, have been somewhat confusing. Jones *et al.* (1976), studying *in vitro* prothrombin synthesis, found that phylloquinone was as active as any of the menaquinones studied, while Friedman and Shia (1976), studying the vitamin K dependent carboxylase system, found that MK-2 and MK-3 were more active than phylloquinone. Both groups found that menadione had little activity in these microsomal preparations. Sadowski (1976) has, however, shown that the reduced form of menadione is an active form of the vitamin in a solubilized (Esmon and Suttie, 1976) preparation of the carboxylase. Whether it is active as such, or is active as a dithiothreitol adduct as suggested by Gardner (1977), has not yet been clarified. The available data suggest that many of the structure–function relationships which have been seen are a function of the interaction of the substituent on the 3-position of the vitamin with the microsomal membrane, and the critical molecular requirement is for a 2-methyl-1,4-naphthoquinone ring, which can be reduced to the hydroquinone.

4.2.2. *Intestinal Absorption*

Vitamin K is absorbed from the intestine into the lymphatic system (Blomstrand and Forsgren, 1968); and, as with other lipid-soluble substances, it requires the presence of both bile and pancreatic juice to allow maximum absorption. The absorption of radioactive phylloquinone has been studied (Shearer *et al.*, 1970, 1974) in patients with impaired fat absorption caused by obstructive jaundice, pancreatic insufficiency, or adult celiac disease. In these patients, as much as 70–80% of the ingested phylloquinone was excreted unaltered in the feces, whereas normal subjects excreted less than 20% of the dose. Treatment of the disease state was shown to drastically improve phylloquinone absorption.

It was demonstrated early in investigations of the role of vitamin K that the inclusion of mineral oil in a diet (Elliott *et al.*, 1940) prevented the absorption of the vitamin, and mineral oil has often been included in vitamin K deficient diets.

The presence of vitamin A also influences vitamin K utilization, and there has been considerable disagreement regarding the nature of the antagonism. Evidence has been presented to both support and refute a systemic effect of vitamin A on vitamin K metabolism. The effect has also been postulated to be at the level of vitamin K absorption, and has been attributed to an effect on bacterial production of vitamin K. These studies have been reviewed by Doisy and Matschiner (1970). The mechanisms by which mineral oil and high vitamin A antagonize vitamin K utilization may not be identical, but it does appear that the development of a vitamin K deficiency, at least in the rat, is enhanced if one of these agents is included in the diet.

A recent series of studies by Hollander and his associates have demonstrated that different forms of vitamin K are absorbed at different sites in the gastro-intestinal tract, and by different mechanisms. In a study utilizing the rat inverted gut sac, Hollander (1973) demonstrated uptake of phylloquinone in proximal gut sac preparations, but relatively poor uptake from distal segments. This uptake was saturable and inhibited by incubation under N_2 or in the presence of 2,4-dinitro-phenol. It was concluded that ingested phylloquinone was absorbed by an energy-dependent process from the proximal portion of the small intestine. In contrast to the results with phylloquinone, Hollander and Truscott (1974b) found that menadione was more effectively absorbed from the distal segment of the small intestine. This uptake was not saturable or energy dependent, and it was concluded that intestinal absorption of ingested menadione proceeded by a bidirectional passive diffusion process which occurred most rapidly in the distal end of the small bowel. It was subsequently shown (Hollander and Truscott, 1974a) that menadione is also absorbed from the colon by a passive process. In contrast to the active transport of phylloquinone observed in inverted gut sacs, Hollander and Rim (1976) found that menaquinone-9 was absorbed from the small intestine by a passive non-carrier-mediated process. The rate of diffusion could, however, be varied by alterations in the lipid and bile salt composition of the incubation medium. The extent to which bacterial synthesis of vitamin K can contribute to the vitamin K requirement of noncoprophagic animals must depend on the ability of the lower part of the intestinal tract, where the bacterial population is greatest, to absorb the vitamin. Uptake of MK-9 from segments of the colon was also studied, and it was shown (Hollander *et al.*, 1976) that the uptake was a passive, nonsatur-able process that showed no evidence of energy dependence or carrier mediation. The results of these studies utilizing rat tissue were interpreted as indicating that vitamin K of bacterial origin could be absorbed from the large bowel of mammals at a sufficient rate to present deficiency symptoms. Shearer *et al.* (1974) have, however, suggested that the concentration of bile salts in the large bowel may not be sufficient to achieve a physiologically significant absorption of bacterial menaquinones in this region of the gut. In any event, the observations available, which are summarized in Table IV, are consistent with the acknowledged difficulty of producing a vitamin K deficiency in most animals by a simple restriction of the vitamin from the diet.

4.2.3. Tissue Uptake and Distribution

Studies of the uptake of vitamin K from the diet and distribution of the vitamin in tissues were hampered by the low specific radioactivity of the vitamin available

Table IV. Gastrointestinal Absorption of Vitamin K[a]

Form of vitamin	Location and mechanism
Phylloquinone	Proximal small intestine (saturable, energy dependent)
Menadione	Distal small intestine (passive)
	Colon (passive)
Menaquinone-9	Small intestine (passive)
	Colon (passive)

[a]Summarized from a series of studies of absorption from isolated intestinal preparations by Hollander and associates. See text for specific references.

until the mid 1960s. Early studies, which have been reviewed by Doisy and Matschiner (1970) and Wiss and Gloor (1966), utilized milligram quantities of vitamin K injected into rats, and they indicated that phylloquinone was specifically concentrated and retained in the liver while menadione was poorly retained in this organ. Menadione was found to be widely distributed in all tissues and to be very rapidly excreted. Losito *et al.* (1968) directly compared the metabolism of menadione and phylloquinone in normal and hepatectomized rats. They used relatively high doses (30–90 μg/100 g body weight) and found a rapid excretion of menadione in the urine, which reached 70% of the injected dose by 24 hr. This excretion was not influenced by hepatectomy. When phylloquinone was used, only about 10% of the radioactivity was found in the urine of normal rats in 24 hr, and a negligible excretion of radioactivity from phylloquinone was observed in hepatectomized rats. The liver has been shown to conjugate menadione to a glucuronide and a sulfate (Losito *et al.*, 1967), and the lack of effect of hepatectomy would suggest that a similar activity might be found in other tissues. These data also suggest that urinary excretion products of phylloquinone are formed only in the liver. More recently, the metabolism of more physiological doses of the vitamin have been studied, and the data obtained (Thierry and Suttie, 1969, 1971) again demonstrate that even at low doses there are significant differences in the tissue distribution of phylloquinone and a water-soluble form of menadione (Table V). Both Taggart and Matschiner (1969) and Thierry *et al.* (1970) found that the major lipophilic product of menadione metabolism formed when low doses of menadione were administered was menaquinone-4.

It is often assumed that a deficiency of vitamin K, like that of other fat-soluble vitamins, would develop slowly because of tissue stores, but this is not supported by the available data. Although about 40% of a 3 μg/100 g body weight dose of phylloquinone is present in the liver at 3 hr after an intravenous injection, this falls to about 10% by 24 hr (Thierry and Suttie, 1971). Additional studies (Thierry *et al.*, 1970) indicated that after a rapid drop during the first few hours following injection, radioactive phylloquinone was removed from rat liver with a half-life of about 17 hr. The inability to rapidly develop a vitamin K deficiency in most species results from the difficulty in preventing any absorption of the vitamin from the diet or from intestinal synthesis rather than from a significant storage of the vitamin.

Konishi *et al.* (1973) have studied vitamin K distribution by the technique of whole body radiography following administration of radioactive menadione, phylloquinone, or menaquinone-4. They found essentially equal distribution of

Table V. Whole-Body Distribution of Phylloquinone and Menadiol Diphosphate[a]

Tissue	Percent of ingested radioactivity	
	Phylloquinone	Menadiol diphosphate
Liver	47.0	2.0
Plasma	2.3	2.9
Heart	0.1	0.2
Kidney	0.5	4.8
Lung	1.3	0.5
Spleen	1.5	0.3
Skeletal muscle	4.4	11.8
Bone	6.1	8.1
Gastrointestinal tract and contents	18.4	33.8

[a]Male rats were given 10 μg of radioactive phylloquinone intravenously or 2 μg of radioactive menadiol diphosphate intraperitoneally and killed after 3 hr. The data represent the total amount of vitamin and vitamin metabolites in the tissues at that time. From Thierry and Suttie (1969, 1971).

phylloquinone and menaquinone-4 at 24 hr following intravenous injection, but a relatively higher deposition of menaquinone-4 than phylloquinone in intestinal contents at early periods. This is consistent with a much more rapid biliary excretion of menaquinone-4 than phylloquinone which was also demonstrated in this study. When radioactive menadione was administered, it was spread over the whole body much faster than the other two compounds, but the amount retained in the tissues was low. Whole body radiography also confirmed that vitamin K is concentrated by organs other than the kidney, and the highest activity was seen in the adrenal glands, lungs, bone marrow, kidneys, and lymph nodes. Whether this distribution indicates any physiological function of the vitamin in soft tissues other than the liver cannot be determined from these studies.

4.2.4. Form of Vitamin K in Tissues

The amount of vitamin K in animal tissues is low, and the various forms of the vitamin which are present in tissues from different species have been difficult to determine. It was recognized early in the studies of the vitamin that phylloquinone is the form of the vitamin found in plant material and that bacterial cells contain a series of menaquinones. The low concentration of the vitamin in animal tissues prevented, however, an assessment of the form of the vitamin in animal tissues. A series of studies by Martius and his co-workers utilized radioactive menadione to establish that it could be converted to a more lipophilic compound which on the basis of their limited characterization (Martius and Esser, 1958) appeared to be menaquinone-4. They also demonstrated (Billeter and Martius, 1960, 1961; Billeter *et al.*, 1964) that if phylloquinone or long-chain menaquinones were administered orally, a considerable portion of the naphthoquinone nucleus ended up as menaquinone-4 in the tissues. The existence of an enzyme system in liver, and possibly in extrahepatic tissues that will alkylate menadione, has been well established (see Section 4.3.1.2). Although these studies have established that menadione can be alkylated, and that intestinal flora or tissues can dealkylate lipophilic forms of the vitamin, they have contributed little to establish the identity of the forms of vitamin normally present in the liver. Because of the limited data available, and the demonstration that liver tissue can form menaquinone-4 from menadione, it has often been assumed that this was the major form of the vitamin in animal tissues. The available data, largely from Matschiner's laboratory, have been reviewed by Matschiner (1970, 1971), and they indicate that phylloquinone is found in the liver of those species ingesting plant material, and that, in addition to this, menaquinones containing from 6 to 13 prenyl units in the alkyl side chain are found in the liver of most species. Some of these long-chain vitamins appear to be partially saturated. It is of interest that menaquinone-4, which is the major metabolite formed when menadione is ingested, has not been identified as a normal liver form of the vitamin. The data in Table VI would therefore suggest that, in addition to some dietary phylloquinone, the liver of most species contains a number of menaquinones which are presumably of bacterial origin.

4.2.5. Cellular Distribution

Early studies of vitamin distribution within the liver (Green *et al.*, 1956) depended on bioassay of isolated fractions and indicated that the vitamin was

Table VI. Various Forms of Vitamin K in Liver[a]

	Species of liver studied						
	Beef[b]	Rabbit[c]	Chicken[c]	Pig[c,d]	Dog[d]	Horse[e]	Human[f,g]
Vitamin Content							
Total K_1 equivalent (μg/g)	1.2	—	—	0.4	0.6	0.2	0.06
Number of forms found	4	2	2	13	19	1	10
Form identified							
Phylloquinone	X[h]	X[h]	X	X[i]		X	X
MK-4 (H_4)[k]		X	X	X[i]			
MK-6					X[j]		
MK-7				X[j]	X[j]		X
MK-8				X[j]	X[j]		X
MK-9				X[j]	X[j]		X[j]
MK-10	X			X[j]	X[j]		X[j]
MK-11	X				X		X
MK-12	X				X		
MK-13					X		

[a] Adapted from Matschiner (1971). The presence of a particular form (indicated by X) does not mean that it represents a significant amount of the total activity, and the original data should be consulted.
[b] Matschiner and Amelotti (1968).
[c] Hamilton and Dallam (1968).
[d] Duello and Matschiner (1971*a,b*).
[e] Duello and Matschiner (1970).
[f] Duello and Matschiner (1972).
[g] Rietz *et al.* (1970).
[h] Observed by Duello and Matschiner (personal communication).
[i] Proposed by Hamilton and Dallam (1968), not found by Duello and Matschiner (1971*a,b*).
[j] Partially saturated forms of these menaquinones also identified.
[k] 2-Methyl-3-Δ^6-dehydrophytyl-1,4-naphthoquinone.

distributed in all liver subcellular fractions. Bell and Matschiner (1969*b*) injected 0.02 or 3 μg of phylloquinone to vitamin K deficient rats and found that over 50% of the liver radioactivity was recovered in the microsomal fraction and that substantial amounts were found in the mitochondria and cellular debris fractions. Thierry and Suttie (1971) studied the specific activity (picomoles vitamin K/mg protein) of injected radioactive phylloquinone and found that only the mitochondrial and microsomal fractions had a specific activity which was enriched over that of the entire homogenate and that the highest specific activity was in the mitochondrial fraction. As neither group expressed their data in terms of both specific activity and total radioactivity in various fractions, and the amounts injected were not the same, it is not possible to determine if the apparent discrepancy is real between the two sets of data in the degree to which the vitamin is concentrated in microsomal or mitochondrial membranes. A subsequent study (Matschiner and Bell, 1972) utilizing the more biologically active *cis* isomer of phylloquinone indicated that the largest fraction of liver radioactivity was in the microsomal fraction but that the mitochondrial fraction had the highest specific activity. Nyquist *et al.* (1971) studied the distribution of radioactive phylloquinone in more detail and found the highest specific activity (protein basis) to be in the Golgi and smooth microsomal membrane fractions. Although there has been some variation in the results of different studies, it is clear that the vitamin localizes in various cellular membranes and that significant amounts of the vitamin are probably found in nontarget membranous fractions. A number of factors probably influence the intracellular distribution of the vitamin. Knauer *et al.* (1975) have shown that there are variations in the subcellular distribution of the *cis* and *trans* forms of phylloquinone, and Thierry and Suttie (1971) have demonstrated that the vitamin K antagonist 2-chloro-3-phytyl-1,4-naphthoquinone has a significant influence on both the whole body and subcellular distributions of radioactive phylloquinone. The limited number of data which are available (Knauer *et al.*, 1976) would suggest that the amount of vitamin K in the cytoplasmic fraction of the cell is depleted rather rapidly during the development of a deficiency state and that the various membrane fractions lose the vitamin more slowly and at roughly the same rate.

4.3. Metabolism

4.3.1. Biosynthesis

4.3.1.1. Biosynthesis of the Naphthoquinone Ring

Although animals are not able to synthesize the 2-methyl-1,4-naphthoquinone ring needed for vitamin K, the bacterial synthesis of this ring has been studied, mainly in *Mycobacterium phlei* and *Escherichia coli*. These studies have resulted in the identification of the biochemical source of all of the carbons of the ring (Glover, 1970) and an identification of some of the intermediates involved (Campbell *et al.*, 1971; Baldwin *et al.*, Bentley, 1975*a,b*). As would be expected, the carbons of the aromatic ring are furnished by shikinic acid, and the remaining three carbons of the naphthalene ring are furnished by α-ketoglutarate. The intermediate in this reaction, following the decarboxylation and additions of the remaining four carbons of α-ketoglutarate to the shikinic acid acceptor, would be *O*-succinyl-benzoic acid, which would be converted to 1,4-dihydroxy-2-naphthoic acid. The pathway from this compound to the menaquinones may depend on the organism. It has been shown (Baldwin *et al.*, 1974) that, at least in the case of *M. phlei*, this compound is not decarboxylated to a symmetrical intermediate, but rather is methylated and then decarboxylated and prenylated prior to oxidation to yield the menaquinone structure. These two steps might be closely linked and naphthoquinol would not necessarily be a free intermediate in this pathway. This pathway is outlined in Fig. 5. Some organisms produce a large amount of demethylmenaquinones, which can be converted to menaquinones. Menaquinone biosynthesis has also been studied in extracts of *E. coli* (Shineberg and Young, 1976; Bryant and Bentley, 1976), and in this system it appears that decarboxylation and prenylation result in demethylmenaquinone formation. This is presumably a concerted reaction as there is no evidence for the presence of the symmetrical naphthoquinone or naphthoquinol intermediate. A subsequent step would then be the methylation reaction catalyzed by these organisms to form the menaquinones (Fig. 5).

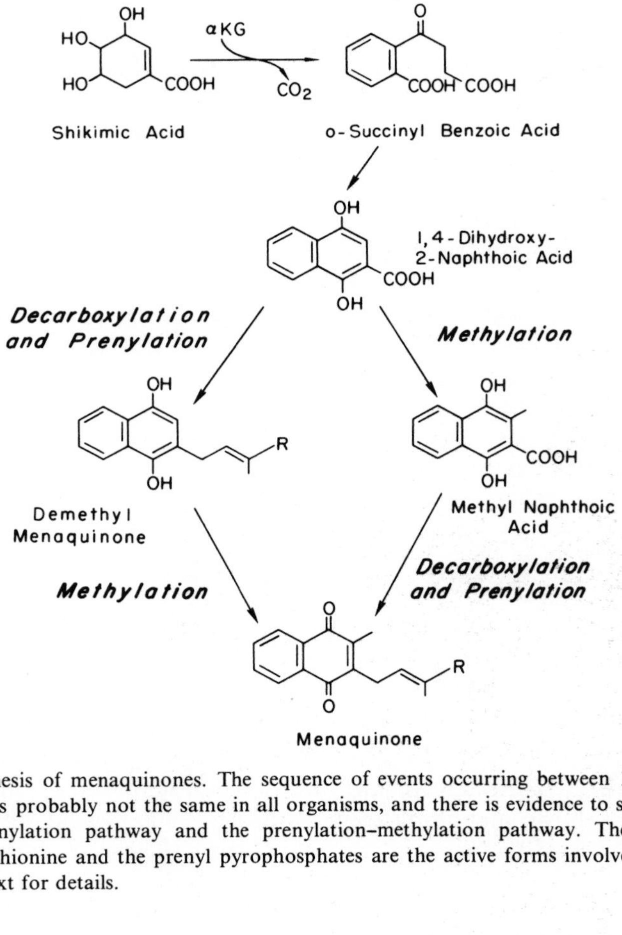

Fig. 5. Biosynthesis of menaquinones. The sequence of events occurring between 1,4-dihydroxy-2-naphthoic acid is probably not the same in all organisms, and there is evidence to support both the methylation–prenylation pathway and the prenylation–methylation pathway. The methyl group comes from methionine and the prenyl pyrophosphates are the active forms involved in the alkylation step. See text for details.

4.3.1.2. Alkylation of Menadione

Animals cannot synthesize the naphthoquinone ring, but they are able to alkylate menadione to menaquinones, chiefly menaquinone-4. The transformation of menadione to menaquinone-4 was first observed by Martius and Esser (1958), and Martius subsequently presented evidence that administered phylloquinone or other alkylated forms could be converted to menaquinone-4. Although it was originally believed that the dealkylation and subsequent realkylation with a geranylgeranyl side chain occurred in the liver, it was subsequently shown (Billeter *et al.*, 1964) that phylloquinone was not converted to menaquinone-4 unless it was administered orally. This suggested that intestinal bacterial action was required for the dealkylation step. It is, however, probable that some dealkylation does occur in animal tissue (Martius, 1967) and that some interconversion of phylloquinone and menaquinones can occur without bacterial action.

Taggart and Matschiner (1969) demonstrated that the rat could convert injected menadione to a lipophilic form, presumably menaquinone-4, and the ability of chick liver to form menaquinone-4 from injected menadione was unequivocally demonstrated by Dialameh *et al.* (1971), who were able to isolate sufficient menaquinone-4 from the liver of deficient chicks injected with menadione to characterize it by mass spectrometry. Martius (1961) demonstrated that menadione could be converted to menaquinone-4 by an *in vitro* incubation of chick liver homogenates with geranylgeranyl pyrophosphate. The highest alkylating activity was found in the mitochondrial fraction. These studies were extended by Dialameh *et al.* (1970), who demonstrated that chick liver homogenates could use geranyl pyrophosphate, farnesyl pyrophosphate, or geranylgeranyl pyrophosphate as alkyl donors for the synthesis of menaquinones from menadione. The corresponding free alcohols were inactive, the highest activity was found in microsomes, and the system was inhibited by O_2. It was shown that the highest alkylation rate was observed with the nonphysiological donor, farnesyl pyrophosphate, and that the activity of the alkylating system was about 6–7 times higher in chick liver microsomes than in rat liver microsomes.

These studies have demonstrated that animal tissues can convert ingested menadione to a biologically active form. There is no evidence from *in vivo* studies that menadione can function without alkylation, and although there are undoubtedly small amounts of other menaquinones formed, menaquinone-4 appears to be the major metabolite. It is not, however, a major form in the absence of menadione administration, and the higher isoprenalogs and absorbed phylloquinone predominate in most species.

4.3.2. Metabolic Degradation

4.3.2.1. Menadione

The metabolism of menadione has been more completely studied than that of phylloquinone or the menaquinone isoprenalogs. Early studies of menadione metabolism in the rat (Hoskin *et al.*, 1954; Hart, 1958) led to the identification of three different urinary conjugates—the phosphate, sulfate, and glucuronide of menadiol. These did not account for all of the possible metabolites; and when menadione metabolism was studied in an isolated perfused rat liver (Losito *et al.*, 1957), it was found that 93% was metabolized in 5 hr. The major product identified was the glucuronide, which was excreted in the bile. The major plasma metabolite that was characterized was the sulfate conjugate of menadiol, but about 70% of the radioactivity was in the form of metabolites that could not be identified. The simultaneous administration of Warfarin had no influence on the excretion pattern observed. In a subsequent study (Losito *et al.*, 1968), it was shown that 70% of a physiological dose of menadione was excreted in the urine of both normal and hepatectomized rats in 24 hr. Although the relative amounts did differ, the compounds identified as the glucuronide and sulfate were found in both cases, indicating that extrahepatic conjugation was a significant process.

4.3.2.2. *Alkylated Forms of the Vitamin*

Metabolic products of phylloquinone and the menaquinones have been less extensively studied. Taylor *et al.* (1956) demonstrated that the major route of excretion of intravenously administered radioactive phylloquinone was the feces. Only a small percentage of fecal or biliary radioactivity was present as unmetabolized phylloquinone, but no information was obtained to identify the metabolites of phylloquinone which were present in the bile and feces. Wiss and Gloor (1966) observed that the side chains of phylloquinone and menaquinone-4 were shortened to seven carbon atoms, yielding a carboxylic acid group at the end which cyclized to form a γ-lactone. This lactone was excreted in the urine, presumably as a glucuronic acid conjugate. The metabolism of radioactive phylloquinone has also been studied in man by Shearer *et al.* (1972). When a dose of either 1 mg or 45 μg of vitamin K was injected, it was found that about 20% of the dose was excreted in the urine in a 3-day study and 40–50% was excreted in the feces via the bile. In contrast to the lack of effect on menadione metabolism, Warfarin administration greatly increased urinary excretion and decreased fecal excretion of phylloquinone (Shearer *et al.*, 1973). The chemical nature of the urinary components was studied (Shearer and Barkhan, 1973; Shearer *et al.*, 1974), and it was shown that most of the metabolites were present as glucuronides. Two different aglycones of phylloquinone were identified (Fig. 6), and it was concluded that the γ-lactone previously identified is an artifact formed by the acidic conditions used in previous studies. The urinary metabolites of phylloquinone are substantially altered by Warfarin administration, and although they have not been positively identified, they appear to be mainly glucuronides of degradation products of phylloquinone-2,3-epoxide (Shearer *et al.*, 1974). The biliary and fecal metabolites of phylloquinone in man have not been identified, but they have been shown to include a major metabolite which is less polar than the urinary metabolites shown in Fig. 6, and which may be an esterified form of these aglycones (Shearer *et al.*, 1974).

It seems likely that a number of excretion products of vitamin K have not been identified. Matschiner (1970) fed doubly labeled phylloquinone (a mixture of [6,7-^{3}H]phylloquinone and [phytyl-U^{14}C]phylloquinone with a ^{3}H/^{14}C ratio of 66) to deficient rats and recovered radioactivity in the urine with a ^{3}H/^{14}C ratio

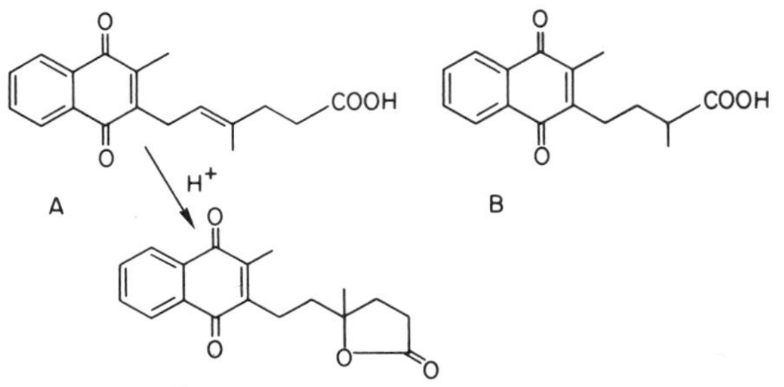

Fig. 6. Urinary metabolites of phylloquinone. Compounds A, 2-methyl-3-(5′-carboxy-3′-methyl-2′-pentenyl)-1,4-naphthoquinone, and B, 2-methyl-3-(3′-carboxy-3′-methylpropyl)-1,4-naphthoquinone, are the major aglycones of phylloquinone which can be obtained from urine after treatment with β-glucuronidase. Compound C, 2-methyl-3-(5′-carboxy-3′-methylpentyl)-1,4-naphthoquinone lactone, is probably formed during the isolation of these compounds due to the acidic conditions which are used. From Shearer *et al.* (1974).

of 264. This ratio is higher than would be expected for the excretion of a lactone with seven carbons remaining in the side chain. The ratio is more nearly that which would be expected from the excretion of the five-carbon acid derivatives identified by Shearer and Barkhan (1973) and suggests that some even more extensively degraded metabolites of phylloquinone were present. An exchange of the phytyl side chain of phylloquinone for the tetraisoprenoid chain of menaquinone-4 before degradative metabolism would also be expected to contribute to an increase in the $^3H/^{14}C$ ratio. Although the current data suggest that this exchange does occur, it is doubtful that it is extensive enough to account for the relative increase in tritium activity in the urinary products.

It is clear from the limited studies available that menadione is rapidly metabolized and excreted, and that only a relatively small portion gets converted to biologically active menaquinone-4. The metabolism of phylloquinone and the menaquinones is much slower, and, although the major products may have been identified, there are a number of urinary and biliary products not yet characterized. There is, however, no evidence to indicate that the isoprenoid forms of the vitamin must be subjected to any metabolic transformation before they are biologically active.

4.3.3. Vitamin K Epoxide

From a quantitative standpoint, the most significant metabolite of vitamin K is undoubtedly its 2,3-epoxide, commonly called "vitamin K oxide." This metabolite was discovered by Matschiner *et al.* (1970), who were following up their observation that there was more radioactivity from tritium labeled phylloquinone in the liver of Warfarin-treated rats than control rats. The increased radioactivity was shown to be due to the presence of a significant amount of a metabolite more polar than phylloquinone which was isolated and characterized as phylloquinone 2,3-eposide (XVII). Further studies of this compound (Bell *et al.*, 1972) revealed that about 10% of the vitamin K in the liver of a normal rat is

XVII

present as the oxide and that this can become the predominant form of the vitamin following treatment with coumarin anticoagulants. The formation of this metabolite is not confined to the rat, and Shearer *et al.* (1974) have shown that injected vitamin K is metabolized to the oxide in man and that under conditions of Warfarin therapy the major form of the vitamin circulating in the plasma is the epoxide. The recent development of a high-pressure liquid chromatographic method (Elliott *et al.*, 1976) for separating vitamin K and the epoxide will greatly facilitate studies of interrelationships of these compounds. The possible significance of vitamin K epoxide in the action of the vitamin or in the action of the coumarin anticoagulants will be discussed in Sections 4.6.9, 4.6.11.3, and 4.6.11.4.

4.4. Dietary Requirement

4.4.1. Requirement for Animals

The establishment of the dietary level of vitamin K which is required for various species has not been an easy task. Although all species of animals presumably need vitamin K to synthesize the essential clotting factors, it is difficult to demonstrate dietary requirement in many species. The difficulties presumably come from the varying degrees to which different species can utilize the large amount of vitamin K synthesized by intestinal bacteria and the degree to which different species practice coprophagy. A spontaneous deficiency of vitamin K was first noted in chicks, and poultry are much more likely to develop symptoms of a dietary deficiency than any other species. It is usually assumed that the site of vitamin K synthesis in the chick is too close to the distal end of the tract to permit absorption, although it is also possible that the rapid rate of passage of material through the chick intestine and the relatively short length of large intestine in this species contribute to the problem.

There appears to be little evidence that ruminants, with the vitamin K synthesized in their large bacterial population available for subsequent utilization, need a source of vitamin in the diet. It has, however, been possible to produce a deficiency in most monogastric species. Values for vitamin K requirement from different studies are difficult to compare because of different forms of the vitamin that were used, and different methods that were employed to establish the requirement. Studies have been carried out using either a curative or a preventive type of assay. In the former case, animals are first made hypoprothrombinemic by feeding a vitamin K deficient diet. Following this, either they are injected with the vitamin and prothrombin levels are assayed after a few hours or 1 day, or they are fed diets containing various amounts of the vitamin for a number of days and the response in clotting factor synthesis is noted. In a preventive or prophylactic assay, an attempt is made to determine the minimum concentration of the vitamin which must be present in a diet to maintain normal clotting factor levels.

Although phylloquinone has usually been used as the source of vitamin in experimental nutrition studies, other forms have often been used when more practical diets have been considered. Menadione is usually considered to be from 20% to 40% as effective as phylloquinone on a molar basis, but this depends a great deal on the type of assay used. Menadione is rather ineffective in a curative assay where the rate of its alkylation to a menaquinone is probably the important factor, but often shows activity nearly equal to that of phylloquinine in a long-term preventive assay. Practical nutritionists have often preferred to add a water-soluble form of menadione to rations, and the most commonly used compound has been the menadione sodium bisulfite complex (MSBC). This compound appears to be about as active on a molar basis as phylloquinone in poultry rations, and the activities of menadione, MSBC, and phylloquinone are therefore roughly equal on a weight basis, at least in poultry rations.

More detailed discussions of the vitamin K requirements of various species are available in articles by Scott (1966), Doisy and Matschiner (1970), and Griminger (1971). The data indicate that the requirement for most species falls in a range of from 2 to 200 μg vitamin K/kg body weight/day. The data in Table VII, which have been adapted from a table presented by Griminger (1971), give an indication of the magnitude of the requirement for various species. It should be remembered that this requirement can be altered by variations in age, sex, or strain and that any condition influencing lipid absorption or conditions altering intestinal flora will have an influence on these values. The effect of sex on vitamin K requirement has been studied in some detail (Matschiner and Bell, 1973; Matschiner and Willingham, 1974), and the increased vitamin requirement of male rats may be related to both the metabolism of the vitamin and required cellular concentrations. The basis for variation in requirement for different species is not known, but Knauer *et al.* (1976) have presented data which suggest that differences in the requirement for vitamin K between sexes and strains of rats are due principally to different required concentrations of vitamin K in the liver, and not differences in absorption or turnover of the vitamin. The relative importance

Table VII. Vitamin K Requirement of Various Species[a]

Species	Body weight basis (μg/kg/day)	Dietary basis (μg/kg diet)
Dog	1.25	60
Pig	5	50
Rhesus monkey	2	60
Rat, male	11–16	100–150
Chicken	80–120	530
Turkey poult	180–270	1200

[a]Requirement based on the amount of vitamin needed to prevent the development of a deficiency, with no correction for any difference in potency of different forms of the vitamin on a weight basis. The data have been summarized from a more extensive table (Griminger, 1971).

of each of these factors in explaining the requirement of different species has not been determined.

4.4.2. Human Requirement

The requirement of the adult human for vitamin K is extremely low, and there is no possibility of a simple dietary deficiency developing in the absence of complicating factors. The low requirement and the relatively high levels of vitamin K found in most diets prevented an accurate assessment of the requirement until recent years. Frick *et al.* (1967) studied the vitamin K requirement of starved, debilitated patients given antibiotics to decrease intestinal vitamin K synthesis and fed intravenously. They determined that 0.1 $\mu g/kg/day$ was not sufficient to maintain normal prothrombin levels and that 1.5 $\mu g/kg/day$ was sufficient to prevent any decreases in clotting factor synthesis. Their data would indicate that the requirement was on the order of 1 $\mu g/kg/day$. Doisy (1971) fed a chemically defined diet providing less than 10 μg of vitamin K per day to two normal subjects and was able to deplete prothrombin concentrations to less than 50% by about 20 weeks. Mineral oil and antibiotics were administered during a portion of this period to decrease vitamin absorption and synthesis. These patients responded to the administration of about 0.5 μg vitamin K/kg/day by rapidly restoring clotting activity to normal. It was concluded from this study that about 1 μg vitamin K/kg/day was sufficient to maintain normal clotting factor synthesis in the normal adult human. O'Reilly (1971) maintained four normal volunteers on a diet containing about 25 μg vitamin K/day and administered antibiotics to decrease intestinal synthesis. Prothrombin activity was maintained in a normal range of from 70% to 100% of normal during a 5-week period, with lower values observed near the end of the study. These data would suggest that prothrombin concentrations can be maintained near the low end of the normal range on a diet containing about 0.5 μg vitamin K/kg/day. The limited studies available would therefore suggest that the vitamin requirement of the human is in the range of 0.5–1.0 μg vitamin K/kg/day. This amount of vitamin is exceeded in almost any diet which is in other respects nutritionally adequate, and a simple dietary deficiency appears to be of little concern.

This does not mean that vitamin K responsive hypoprothrombinemia cannot be a clinically significant human problem. O'Reilly (1976) has reviewed the potential problem areas and has pointed out that there are basic factors needed to prevent a vitamin K deficiency: (1) a normal diet containing the vitamin, (2) the presence of bile in the intestine, (3) a normal absorptive surface in the small intestine, and (4) a normal liver. It is therefore not surprising that cases of insufficient vitamin K have been noted in both breast-fed infants and infants receiving an artificial formula deficient in vitamin K and also in adults on protracted antibiotic treatment or receiving long-term parenteral hyperalimentation without vitamin K supplementation. Malabsorption of vitamin K has also occurred as a result of obstructive jaundice, biliary fistula, pancreatic insufficiency, steatorrhea, or chronic diarrhea. Hepatocellular disease or hepatic immaturity in premature infants can also result in a failure to utilize sufficient amounts of vitamin K that may be in the diet. It is therefore clear that cases of at least an acquired vitamin K deficiency do occur in the adult population, and, although they are relatively rare, they do present a significant problem for some individuals. It has also been suggested (Hazell and Baloch, 1970) that a relatively high percentage of the older adult population may have a hypoprothrombinemia that responds to administration of oral vitamin K.

A hemorrhagic disease of the human newborn has been a long-recognized syndrome and one which is at least in part responsive to vitamin K therapy (Owen, 1971). The condition is at its worst a few days after birth, and it appears to be associated almost exclusively with breast-fed infants. It can be effectively prevented by parenteral administration of 100 μg of menadione sodium bisulfite at birth (Keenan *et al.*, 1971).

4.5. Antagonists of Vitamin Action

4.5.1. Coumarin Derivatives

The first antagonist of the action of vitamin K was actively being investigated during the period that Dam was involved in establishing the essentiality of vitamin K in the diet. The history of the discovery of the anticoagulant action of coumarin derivatives has been documented and discussed by Link (1959). A hemorrhagic disease of cattle was described in Canada and the American midwest in the 1920s, and, as it was traced to the consumption of improperly cured sweet clover hay, the disease was commonly referred to as "sweet clover disease." Early investigators recognized that the disease could be cured by the substitution of good hay for the spoiled hay and that if serious hemorrhages did not develop the animals could be aided by transfusion with whole blood from healthy animals. By the early 1930s, it was established that the cause of the prolonged clotting times was a decrease in the prothrombin activity of blood. The compound present in spoiled sweet clover responsible for this disease had been studied by a number of investigators but was finally isolated and characterized by Link's group during the period from 1933 to 1941. A reliable biological assay was developed, and the active hemorrhagic agent was crystallized (Campbell and Link, 1941) and characterized as 3,3'-methyl-*bis*-(4-hydroxycoumarin) (Stahmann *et al.*, 1941); it was given the trade name Dicumarol (XVIII).

Dicumarol was successfully used as a clinical agent for anticoagulant therapy rather than heparin in some early studies, and a large number of substituted 4-hydroxycoumarins were synthesized both in Link's laboratory and elsewhere. The most successful of these, both clinically for long-term lowering of the vitamin K dependent clotting factors and as a rodenticide, have been Warfarin, 3-(α-acetonylbenzyl)-4-hydroxycoumarin (XIX), or its sodium salt; Marcoumar, 3-(1-phenylpropyl)-4-hydroxycoumarin (XX); and Tromexan, 3,3'-carboxymethylene-*bis*-(4-hydroxycoumarin) ethyl ester (XXI). The various drugs which have been used differ in the degree to which they are absorbed from the intestine and in their plasma half-life and presumably in their effectiveness as a vitamin K antagonist in the active site. Because of this, their clinical use differs. Much of the information on the structure–activity relationships of the 4-hydroxy-

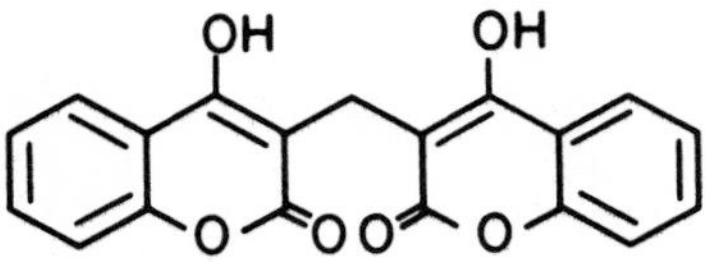

XVIII 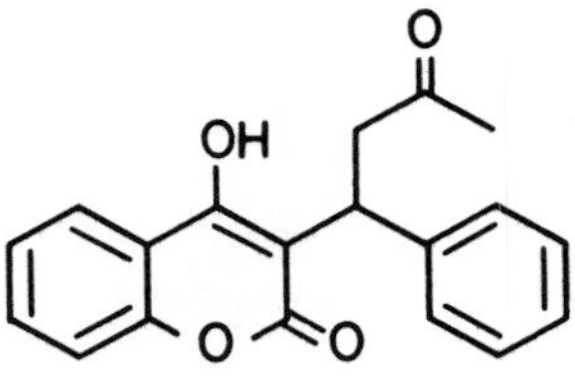   XIX

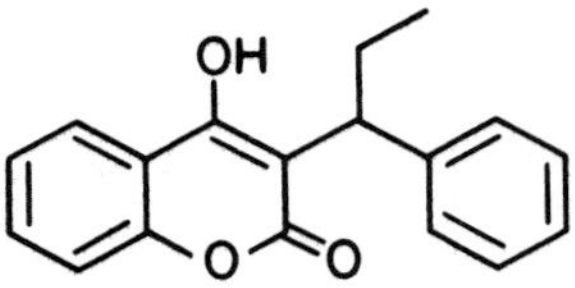

XX 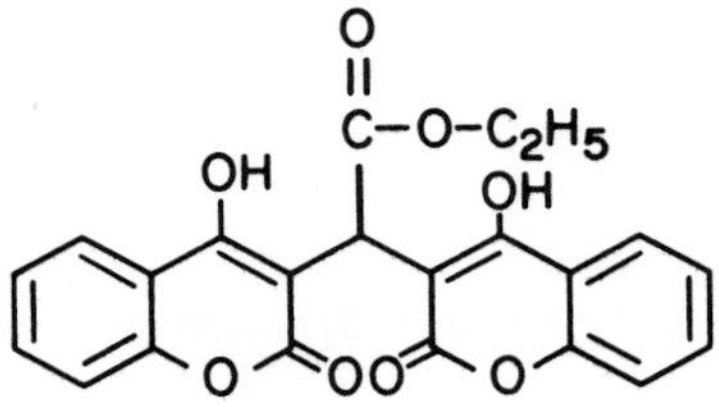   XXI

coumarins has been reviewed by Renk and Stoll (1968). These drugs are synthetic compounds, and, although the clinically used compound is the racemic mixture, studies of the two optical isomers of Warfarin have shown that they differ both in their effectiveness as anticoagulants and in the influence of other drugs on their metabolism. The clinical uses of these compounds and many of their pharmacodynamic interactions have been reviewed by O'Reilly (1976).

Warfarin not only is used as a clinical anticoagulant but also is widely used as a rodenticide. Concern has been expressed in recent years because of the identification of anticoagulant-resistant rat populations, first in Northern Europe (Boyle, 1960; Lund, 1964; Drummond, 1966) and subsequently in the United States (Jackson and Kaukeinen, 1972). These rats have both an increased resistance to warfarin and an increased requirement for vitamin K (Hermodson et al., 1969) and appear to have the same genetic alteration as that described by O'Reilly (1971) in human patients. Concern over the spread of the anticoagulant-resistant rat population (Jackson et al., 1975) has led to renewed interest in synthesis of coumarin derivatives, and a study by Hadler and Shadbolt (1975) has resulted in the synthesis of a number of compounds that not only are effective (Hadler et al., 1975; Rennison and Hadler, 1975) in the Warfarin-resistant rat but also are much more active in normal animals than most compounds synthesized in the past. Because of their widespread use, there has been a continued interest in studies of the mechanism by which the coumarin anticoagulants exert their effect. This problem will be considered in detail in Section 4.7.

4.5.2. 1,3-Indandiones

A second class of chemical compounds having anticoagulant activity which can be reversed by vitamin K administration was first identified by Kabat et al.

XXII   XXIII

(1944) as 2-substituted-1,3-indandiones. A large number of these compounds have also been synthesized, and two of the more commonly used members of the series have been 2-phenyl-1,3-indandione (XXII) and 2-pivalyl-1,3-indandione (XXIII). These compounds have had some commercial use as rodenticides, but because of the potential for hepatic toxicity (O'Reilly, 1976) they are no longer used clinically. Studies on the mechanism of action of these compounds have not been so extensive as those on the 4-hydroxycoumarins, but the observations that Warfarin-resistant rats are also resistant to the indandiones and the ability of these compounds to inhibit the conversion of the 2,3-epoxide of vitamin K to the vitamin (Ren *et al.*, 1974) would suggest that the mechanism of action of the indandiones is similar to that of the 4-hydroxycoumarins.

4.5.3. 2-Halo-3-phytyl-1,4-naphthoquinones

During the course of a series of investigations into the structural requirements for vitamin K activity, Lowenthal *et al.* (1960) found that the replacement of the 2-methyl group of phylloquinones by a chlorine atom to form 2-chloro-3-phytyl-1,4-naphthoquinone (XXIV) or a bromine atom to form 2-bromo-3-phytyl-1,4-naphthoquinone resulted in compounds which were potent antagonists of vitamin K. These compounds can be synthesized (Lowenthal and Chowdhury, 1970) by alkylation of the corresponding 2-halo-1,4-naphthoquinone by essentially the same procedures used to synthesize phylloquinone from mena-

XXIV   XXV

XXVI   XXVII

dione. The most active of these two compounds is the chloro derivative (commonly called "Chloro-K"), and in a series of publications Lowenthal (1970) has shown that, in contrast to the coumarin and indandione derivatives, Chloro-K acts as if it were a true competitive inhibitor of the vitamin at its active site(s). It has also been shown (Parmer and Lowenthal, 1962) that oxidative phosphorylation of isolated rat liver mitochondria is not influenced by prior treatment of the animals with Chloro-K, which would support the hypothesis that it is a direct antagonist of vitamin K at its active site rather than being a nonspecific quinone antagonist. Chloro-K may also interfere with those enzymes responsible for metabolizing vitamin K, as Thierry and Suttie (1971) found significant alterations in the tissue distribution of radioactive vitamin K when it was administered at the same time as Chloro-K. Because of its distinctly different mechanism of action than the commonly used coumarin anticoagulants, Chloro-K has been used as a probe of the mechanism of action of vitamin K, and as it is an effective anticoagulant in coumarin anticoagulant-resistant rats (Suttie, 1973b), it has been suggested as a possible rodenticide.

4.5.4. Other Antagonists

Lowenthal's studies of possible agonists and antagonists of vitamin K indicated (Lowenthal and MacFarlane, 1965) that some of the *para*-benzoquinones have biological activity. The benzoquinone analogue of vitamin K_1 2,5,6-trimethyl-3-phytyl-1,4-benzoquinone has weak vitamin K activity, and 2-chloro-5,6-trimethyl-3-phytyl-1,4-benzoquinone (XXV) is an antagonist of the vitamin. When these compounds were modified to contain shorter isoprenoid side chains at the 3-position, they were neither agonists nor antagonists. Because of the similarities in structure of vitamin K and coenzyme Q, it is not unreasonable that some compounds could act as inhibitors of both, and Combs *et al.*(1976) have determined that 2-hydroxy-3-*n*-dodecylmercapto-1,4-naphthoquinone is both an inhibitor of mitochondrial coenzyme Q reactions and an anticoagulant.

Two compounds that appear to be rather unrelated to either vitamin K or the coumarins that have anticoagulant activity have recently been described. The first 2,3,5,6-tetrachloro-4-pyridinol (XXVI), has been shown (Marshall, 1972a,b) to have anticoagulant activity, and on the basis of its action in Warfarin-resistant rats (Ren *et al.*, 1974) it would appear that it functions as a direct antagonist of the vitamin, as does Chloro-K. A second series of compounds even less structurally related to the vitamin are the 6-substituted imidazole [4,5-b]pyrimidines. These compounds have been described by Bang *et al.* (1975) as antagonists of the vitamin, and the action of 6-chloro-2-trifluoromethyl-imidazo[4,5-b]pyrimidine (XXVIII) in Warfarin-resistant rats would suggest that they are functioning in the same way as a coumarin or an indandione type of compound.

A hypoprothrombinemia can also be produced in some species by feeding animals sulfa drugs and antibiotics. There is little evidence that these compounds are doing anything other than decreasing intestinal synthesis of the vitamin by altering the intestinal flora. They should therefore not be considered as antagonists of the vitamin.

4.6. Metabolic Role of Vitamin K

4.6.1. Historical Development

Although it was clear soon after Dam's discovery of vitamin K in the mid to late 1930s that the vitamin was involved in the biosynthesis of prothrombin, knowledge of the cellular events responsible for the production and the regulation of these metabolic events has come largely from investigations spanning only the last 10–15 years. There was a 25-year period following the discovery of the vitamin during which a great deal was learned about the biological activity of various forms of both the vitamin and its antagonists, and about the nutritional and clinical significance of the vitamin, but little information regarding its mechanism of action was obtained. Anderson and Barnhart (1964a,b) conclusively demonstrated that prothrombin is produced in the liver, but the lack of a general understanding of the mechanism of protein biosynthesis prevented serious experimental approaches to the cellular or molecular mechanisms involved until the last decade.

A number of theories to explain the role of vitamin K in prothrombin synthesis which were eventually shown to be incorrect were, however, proposed during the last 40 years. Dam *et al.* (1936) originally proposed that the vitamin, or at least a portion of it, was a part of the prothrombin molecule, but this observation could never be confirmed. Martius postulated that the vitamin had a function in mammalian electron transport such that a deficiency would lead to a defect in oxidative phosphorylation and a low cellular ATP level. This would then result in a decrease in the concentration of a protein with such a rapid turnover rate as prothrombin. He cited evidence (Martius and Nitz-Litzow, 1954) of low P:O ratios in vitamin K deficient chicks which were not supported by other investigators (Paolucci *et al.*, 1963; Wosilait, 1966). In the mid 1960s, Olson (1964) suggested that the rate of prothrombin production is regulated by an effect of vitamin K on DNA transcription. This hypothesis was refuted (Suttie, 1967; Hill *et al.*, 1968; Lowenthal and Simmons, 1967; Polson and Wosilait, 1969), and subsequent investigations centered around two alternate hypotheses: (1) that the vitamin acts at a ribosomal site to regulate the *de novo* rate of prothrombin synthesis or (2) that it functions postribosomally in a metabolic step that converts a precursor protein, which can be produced in the absence of the vitamin, to active prothrombin. The subsequent discussion will follow the development of the precursor hypothesis to the point where it appears that this is the only tenable explanation of the available data. Various aspects of these studies have been reviewed in recent years (Suttie, 1974; Suttie *et al.*, 1974; Stenflo and Suttie, 1977; Suttie and Jackson, 1977). It was possible, however, to explain many of the early experimental observations by either theory of vitamin K action, and a review by Olson (1974) traces the development of the *de novo* theory for control of prothrombin biosynthesis in his laboratory.

4.6.2. Indirect Evidence for a Prothrombin Precursor Protein

The possibility that a precursor protein was involved in the formation of prothrombin was probably first clearly stated by Hemker *et al.* (1963), who noted a clotting time abnormality in plasma from patients receiving anticoagulant therapy and postulated that it was due to the presence of an inactive form of prothrombin in these patients. The time course of prothrombin appearance in the plasma which was observed by a number of investigators (Bell and Matschiner, 1969a; Hill *et al.*, 1968; Pyörälä, 1965; Suttie, 1970) when vitamin K was administered to severely hypoprothrombinemic rats also suggested the presence of a significant pool of a precursor protein which could be converted to prothrombin following vitamin administration. There was a delay in the appearance of plasma prothrombin which was somewhat dependent on the dose of vitamin K, but which usually lasted 30–60 min after vitamin K administration, followed by a burst of synthesis. Both Pyörälä (1965) and Bell and Matschiner (1967a) clearly pointed out that the rate of prothrombin synthesis observed during this initial period exceeded the theoretical induction curve based on the experimentally determined half-life of prothrombin. Dulock and Kolmen (1968) observed a similar response in plasma prothrombin when vitamin K was administered to dogs previously given Warfarin. It was subsequently shown (Shah and Suttie, 1972) that the appearance of plasma prothrombin in vitamin K treated hypoprothrombinemic rats was preceded by a transient increase of prothrombin in rat liver microsomal preparations. Microsomal prothrombin peaked about 10 min after vitamin K was administered and then fell as prothrombin appeared in the plasma, suggesting that a liver precursor in the hypoprothrombinemic rat was being converted to prothrombin in a vitamin-dependent step and that following depletion of this pool the rate of prothrombin synthesis slowed and became dependent on the rate of synthesis of the precursor.

It was demonstrated (Bell and Matschiner, 1969a; Hill *et al.*, 1968; Suttie, 1970) that the vitamin K stimulated initial burst of prothrombin in the hypoprothrombinemic intact rat was decreased only slightly by prior administration of an amount (Polson and Wosilait, 1969) of the protein synthesis inhibitor cycloheximide sufficient to block prothrombin synthesis in a normal rat. It was also shown (Suttie, 1970) that the increase in plasma prothrombin observed between the first and second hours following vitamin K administration was blocked by this dose of cycloheximide. These studies strongly suggested that protein synthesis was not involved in the vitamin K dependent step of prothrombin synthesis, but did not offer final proof. It was always possible that, although the amount of protein synthesis inhibitor used was sufficient to block the synthesis of most proteins in the system studied, it was for some reason not blocking the formation of prothrombin. Direct evidence of the presence of a liver precursor protein was obtained when Shah and Suttie (1971) demonstrated that the prothrombin produced when hypoprothrombinemic rats were given vitamin K and cycloheximide was <u>not</u> radiolabeled if radioactive amino acids were administered at the same time as the vitamin. If the vitamin had initiated *de novo* synthesis of prothrombin, and for some reason prothrombin synthesis was not blocked by cycloheximide, the newly formed prothrombin should have contained a high level of radioactivity. This study further indicated that administration of radioactive amino acids to hypoprothrombinemic vitamin K deficient rats prior to cycloheximide and vitamin K administration resulted in the formation of radioactive plasma prothrombin. These observations were consistent with the presence of a precursor protein pool in the hypoprothrombinemic rat that was rapidly being synthesized and that could be converted to prothrombin in a step which did not require protein synthesis.

Although the observations in intact animals rather consistently supported the existence of a precursor to prothrombin, experiments in perfused livers were less conclusive. Puromycin was reported to be both effective (Suttie, 1967) and ineffective (Olson *et al.*, 1966) in blocking a clotting factor response to the vitamin, and Kipfer and Olson (1970) reported that vitamin K was able to specifically reverse the effect of cycloheximide on those ribosomes synthesizing prothrombin but not on the general ribosomal population of an isolated perfused liver. Some of the confusing data might be related to observations (Olson, 1974) that vitamin K administration in the perfused liver results in excretion of incompleted forms of prothrombin which do not have biological activity. A cultured hepatoma cell system (Munns *et al.*, 1976) that produces prothrombin in response to vitamin K has recently been described, and it may prove useful in studying these responses. Plasma clotting Factors VII, IX, and X also depend on vitamin K for their synthesis, and studies of their formations have also been reported. Studies of Factor VII formation in liver slices or isolated liver cells (Babior, 1966; Lowenthal and Simmons, 1967; Pool and Borchgrevink, 1964; Prydz, 1964; Ranhotra and Johnson, 1969) have not contributed a great deal to an understanding of the system. More recent studies (Babior and Kipnes, 1970; Rez and Prydz, 1971) of Factor VII formation in cell-free systems have also been inconclusive but have tended to support the hypothesis that there is a liver precursor to Factor VII which can be converted to Factor VII without the need of additional protein synthesis.

4.6.3. Immunochemical Evidence for a Prothrombin Precursor

The hypothesis that there was a liver precursor to prothrombin was strengthened by direct observations that the plasma of man or animals treated with coumarin anticoagulants contained a protein which was in many ways similar to prothrombin. A protein which was antigenically similar to prothrombin but which lacked biological activity was first demonstrated in the plasma of human patients receiving anticoagulant therapy by Ganrot and Nilehn (1968), and an inactive prothrombin species in the patients was detected by the use of staphylocoagulase by Josso *et al.* (1968, 1970). Other workers (Brozovic and Gurd, 1973; Cesbron *et al.*, 1973; Denson, 1971; Hemker *et al.*, 1970) subsequently confirmed the presence of this protein which could presumably represent a plasma form of the hypothesized liver precursor.

A similar protein was first demonstrated in bovine plasma by Stenflo (1970) and subsequently confirmed by other workers (Malhotra and Carter, 1971; Nelsestuen and Suttie, 1972*a*; Reekers *et al.*, 1973; Wallin and Prydz, 1975). The presence of such a protein in the plasma of other species administered oral anticoagulants has been somewhat controversial, and evidence for its existence has been sought by both immunochemical methods and by the assay of thrombin generation by nonphysiological activators. There are reports that plasma from rats administered anticoagulants does contain an appreciable amount of an abnormal form of prothrombin, while other investigations (Carlisle *et al.*, 1975; Olson, 1974) have found very little of this species. Both Olson (1974) and Carlisle *et al.* (1975) observed some abnormal prothrombin in the plasma of anticoagulant-treated chicks, while the latter group has reported that this protein appeared to be missing in plasma from anticoagulant-treated mice, hamsters, guinea pigs, rabbits, and dogs. Data to support the presence of abnormal forms of the other vitamin K dependent clotting factors in the plasma of anticoagulant-treated human patients or animals have recently been received (Suttie and Jackson, 1977).

4.6.4. *Isolation and Characterization of the Abnormal Prothrombin*

The identification of these new proteins in the plasma of anticoagulant-treated animals provided the stimulus for a series of investigations that culminated in an understanding of the chemical nature of the postribosomal modification of prothrombin. Only recently (Guillin *et al.*, 1977; Elion *et al.*, 1976) has the human abnormal prothrombin been subjected to chemical characterization. The protein from bovine plasma has been purified by Stenflo and Ganrot (1972), Nelsestuen and Suttie (1972*a*), Malhotra and Carter (1971), Hemker and Reekers (1974), and Wallin and Prydz (1975), but the majority of the information regarding its structure was obtained by the first two groups. This protein has been given different names by various investigators: "protein induced by vitamin K absence" (PIVKA), "abnormal prothrombin," "isoprothrombin," "Dicumarolized prothrombin," "Dicumarol-induced prothrombin," "acarboxyprothrombin," or "atypical prothrombin." Each of these names has some merit, but, for the sake of consistency, the plasma protein produced in the bovine administered an oral anticoagulant will be called "abnormal prothrombin" in the subsequent discussions.

The initial studies of this protein (Nelsestuen and Suttie, 1972*a*; Stenflo, 1972; Stenflo and Ganrot, 1972) indicated that it appeared to have the same molecular weight and amino acid composition as normal prothrombin, and that it did not adsorb to insoluble barium salts as did normal prothrombin. The lack of barium salt adsorption and the calcium-dependent electrophoretic and immunochemical properties suggested a difference in calcium-binding properties of these two proteins which was directly demonstrated by Nelsestuen and Suttie (1972*b*) and confirmed by Stenflo and Ganrot (1973). The difference in calcium binding was shown by Stenflo (1973) to be a property of an amino-terminal peptide (prothrombin fragment 1) which could be derived from the two proteins. A further study of the two proteins (Bjork and Stenflo, 1973) by optical rotatory dispersion and circular dichroism revealed that the presence of calcium ions caused conformational changes in prothrombin that were not found in the abnormal prothrombin. The observation (Nelsestuen and Suttie, 1972*a*) that the abnormal prothrombin could yield thrombin when treated with trypsin or snake venoms which contain prothrombin activators indicated that this portion of the molecule was normal and that the critical difference in the two proteins was the inability of the abnormal protein to bind to calcium ions which are needed for the phospholipid-stimulated activation by Factor X_a. Although the phospholipid interactions were not investigated during the early studies of the abnormal prothrombin, it has now been shown (Esmon *et al.*, 1975*c*) that the abnormal prothrombin will not bind to a phospholipid surface in the presence of calcium ions, and that the addition of phospholipid, which drastically stimulates the $[X_a, Ca^{2+}]$ activation of prothrombin, has no effect on the rate of activation of abnormal prothrombin.

The early studies of abnormal prothrombin indicated that its carbohydrate content and structure were probably similar to those of normal prothrombin (Nelsestuen and Suttie, 1972c), and offered final refutation of earlier claims (Johnson *et al.*, 1971, 1972; Pereira and Couri, 1971) that the vitamin K dependent step in the formation of prothrombin involved glycosylation of the protein. Later studies by Pereira and Couri (1972) and the observation (Henriksen *et al.*, 1976; Nelsestuen and Suttie, 1971) that asialo- and aglycoprothrombin retain biological activity and still adsorb to barium salts also made this hypothesis unlikely. It has been claimed (Morrison and Esnouf, 1973) that the abnormal prothrombin from human plasma differs from normal plasma prothrombin in its carbohydrate content. If so, it would differ appreciably from the major species of bovine abnormal prothrombin that have been characterized. There may, however, be other minor species in the plasma, and the studies of Malhotra and Carter (1971, 1972; Malhotra, 1972a,b) suggest that different preparations of abnormal bovine prothrombin may be obtained from plasma, depending on the initial steps in the isolation. The properties of the abnormal bovine prothrombin are compared to those of normal prothrombin in Table VIII, and a study of the abnormal human prothrombin (Guillin *et al.*, 1977) indicates that it has similar properties.

The initial studies of the abnormal prothrombin clearly implicated the calcium-binding region of prothrombin as the vitamin K dependent region, but provided no evidence of the chemical nature of this region. Nelsestuen and Suttie (1973) isolated an acidic peptide from a tryptic digest of normal bovine prothrombin which would adsorb to insoluble barium salts and which bound cal-

Table VIII. Properties of Abnormal Bovine Prothrombin[a]

Property	Comparison to prothrombin
Molecular weight	Indistinguishable
Amino acid composition[b]	Apparently identical
End-terminal residues	Apparently identical
Carbohydrate composition	Apparently identical
Immunochemical determinants	Similar or identical
Electrophoretic mobility	Similar without Ca^{2+}
Hydrodynamic properties	Indistinguishable
Circular dichroism spectra	Indistinguishable
Adsorption to Ba salts	Very low
Ca^{2+} binding	Very low
Ca^{2+}-dependent phospholipid binding	Lacking
Biological activity[c]	Lacking or very low
Activation by trypsin	Apparently identical
Activation by *E. carinatus* venom	Apparently identical

[a] Properties determined from various studies; see text for references.
[b] Based on analysis of acid hydrolyzates.
[c] Activation with $[X_a, V, PL, Ca^{2+}]$.

cium ions in solution. This peptide, which was a portion of the fragment 1 region of prothrombin, contained a high proportion of acidic amino acid residues and had an anomalously high apparent molecular weight as determined on molecular sieve columns. Stenflo (1974) later isolated two acidic peptides from the fragment 1 region by different methods, and both groups postulated the existence of some unknown acidic, nonpeptide, prosthetic group attached to this portion of the molecule. The peptides isolated by these two groups could not be obtained when similar isolation procedures were applied to preparations of abnormal prothrombin. Abnormally acidic peptides which adsorbed to barium salts were subsequently isolated from the fragment 1 portion of prothrombin (Benson and Hanahan, 1975; Magnusson, 1973; Skotland *et al.*, 1974).

4.6.5. Characterization of γ-Carboxyglutamic Acid

All of the investigators involved in attempts to characterize the calcium-binding portion of prothrombin appeared to be looking for some acidic group(s) attached to the glutamic acid residues of prothrombin, presumably through an ester linkage. The vitamin K dependent modification was, however, much simpler. Stenflo *et al.* (1974) succeeded in isolating a tetrapeptide (residues 6–9 of prothrombin) which had an apparent sequence of Leu-Glu-Glu-Val and demonstrating by a combination of mass fragmentation, NMR spectra, and chemical synthesis that the glutamic acid residues of this peptide were modified so that they were present as γ-carboxyglutamic acid (3-amino-1,1,3-propanetricarboxylic acid) residues (Fig. 7). Independently, Nelsestuen *et al.* (1974), by rather similar methods, characterized γ-carboxyglutamic acid (Gla) from a dipeptide (residues 33 and 34 of prothrombin) which appeared originally to be Glu-Ser. These characterizations of the modified glutamic acid residues in prothrombin were confirmed by Magnusson *et al.* (1974), who have shown that all ten of the first 33 Glu residues in prothrombin are modified in this fashion. Further details of these characterizations of γ-carboxyglutamic acid have now been published (Fernlund *et al.*, 1975; Morris *et al.*, 1976).

Factor X is also a calcium-binding vitamin K dependent clotting factor. The amino-terminal region of the light chain of Factor X is homologous with the amino-terminal region of prothrombin, and it has been shown (Enfield *et al.*, 1975; Howard and Nelsestuen, 1975) to also contain these modified glutamic acid

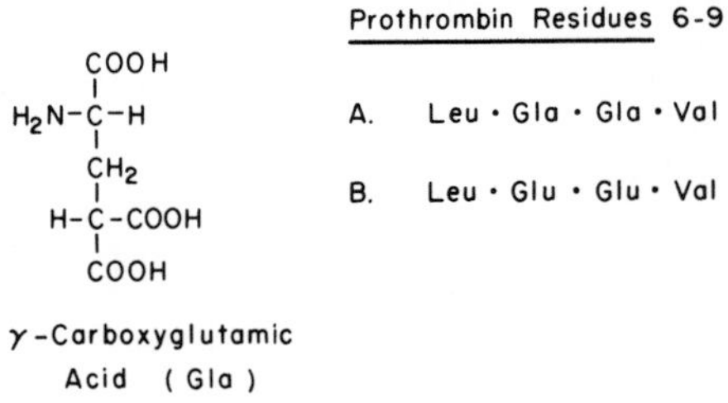

Fig. 7. Structure of γ-carboxyglutamic acid. This carboxylated glutamic acid residue was identified by a comparison of the properties of the tetrapeptide A, obtained from a digest of bovine prothrombin, and the tetrapeptide B, obtained from a similar digest of the abnormal form of prothrombin which is produced in the coumarin-treated bovine.

residues. The amino-terminal region of Factor IX has a considerable amount of homology with prothrombin and Factor X, and it, as well as another vitamin K dependent plasma protein, has been shown (Bucher *et al.*, 1976) to contain γ-carboxyglutamic acid. The position of the Gla residues in prothrombin and the homology to the other vitamin K dependent proteins are shown in Fig. 8.

The original preparations of abnormal bovine prothrombin studies by Stenflo and Suttie were isolated from plasma treated with barium salts to remove any normal prothrombin. Prowse *et al.* (1976) have reported the presence of a prothrombin of low specific activity which will adsorb to barium salts in the plasma of cattle treated for a prolonged period. They have subsequently (Esnouf and Prowse, 1977) reported that this protein contains seven of the ten Gla residues in normal prothrombin and that the prothrombin that does not adsorb to barium salts contains four of the ten Gla residues. These studies strongly suggest that carboxylation is not an "all or nothing" event, and that a gradation of molecules exists when the action of vitamin K is partially blocked. Whether the original preparations of abnormal bovine prothrombin which were studied contained some Gla residues is not yet clear.

4.6.6. *Metabolism of γ-Carboxyglutamic Acid*

The discovery of this new amino acid has opened the question of its metabolic fate. Fernlund (1976) has detected this amino acid in human urine and has determined that the excretion is on the order of 30 μmol/day. This amount is only slightly more than the amount of γ-carboxyglutamic acid which would be available from the normal turnover and degradation of plasma prothrombin. Shah *et al.* (1978) studied the metabolism of radioactive DL-γ-carboxyglutamic acid in the rat and were unable to detect any oxidative degradation of the compound. The amino acid was very rapidly cleared from the plasma and excreted in the urine. Although injected γ-carboxyglutamic acid was found in skeletal

γ-Carboxyglutamic Acid Residues in Plasma Proteins														
Prothrombin	Ala	Asn	Lys	Gly	Phe	Leu	Gla	Gla	–	Val	Arg	Lys	Gly	Asn	Leu
Factor IX	Tyr	Asn	Ser	Gly	Lys	Leu	Gla	Gla	Phe	Val	Arg	–	Gly	Asn	Leu
Factor X	Ala	Asn	Ser	–	Phe	Leu	Gla	Gla	–	Val	Lys	Gln	Gly	Asn	Leu
Factor VII	Ala	Asn	–	Gly	Phe	Leu	(Gla)	(Gla)	Leu	Leu	–	Pro	Gly	Ser	Leu
Protein C	Ala	Asn	Ser	–	Phe	Leu	Gla	Gla	–	Leu	Arg	Pro	Gly	Asn	Leu
Protein S	Ala	Asn	Ser	?	–	Leu	(Gla)	(Gla)							

Fig. 8. Amino-terminal sequences of the vitamin K dependent plasma proteins. The sequences are for the bovine protein except for protein S, which has been isolated only from human plasma. The Factor X and protein C sequences are for the light chain of these two-chain proteins. Dashes refer to spaces that have been inserted to bring the proteins into alignment for better homology. The ? indicates that this residue has not been identified. The residues indicated as (Gla) are presumably Gla residues based on their behavior during sequencing, but have not yet been positively identified. See Bucher *et al.* (1976) and DiScipio *et al.* (1977) for detailed references.

tissue and various organs, it ws there only transiently and was quantitatively excreted in the urine within 48 hr. These two reports would suggest that there is no degradative pathway for γ-carboxyglutamic acid in mammalian tissue and that, as it is formed from the degradation of vitamin K dependent proteins, it is excreted in the urine.

4.6.7. Isolation of Liver Prothrombin Precursor Proteins

Studies of the abnormal prothrombin demonstrated that a thrombinlike activity could be generated from it by treatment with *Echis carinatus* venom. Suttie (1973a) then demonstrated that thrombin activity was generated when microsomes were isolated from Warfarin-treated rats, solubilized in detergent, and treated with *Echis carinatus* venom. A similar increase in this activity was seen when rats were made vitamin K deficient or injected with Chloro-K. Further study (Shah and Suttie, 1973) demonstrated that this activity decreased rapidly when vitamin K was injected, and as its level fell the amount of microsomal pro-thrombin increased and then fell as it moved out of the liver into the plasma.

Based on this assay, Esmon *et al.* (1975a) isolated a protein from the liver of Warfarin-treated rats which had the properties predicted for the prothrombin precursor. The purified precursor was a glycoprotein which was immunochem-ically similar to prothrombin and had a molecular weight indistinguishable from that of rat prothrombin. Electrophoretic and isoelectric focusing analyses indi-cated that the precursor was less negatively charged ($pI = 5.8$) than prothrombin ($pI = 5.0$). Specific proteolysis of the precursor by thrombin, Taipan snake venom, or clotting Factor X_a yielded fragments indistinguishable from those formed by similar proteolysis of rat prothrombin (Grant and Suttie, 1976b). This protein did not adsorb to $BaSO_4$ and its rate of activation to thrombin by Factor X_a and Ca^{2+} was not stimulated from the addition of phospholipid. This protein (precursor I) differed from prothrombin in that it did not contain sialic acid resi-dues or γ-carboxyglutamic acid. It differed from the bovine plasma abnormal prothrombin in that it lacked the sialic acid residues which the plasma protein contains. A second protein with properties very similar but with an isoelectric point of 7.2 has now been isolated (Grant and Suttie, 1976a) from the same microsomal preparations. The increased basic nature of this protein is a prop-erty of the amino-terminal region of the molecule, but the chemical alteration responsible for the shift in pI has not been determined. Which of these proteins, if either, is the physiological precursor of prothrombin has not been determined. The properties of these two proteins are compared to those of rat prothrombin in Table IX. It is possible that the precursor species which is the substrate for the vitamin K dependent carboxylase might be substantially different than the pro-teins which have been purified.

Precursorlike activity has been detected (Carlisle *et al.*, 1975) in liver micro-somal preparations of species other than the rat, but none of these activities has been purified. Morrissey *et al.* (1973) have isolated a protein from liver micro-somes of Warfarin-treated rats which has properties similar to those of the pro-tein isolated by Esmon *et al.* (1975a) which they called "isoprothrombin." This protein was obtained from an antibody affinity column, and sufficient amounts of it have not been available to determine any of its chemical properties.

Table IX. *Properties of Rat Plasma Prothrombin and Rat Liver Precursors*[a]

Property	Prothrombin	Precursor I	Precursor II
Molecular weight (SDS gel)	85,000	85,000	85,000
pI	5.0	5.8	7.2
Neutral sugars	Yes	Yes	Yes
Sialic acid	Yes	No	No
Phospholipid binding (with Ca^{2+})	Yes	No	No
Adsorbs to $BaSO_4$	Yes	No	No
Thrombin generation [X_a, V, PL, Ca^{2+}]	Yes	No	No
Thrombin generation (snake venoms)	Yes	Yes	Yes

[a]Data summarized from Esmon *et al.* (1975*a*) and Grant and Suttie (1976*a,b*).

4.6.8. Prothrombin Production—Vitamin K Dependent Carboxylation

4.6.8.1. In Vitro Systems

A number of *in vitro* cell-free systems which showed an increase in Factor VII or Factor X activity upon incubation have been described over the last 10 years (Suttie and Jackson, 1977). Only recently (Lowenthal and Jaeger, 1977) has it been shown that any of these systems responded to the *in vitro* addition of vitamin K. They have therefore been of limited value in determining the metabolic role of vitamin K. Johnston and Olson (1972*a,b*) described an *in vitro* system that produces radioimmunologically detectable amounts of prothrombin upon incubation. The amount of prothrombin produced in this system is dependent on the prior vitamin K status of the rats used, but it does not respond to the *in vitro* addition of the vitamin. More recently, the synthesis of an immune reactive prothrombin species has been demonstrated in a heterologous system using rat liver mRNA and rabbit reticulocytes (Nardacci *et al.*, 1975). The material produced can be degraded by *E. carinatus* venom and appears to have a molecular weight of about 75,000. Presumably this material corresponds to the primary gene product (precursor protein) rather than to prothrombin, but it has not yet been completely characterized.

The first vitamin K dependent *in vitro* system which produced prothrombin was that described by Shah and Suttie (1974). In this system, postmitochondrial supernatants from vitamin K deficient rats were shown to respond to the addition of vitamin K by producing a significant amount of prothrombin as assayed by the standard two-stage assay. Prothrombin was dependent on an energy supply and was inhibited by antagonists of vitamin K but not by inhibitors of polypeptide chain biosynthesis.

4.6.8.2. Vitamin K Dependent Carboxylation

After the vitamin K dependent step in prothrombin synthesis was shown to be the formation of γ-carboxyglutamic acid residues, the same system was used to demonstrate (Esmon *et al.*, 1975*b*) that the addition of vitamin K and $H^{14}CO_3^-$ promoted the carboxylation of microsomal proteins (Fig. 9). This carboxylation had the same requirements as the *in vitro* prothrombin-synthesizing system. It

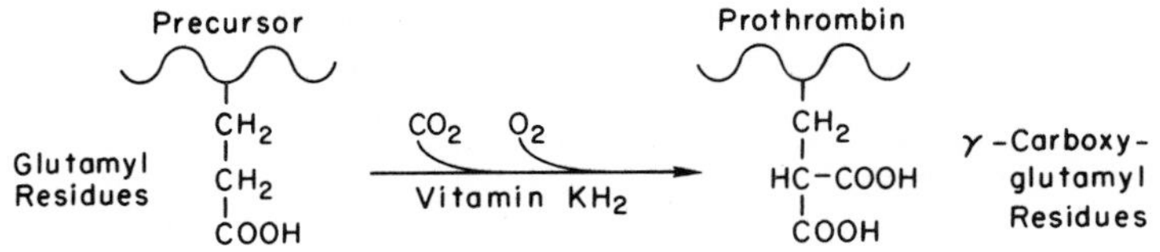

Precursor   Prothrombin

Glutamyl CH_2 CO_2 O_2 CH_2 γ-Carboxy-
Residues CH_2 $\xrightarrow{\text{Vitamin KH}_2}$ $HC-COOH$ glutamyl
                  $COOH$ $COOH$ Residues

Fig. 9. The vitamin K dependent reaction. The reaction requires the reduced form of vitamin K (KH_2) and O_2.

was possible to isolate radioactive prothrombin from this system following incubation and show that essentially all of the incorporated radioactivity was present as γ-carboxyglutamic acid residue in the fragment 1 region of prothrombin. These observations would appear to offer final proof of the role of the vitamin in the biosynthetic process. An *in vivo* demonstration of $H^{14}CO_3^-$ incorporation into prothrombin has also been claimed by Girardot *et al.* (1974).

The vitamin K dependent carboxylase has been studied (Friedman and Shia, 1976; Girardot *et al.*, 1976; Jones *et al.*, 1976; Sadowski *et al.*, 1976) in washed microsomes, where the activity requires the presence of the precursor, O_2, vitamin K, and HCO_3^-, and is stimulated by an energy source and factor(s) present in the postmicrosomal supernatant. A major factor in the supernatant is a protein(s) acting as a NAD^+ ($NADP^+$) reductase. This requirement can be replaced with the addition of the reduced form of the nucleotides (NADH or NADPH) to the system, and the requirement for reducing equivalents from pyridine nucleotides in the systems has now been shown to be largely a requirement for the reduced form of vitamin K. It has also been shown (Friedman and Shia, 1976) that dithiothreitol (DTT) can be used as the source of reducing equivalents for this reaction and that this reducing agent might also be functioning to protect an essential sulfhydryl group in the enzyme system. Studies of the relative activity of various homologues of vitamin K in these systems (Friedman and Shia, 1976; Jones *et al.*, 1976) have not been consistent, and the structural requirement for activity remains open. The carboxylase activity in this microsomal preparation can be inhibited by Warfarin, and this inhibition can be overcome by high concentrations of the vitamin (Sadowski *et al.*, 1976).

The vitamin K dependent carboxylase activity has now been solubilized (Esmon and Suttie, 1976; Girardot *et al.*, 1976; Mack *et al.*, 1976), and the solubilized preparation retains the basic requirement for reduced vitamin K and O_2 of the membrane-associated system. In contrast to the microsomal system, the solubilized system is not inhibited by Warfarin but is still sensitive to a direct vitamin K antagonist such as 2-chloro-3-phytyl-1,4-naphthoquinone and is inhibited by agents which interfere with electron transfer reactions. The solubilized system has been particularly useful in clarifying the need for ATP in the system. Incubation in the absence of ATP and the presence of ATP inhibitor does inhibit the membrane-bound (microsomal) carboxylase but not the solubilized system (Esmon and Suttie, 1976). These data suggest that the energy to drive the carboxylation does not come from ATP, but may come from the reoxidation of the reduced vitamin in the system. A recent study of Jones *et al.* (1977) has indicated that the active species in the carboxylation reaction is CO_2 rather than HCO_3^-. As the biotin-mediated carboxylase requires HCO_3^-, this finding would

Table X. *Properties of the Vitamin K Dependent Carboxylase*[a]

A. Absolute requirements	B. Known inhibitors
Vitamin K and NAD(P)H or	Chloro-K
vitamin KH$_2$	Warfarin (or other coumarins)[b]
O$_2$	p-Hydroxymercuribenzoate
HCO$_3^-$(CO$_2$)	Spin-trapping agents (high concentration)
Presence of precursor	Ethanol
B. Noninhibitory conditions	D. Stimulatory conditions
ATP analogue AMPP(NH)P[c]	Dithiothreitol
Avidin	Additional substrate (peptide)
Cyt P$_{450}$ inhibitors	High salt concentration
EDTA	

[a]There is not complete agreement in the published literature on all points; the properties assigned represent the author's evaluation of the consensus of the published literature.
[b]Only when intact microsomes are present.
[c]Some inhibition when intact microsomes are present.

be consistent with the inability of workers in the field to demonstrate any effect of biotin deficiency or the *in vitro* addition of avidin on this system.

The currently established properties of the vitamin K dependent carboxylase are presented in Table X. Studies of the mechanism of this carboxylase should be facilitated by the observation (Suttie *et al.*, 1976) that the pentapeptide Phe-Leu-Glu-Glu-Val will serve as a substrate for the carboxylase. A solubilized microsomal system which carries out a vitamin K dependent conversion of isolated bovine abnormal prothrombin to prothrombin has also been described (Vermeer *et al.*, 1976). This system appears to have many of the same properties as the vitamin K dependent carboxylase system, but it has not yet been fully characterized. A recent report by Lowenthal and Jaeger (1977) suggests that the role of vitamin K in an *in vitro* system that produces Factor VII is not to promote the carboxylation of precursor glutamyl residues but rather is some other as yet unidentified reaction. The report does not indicate what this step might be, nor does it present any evidence that the carboxylase reaction is not the vitamin-dependent step.

4.6.9. Molecular Role of Vitamin K

Although in one sense the efforts of the past few years have been successful in determining the mechanism of action of vitamin K, in another sense they have only opened the field. The preponderance of evidence would indicate that the vitamin functions to promote the carboxylation of a microsomal precursor to prothrombin and of the other vitamin K dependent proteins. However, there are essentially no data to establish what its role is in this reaction. It appears (Fig. 10) that there are three generalized possibilities: it may function to activate (or transfer) CO$_2$ for this carboxylation; it may function to labilize the hydrogen at the γ-carboxyl of the precursor so that it may accept the CO$_2$; or it may function as an activator of one of the enzymes in this reaction. The last possibility is probably least likely, but there is little information which would indicate which of the other two possibilities is most likely. A role of the vitamin as a membrane-

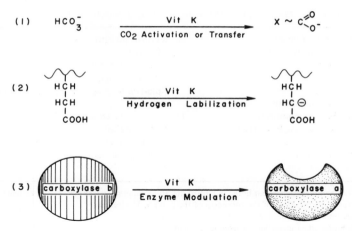

Fig. 10. Possible roles of vitamin K in the carboxylation reaction. The first possibility suggests that the vitamin is involved in either activating or acting as a carrier for CO_2. The "X" shown in the figure could be either vitamin K itself or some unknown carrier. The second possibility suggests that the vitamin is in some unknown way used to labilize a hydrogen on the γ-carbon of the glutamyl residue so that CO_2 can attack this carbon. Although shown in the figure as the formation of a formal carbanion, such an extensive charge separation would not be required. The third possibility suggested is that the vitamin in some unknown way is required to keep the enzyme in an "active" state.

associated lipid-soluble CO_2 carrier analogous to the action of biotin in cytoplasmic systems is attractive, but there is no experimental evidence to support such a role. Likewise, it seems possible that the reoxidation of the reduced form of the vitamin could be coupled to a labilization of a hydrogen on the γ-carbon of the glutamyl residues of the precursor to allow a direct attack of CO_2, but there is no experimental evidence to support this. Advances in this field have been rapid in the last few years but are now being hampered by the crudeness of the carboxylase system. It is likely that definitive studies of the molecular role of the vitamin will be difficult until some success is achieved in purifying the carboxylase activity from liver microsomes.

4.6.10. *Vitamin K Epoxidase and Epoxide Reductase*

Any theory of the mechanism of action of vitamin K must take into consideration the possibility that the formation of vitamin epoxide is involved in the reaction. There has been an extensive investigation of the metabolism of the 2,3-epoxide of vitamin K (K-oxide) over the past few years. Bell and Matschiner (1970, 1972) demonstrated that Warfarin administration blocks the action of a liver enzyme which reduces vitamin K oxide to the vitamin and postulated that Warfarin exerts its anticoagulant effect through a buildup of this metabolite, which then blocks the action of the vitamin. This theory, which has now been shown not to be tenable, is discussed below. More recently, Willingham and Matschiner (1974) have postulated that the formation of the epoxide ("epoxidase" activity) is an obligatory step in the action of the vitamin in promoting prothrombin biosynthesis (Fig. 11). This hypothesis was originally based on observations that "epoxidase" activity increased in liver under various conditions

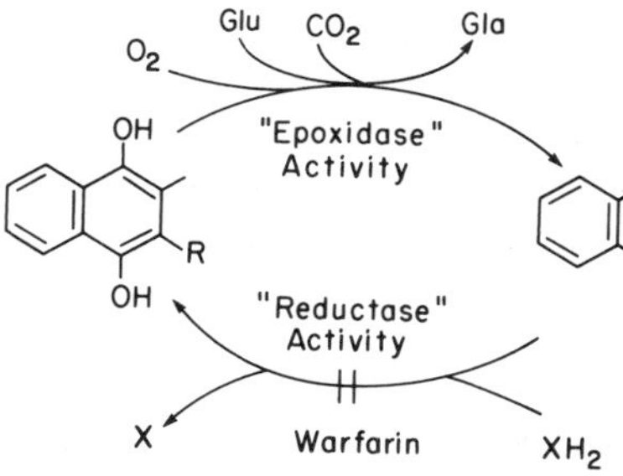

Fig. 11. Action of vitamin K "epoxidase" and "epoxidase reductase." Both of these enzymatic activities are present in the same microsomal preparations that will catalyze the vitamin K dependent carboxylase. The reductase activity is inhibited by Warfarin and other coumarin anticoagulants. The physiologically important reductant in this reaction is not known, but the reaction is most efficiently driven in *in vitro* preparations by dithiothreitol. The substrate for epoxide formation is the reduced form of the vitamin, and it would appear that the reducing equivalents needed to drive the epoxidation are furnished by the reduced vitamin itself. There is no conclusive evidence that the conversion of Glu residues to Gla residues is coupled in some obligatory manner to the epoxidation events, and it may be that carboxylation and epoxidation are two separate events with similar properties which are located in the same preparation.

in a manner that paralleled concentrations of the prothrombin precursor. The theory is also supported by observations on the effects of various anticoagulants in normal and Warfarin-resistant rats (Bell *et al.*, 1976; Willingham *et al.*, 1976). The initial studies of the epoxidase (Willingham and Matschiner, 1974) indicated that the reaction required microsomes, soluble protein, a heat-stable factor, and O_2. Subsequent studies (Willingham *et al.*, 1976) indicated that epoxidase activity was inhibited by tetrachloro-4-pyridinol and Chloro-K. The reaction has now been studied by Sadowski *et al.* (1977), who have shown that the cytoplasmic requirement can be replaced by NADH, and, as with the vitamin K dependent carboxylase, no additional source of reducing equivalents is needed if the hydroquinone of vitamin K is used as a substrate. It was also demonstrated that the oxygen which is incorporated into the epoxide ring arises from molecular oxygen rather than water. In general, those conditions which favor epoxide formation in these microsomal preparations also are conditions which stimulate vitamin K dependent carboxylation. One exception is the lack of dependence of the epoxidase reaction on the addition of HCO_3^- (CO_2). The available evidence is far from conclusive, but it does suggest that both of these reactions involve some components of a microsomal redox system and that they might somehow be coupled.

Properties of the vitamin K epoxide reductase have been less extensively studied. The activity, which is membrane bound and stimulated by thiol compounds, has been studied by Matschiner *et al.* (1974), and by Zimmerman and Matschiner (1974) in crude liver homogenates, and in 10,000g pellets suspended in cytosol. In this system the reduction of the epoxide to the vitamin was stimulated by dithiothreitol (DTT) and inhibited by Warfarin. Preparations from Warfarin-resistant rats were stimulated less by DTT, and were less sensitive to Warfarin.

This reaction has been studied in washed microsomal preparations by Sadowski *et al.* (1977). In this system the reaction is almost completely dependent on the addition of DTT and is blocked by Warfarin. The physiological reducing agent has not been determined, but it has been shown that lipoic acid will function nearly as well as DTT. The reductase activity in Warfarin-resistant rat microsomes is much less sensitive to Warfarin, but is sensitive to Difenacoum, a coumarin anticoagulant that is effective in Warfarin-resistant rats.

4.6.11. Mechanism of Coumarin Action

Other than studies of the role of vitamin K itself, there have been few areas of research that have sustained as much interest as investigations of the mechanism of action of the coumarin anticoagulants. As there have been in the past a number of hypotheses concerning the mechanism of action of vitamin K, it is understandable that numerous theories regarding the mechanism of action of the coumarin anticoagulants have been put forth over a period of years.

4.6.11.1. The Vitamin K Receptor Protein Theory

Although seldom clearly expressed, it has been assumed by many workers that the coumarin anticoagulants are competitive antagonists of vitamin K for a binding site on a receptor protein. This viewpoint is based on the general observation that the actions of one of these compounds can be counteracted by an increased dose of the other. Vitamin K and the coumarins do show some structural similarity to one another, but chemically they are different. This suggests that they might act at a different site on a receptor protein in the same manner as what would now be called a negative allosteric effector of this regulatory protein. This two-site view is based on the apparent chemical dissimilarity of the compounds and observations on the response to mixtures of vitamin K and coumarins. When increasing amounts of an inhibitory ratio of vitamin K and a coumarin are given to rats, the response seen is not the continual inhibition that would be expected from an antagonist and its competitive antagonist, but rather the coumarin inhibition is overcome. This noncompetitive nature of the antagonism was clearly discussed in an early review by Woolley (1947) and has been extensively studied by Lowenthal and his group (Lowenthal, 1970). There has, however, been no experimental determination of what the ratio of the two compounds administered is at any point beyond the injection site, nor is there even any indication that the relative retention of the two compounds in the liver remains the same as the total dose is increased. The general hypothesis of a separate site of Warfarin action was elaborated on and extended by Olson. The observations which suggested that Warfarin and vitamin K both interact with a ribosomal regulatory protein to promote or inhibit the *de novo* synthesis of prothrombin have been reviewed by Olson (1974). This role of Warfarin as an allosteric inhibitor of a vitamin K binding ribosomal protein is, of course, not consistent with the currently accepted role of the vitamin in the microsomal carboxylase reaction.

It has been demonstrated (Thierry *et al.*, 1970; Olson, 1974) that Warfarin is bound to isolated microsomal and ribosomal preparations, and further studies have shown that microsomal Warfarin binding is increased if microsomes are isolated from vitamin K deficient rats, and decreased if animals are given the vitamin just prior to the time they are killed (Lorusso and Suttie, 1972). These data could be interpreted as support for a common binding site for both compounds, or they could indicate that the vitamin caused some conformational change in a protein to alter its binding affinity for Warfarin.

4.6.11.2. The Transport Site Theory

Based largely on experiments where varying ratios of vitamin K and Warfarin or vitamin K and Chloro-K were injected into rats, Lowenthal and his co-workers (Lowenthal and MacFarlane, 1964, 1967; Lowenthal and Birnbaum, 1969) have postulated that Warfarin acts by blocking a normal transport route for vitamin K, while Chloro-K acts as a competitive inhibitor of the vitamin at its physiologically active site. At what membrane this Warfarin-sensitive site functions or what cellular pools of the vitamin it separates is not defined by this theory. It is postulated that high doses of the vitamin overcome the action of coumarins because the vitamin can bypass the specific transport site and reach its receptor protein by an alternate, non-Warfarin-sensitive route after the cellular concentration of the vitamin reaches a certain level. This theory assumes that the coumarin anticoagulants have no effect on whatever metabolic step in the system actually requires the vitamin. The evidence to support the theory is indirect, and the same lack of knowledge of the intracellular concentrations of Warfarin and vitamin K that was mentioned in regard to the receptor protein theory complicates the interpretation. Alternate interpretations of the agonist–antagonist relationship are also consistent with the available data.

4.6.11.3. The Vitamin K Oxide Theory

A theory of coumarin anticoagulant action that received considerable attention was that proposed by Matschiner and Bell following their demonstration (Matschiner *et al.*, 1970) that the 2,3-epoxide of vitamin K (vitamin K oxide) is a normal metabolite of the vitamin in rat liver. They noted (Bell and Matschiner, 1970, 1972; Bell *et al.*, 1972) that the ratio of vitamin K oxide to vitamin K increased when rats were given Warfarin, and postulated that the epoxide acted as a competitive inhibitor of vitamin K at its site of action and that Warfarin was an inhibitor of vitamin K action only to the extent that it increased the cellular ratio of oxide to vitamin. It was also shown that, in addition to the "epoxidase" which forms the vitamin K oxide, rat liver contains an epoxide "reductase" which reduces it back to the vitamin, and that the *in vitro* activity of the reductase was strongly inhibited by Warfarin (Matschiner *et al.*, 1974; Zimmerman and Matschiner, 1974). Much of this theory rested on the observations that when the K-oxide:K ratio in the liver was high, prothrombin production was blocked. Observations (Bell and Caldwell, 1973; Ren *et al.*, 1974; Bell *et al.*, 1976) that rats which have a genetic resistance to Warfarin did not respond to Warfarin administration by increasing the concentration of the epoxide were also consistent with this theory. The data were complicated by the fact that in these experiments Warfarin was always administered to keep the ratio high, and it was difficult to determine if the high oxide level was responsible for the block in prothrombin production, or if it was merely occurring at the same time. It was subsequently shown (Sadowski and Suttie, 1974) that prothrombin synthesis did occur in the presence of a high oxide to vitamin K ratio when a large dose of the epoxide was injected into vitamin K deficient rats in the absence of Warfarin. Goodman *et al.* (1974) also demonstrated that the maintenance of a high liver epoxide to vitamin K ratio in normal animals by continual injection of the epoxide did not cause an inhibition of prothrombin synthesis. These observations and other studies of the correlation between epoxide:K ratios and prothrombin synthesis (Caldwell *et al.*, 1974) appear to rule out a possible role of the epoxide as an inhibitor of vitamin action. They do not, however, invalidate the hypothesis that the inhibition of the epoxide reductase might be involved in the anticoagulant action of Warfarin.

4.6.11.4. Epoxide Reductase Theory

Although it has been convincingly demonstrated that an elevated concentration of vitamin K epoxide does not inhibit the action of vitamin K, it is still possible that the effect of Warfarin on the epoxide reductase is the biologically important action of the drug. If, as has been suggested (Willingham and Matschiner, 1974), the cyclic interconversion of the vitamin to its epoxide and back is required for its action, then an inhibition of this cycle would result in an inhibition of the action of the vitamin. The general observations that were used to support the epoxide inhibition would in a sense support the theory that the vitamin K epoxide reductase is the enzyme sensitive to Warfarin action.

4.6.11.5. Warfarin-Resistant Rats

The availability of strains of wild rats resistant to the coumarin anticoagulants has provided another tool to study the action of the coumarins. These animals are resistant to the coumarins and indandiones but not (Suttie, 1973b) to Chloro-K, and experimental observations made using these animals can be construed to fit all of the theories proposed. The observation that the Warfarin-resistant rats also have a high requirement for vitamin K (Hermodson *et al.*, 1969) suggests that the mutation might well have been at a protein which binds both the vitamin and Warfarin, and it has been observed (Thierry *et al.*, 1970; Lorusso and Suttie, 1972) that microsomes isolated from normal rat liver have a greater Warfarin-binding capacity than those isolated from Warfarin-resistant animals. The available evidence suggests that this binding is a property of a microsomal membrane protein which could be the true vitamin K receptor protein, or it could be a protein involved in the transport of either vitamin K or the coumarins, or even be a Warfarin-metabolizing enzyme. Warfarin metabolism in these rats does not, however, seem to be drastically altered (Pool *et al.*, 1968; Thierry *et al.*, 1970). The lack of Warfarin sensitivity of the "epoxide reductase" in Warfarin-resistant rats suggests that the altered Warfarin-binding protein could be a component of the "reductase" system, and, if this enzyme is involved in the physiological function of the vitamin, an alteration in this enzyme could explain the altered Warfarin sensitivity.

Information from the Warfarin-resistant rats can also be used to support the transport site theory of coumarin action. It has been shown (Shah and Suttie,

1973) that the action of Chloro-K in Warfarin-resistant rats is the same as that observed in normal rats treated with Warfarin. That is, these rats respond to mixtures of Chloro-K and Warfarin just as if the proposed carrier site were blocked by Warfarin. It could then be postulated that the mutation was one that destroyed the transport site, and the Warfarin-binding protein would be assumed to be part of this system. The increased vitamin requirement would be explained as the result of the necessity for all of the vitamin to bypass the transport system and reach its functional site by an alternate route which is active only at high intakes of the vitamin. It should be noted that all of the experimental work reported on Warfarin-resistant rats has been carried out on descendants of a strain trapped in Wales. It is possible that other strains may differ and provide additional information.

4.6.11.6. In Vitro Systems

The development of *in vitro* systems to study the action of vitamin K has provided additional tools to study the mechanism of action of the coumarins. The microsomal vitamin K dependent carboxylase is sensitive to the addition of Warfarin (Sadowski *et al.*, 1976), and when varying ratios of vitamin and Warfarin are administered it responds in the same manner as in an intact rat; that is, at high levels of both Warfarin and vitamin K, the inhibition is overcome. These studies were carried out with intact microsomes, and when the microsomes were solubilized with the detergent Triton X-100 the carboxylase activity was retained, but the system was no longer inhibited by Warfarin (Esmon and Suttie, 1976). These data tend to strongly implicate an interaction of coumarins at a membrane site, and it is significant that the epoxide reductase is not active in these detergent-solubilized preparations.

4.6.11.7. Current Understanding

Some of the theories that have been proposed to explain coumarin action are summarized in Fig. 12. There does not appear to be sufficient evidence at the present time to conclusively support any of these theories, but many can be eliminated by the available data. Although the vitamin K oxide theory generated a great deal of interest, it does not seem tenable at the present time. There also seems little evidence to support the theory that coumarins are allosteric negative effectors of a normal vitamin K binding protein. Much of the support for an allosteric site came from the apparent noncompetitive nature of the antagonism; but, as has been previously discussed, there is no knowledge of what the concentrations of agonist and antagonist are at the critical site. An additional problem with the allosteric site theory is to explain what the normal effector at that site might be. There is certainly no reason to believe that the regulation of clotting factor synthesis should have evolved to include a negative effector site for a compound that would not normally be in the diet and that would be detrimental to the survival of the animals.

Until relatively recently, there were few experimental data that could not be explained by the simple hypothesis that the coumarins interact at the same protein that must interact with vitamin K to promote prothrombin synthesis. The most significant argument against this theory has always been that the antagonism does

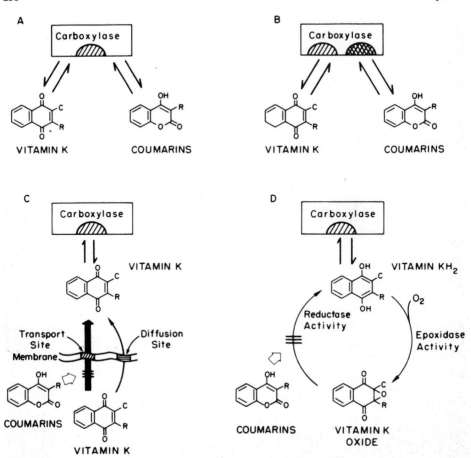

Fig. 12. Theories of coumarin action. The general theory that the coumarin anticoagulants inter-fere with vitamin K action through a direct (A) or an allosteric (B) effect on association of the vita-min with its biochemically important site has received considerable support in the past. It is, how-ever difficult to reconcile this theory with the observations that coumarins are not effective inhibi-tors of the vitamin action in *in vitro* solubilized systems. The vitamin K transport site theory (C) must be considered, but there is still no evidence to indicate the cellular location of the membrane involved. The effectiveness of Warfarin in *in vitro* systems would suggest that it must be the microsomal mem-brane. The vitamin K epoxide reductase theory (D) is consistent with most of the data currently avail-able. The major problem in acceptance of this theory is to explain why the recycling of the vitamin from its epoxide should be so important when there is a large amount of free vitamin in the system. These and other theories of anticoagulant action are discussed in the text.

not appear competitive, but that could be explained as a lack of equilibration of various pools of the vitamin. There is no doubt that coumarins can interact with a number of enzymes that utilize vitamin K. Menadione alkylation is inhibited *in vivo* by Warfarin (Taggart and Matschiner, 1969), coumarin administration *in vivo* increases the conversion of vitamin K to metabolites more polar than the epoxide (Caldwell *et al.*, 1974), coumarins inhibit DT diaphorase (menadione-NADH reductase activity) as well as other vitamin K reductases (Ernster *et al.*,

1973), and, at relatively high concentrations, coumarins and indandiones inhibit the *in vitro* epoxidation of vitamin K (Bell and Stark, 1976). These data suggest that if coumarins can interact with so many other vitamin K binding proteins, they could certainly interact with the vitamin K binding site on the carboxylase, which would suggest that they are better chemical analogues than has usually been assumed. The observations that microsomes from a vitamin K deficient rat will bind more Warfarin than those from a vitamin-sufficient animal would also fit the single-site theory, and the decreased effectiveness of Warfarin and the increased requirement for the vitamin in both the resistant rat and human (O'Reilly, 1971) are consistent with the simple hypothesis that the mutation has been one which lowered the binding affinity of a single protein for both vitamin K and Warfarin. The most significant argument against coumarins acting to directly compete with vitamin K at its site of action is the observation that the solubilized vitamin K dependent carboxylase is not inhibited by the Warfarin. This would seem to suggest that if the vitamin and Warfarin do act at the same site, this is at an enzyme involved in vitamin metabolism rather than the carboxylase system.

There are very few data that cannot be explained by Lowenthal's theory of a Warfarin-sensitive specific transport site for the vitamin. The apparent non-competitive nature of the antagonism, the observations from Warfarin-resistant rats, and the lack of inhibition in solubilized systems would all fit this theory. The mutation which accounts for the resistance could result from alteration of the transport site, and the Warfarin-binding protein could presumably be a component of this site. One problem with this theory is that there is no experimental basis for the hypothesized transport system. Although data (Thierry and Suttie, 1971) on the uptake of vitamin K by the liver compared to other tissues would suggest that there is some specific active transport system or a receptor protein in liver compared to other tissues, it is doubtful that Lowenthal's site could be at the cell membrane. Warfarin does not block the uptake of the vitamin by the liver, nor does it substantially interfere with subcellular localization of the vitamin or its metabolites. There may, however, be intraorganelle pools that are not detected by the relatively crude measurements that have been made. If there is such a pool, it must be very small, as Warfarin acts very rapidly to shut off prothrombin synthesis. If the coumarins were acting only to prevent the transport of the vitamin to a pool where it could then carry out its function, it would appear that synthesis should continue for a limited time while this pool is being depleted. For this theory to be correct, it would seem that the hypothesized transport site would have to be across the microsomal membrane.

The theory that coumarins interfere with the vitamin K epoxide reductase and, therefore, with generation of the reduced vitamin for the carboxylase system is consistent with the observations that the reductase is not active in solubilized systems where the coumarins are inactive. The coumarin sensitivity of this enzyme is also definitely altered in Warfarin-resistant rats. Were it not for this observation, the effect of Warfarin on the reductase would seem to be completely unrelated to the effect of Warfarin on prothrombin synthesis. However, both *in vivo* and *in vitro* this enzyme does appear to be altered in the Warfarin-resistant rats, and it seems rather unlikely that there have been two independent mutational events in these rats, and that both of them involve enzymes that metabolize or interact with vitamin K. The critical question is, however, why is the recycling of the vitamin K epoxide to the vitamin of such importance? Both *in vitro* and *in vivo* it would appear that under conditions of Warfarin administration, where there is a high ratio of epoxide to vitamin, there is still a sufficient amount of vitamin to serve as a substrate for the carboxylase reaction. It is possible that these measurements do not give an indication of the relative concentrations of the important vitamin pool and that continual investigation will be needed to establish if this recycling is of critical importance.

In summary, it appears that data which would definitely select one of these theories of coumarin action over the others are not yet available. However, there are points crucial to each of the theories which are open to experimental attack; and, with an increasing number of data on the action of the vitamin available, and new experimental approaches opening up, a satisfactory explanation for the nature of the antagonisms seems possible in the next few years.

4.7. Non-Clotting-Factor Role of Vitamin K

4.7.1. Vitamin K and Electron Transport

The ability of vitamin K to function in the electron transport chain in certain bacteria has been well documented and has been reviewed in detail (Brodie and Watanabe, 1966; Gel'man *et al.*, 1967; Ramasarma, 1968). Quinone reductases have been demonstrated in a large number of bacterial systems, but in most cases these reductases do not represent a component of the major respiratory chain. The best-studied example of an electron transport chain which appears to specifically require vitamin K is that found in *Mycobacterium phlei* (Brodie *et al.*, 1957). Treatment of extracts from this organism with 360 nm light will cause both the loss of oxidation of NAD^+-linked substrates and the loss of phosphorylation. These activities can be restored by the addition of various forms of vitamin K, but not by the addition of ubiquinone or α-tocopherol (Brodie and Ballantine, 1960*a,b*). Vitamin K in this organism appears to occupy the same position in the respiratory chain as does ubiquinone in mammalian mitochondria, as a link between the two-electron-carrier flavins and the single-electron-carrier cytochromes. An absolute specificity for vitamin K is not common, and a number of bacterial particulate preparations which carry out the oxidation of various substrates have been shown to respond to the addition of either vitamin K or ubiquinone (Gel'man *et al.*, 1967).

Brodie's group (Brodie and Watanabe, 1966) has made an extensive study of oxidative phosphorylation in *M. phlei* in an attempt to demonstrate that a quinol phosphate may be an intermediate in oxidative phosphorylations in this system. Although they have presented evidence for the formation of a quinol phosphate, it cannot be clearly established if this is a product of phosphate transfer from a high-energy phosphate compound in this system, or if it is an obligatory intermediate on the pathway to ATP formation. In addition to their function in the major respiratory chain in a number of organisms, menaquinones have been shown to be associated with the specific reduction of both nitrate (Sasarman *et al.*, 1974) and sulfite (Wagner *et al.*, 1974) in specific bacteria. The evidence that vitamin K has any role in mammalian electron transport (Wosilait, 1961) or mammalian oxidative phosphorylation is even less conclusive. The early claims of Martius and Nitz-Litzow (1953, 1954) that vitamin K deficient animals could not effectively carry out ATP synthesis were not substantiated by further studies (Beyer and Kennison, 1959; Paolucci *et al.*, 1963; Wosilait, 1966; Couri and Wosilait, 1966; Horth *et al.*, 1966; Hill *et al.*, 1966), and it seems unlikely that vitamin K has a function in electron transport and oxidative phosphorylation other than its role as an electron carrier in a limited number of bacteria.

As vitamin K can function as an electron carrier, it is possible to demonstrate a growth requirement for the presence of some type of naphthoquinone in the media of certain bacteria. Such a requirement was first demonstrated by Woolley and McCarter (1940), and a review of the limited amount of information on the subject is available (Koser, 1968). It would appear that relatively few bacteria have an absolute requirement for the vitamin, but that in some other cases the addition of vitamin K to the medium does stimulate growth. Large numbers of bacterial species do synthesize vitamin K active compounds, and a number of studies of the relative distribution of ubiquinone and naphthoquinone in various species are available (Ramasarma, 1968; Gel'man *et al.*, 1967). The data indicate that gram-positive bacteria contain some types of naphthoquinones rather than ubiquinones and that ubiquinones predominate gram-negative organisms. The latter may, however, also contain a form of vitamin K, and *E. coli* contains roughly equal amounts of each. The chemical form of the vitamin in various species has not always been characterized, but menaquinones 7–9 appear to predominate, and both the 2-demethyl analogues and partial hydrogenated members of the menaquinone series have been identified. It is likely that culture conditions can greatly influence the amount of vitamin K present in microorganisms. Variations in oxygen tension have, in particular, been shown to influence both vitamin K and ubiquinone concentrations in some bacteria (Gel'man *et al.*, 1967), and a large increase in menaquinone concentration has been shown to be associated with early stages of sporulation in others (Farrand and Taber, 1974).

4.7.2. *Other Vitamin K Dependent Plasma Proteins*

Although it had been assumed for a number of years that the only vitamin K dependent reactions in animals were those involved in the synthesis of the long-recognized vitamin K dependent plasma clotting factors (II, VII, IX, and X), it is now clear that this is not the case. Through the use of immunochemical methods, Stenflo (1976) demonstrated that there was a fifth vitamin K dependent protein (protein C) in bovine plasma. This protein, in common with the other vitamin-dependent clotting factors, bound to insoluble barium salts and was shown to contain γ-carboxyglutamic acid. It has been demonstrated (Esmon *et al.*, 1976; Kisiel *et al.*, 1976; Bucher *et al.*, 1976) that this protein is an inactive precursor of a serine protease that can be generated by limited trypsin or Factor X_a digestion. Protein C consists of two polypeptide chains, and both the amino acid sequence and the positions of the Gla residues in the amino-terminal region of its light chain show a great deal of homology to the light chain of Factor X, and to the amino-terminal region of prothrombin and Factor IX. The physiological function of this protein is unknown. Current evidence (Esmon *et al.*, 1976; Kisiel *et al.*, 1976) suggests that it is not required in either the extrinsic or intrinsic pathway of blood coagulation as they are currently understood. However, Seegers *et al.* (1976) have reported that protein C is identical to a protein, called "autoprothrombin II-A," involved in epinephrine-induced platelet aggregation which they have previously studied. Further work will be required to establish what role, if any, this protein has in normal hemostatic control. A sixth γ-carboxyglutamic acid-containing plasma protein (protein S) has now been identified in human plasma (DiScipio *et al.*, 1977), which, like protein C, shows considerable structural homology to the long-recognized vitamin K dependent clotting factors. Nothing is known about the physiological role of this protein, but the discovery of these two proteins suggests that other such proteins may be found in the future.

4.7.3. *Nonplasma Vitamin K Dependent Proteins*

The demonstration (Hauschka *et al.*, 1975) that a Gla-containing protein could be isolated from chick bone has raised the possibility that vitamin K dependent proteins might serve important physiological functions in a number of tissues. A similar protein was independently isolated from bovine skeletal tissue by Price *et al.* (1976*a*), and the amino acid sequence of this protein has now been reported (Price *et al.*, 1976*b*). Three of the 49 residues of this small protein have been identified as γ-carboxyglutamyl residues, but there is no apparent structural homology between this protein and the vitamin K dependent plasma proteins. This protein binds tightly to hydroxyapatite crystals, but does not have a strong affinity for amorphous calcium phosphate. The function of this skeletal protein is not known, but its location suggests that it plays some role in either bone calcification or demineralization. The available data are also not sufficient to establish if the proteins that have been isolated are the species that are synthesized, or if they represent a portion of a larger protein that was degraded to leave this low molecular weight protein in the same manner that prothrombin is degraded during activation to leave all of the Gla residues in the F-1 peptide fragment. The presence of hydroxyproline in the bovine protein suggests that it is synthesized in the same cells that synthesize bone collagen rather than being transported to the bone from some other tissue. Preliminary evidence that formation of the bone protein is vitamin K dependent has been presented (Hauschka *et al.*, 1976*a*).

The discovery of a protein(s) in bone, as well as plasma, which contained γ-carboxyglutamic acid has suggested that there might be much more widespread distribution of vitamin K dependent proteins than was once realized. Analyses of a number of calcium-binding proteins from various sources and of barium citrate adsorbing proteins from various plasmas and extracts have indicated (Zytkovicz and Nelsestuen, 1976) the presence of Gla residues only in blood-clotting proteins and vertebrate plasma samples. It has, however, been claimed (Tai and Liu, 1976) that a coagulating enzyme from the *Limulus* crab also contains γ-carboxyglutamic acid. Protein-bound γ-carboxyglutamic acid residues have been identified in some calcium salt containing kidney stones (Lian and Prien, 1976). Neither the physiological significance of these proteins nor their site of formation has yet been determined. The most interesting of the proteins newly recognized to contain Gla residues (Hauschka *et al.*, 1976*b*) is that found in kidney. Kidney cortex contains nearly one residue of Gla for each 10^4 amino acid residues, and, as kidney microsomal preparations are able to carry out a vitamin K dependent incorporation of $H^{14}CO_3^-$ into protein(s), it is doubtful that the Gla residues can be accounted for by fragments of clotting factors being cleared from the plasma. The kidney plays an important role in Ca^{2+} homeostasis, but whether this protein is involved in this process or has some other function has not yet been determined.

In considering the physiological role of these non-clotting-factor vitamin K dependent proteins, the widespread use of vitamin K antagonists as clinical anticoagulants should be considered. Large numbers of patients are routinely given sufficient amounts of coumarin anticoagulants to depress vitamin K dependent clotting factor levels to from 15% to 35% of normal. If vitamin K is required for the synthesis of other proteins with important physiological functions, it might be expected that a number of problems unrelated to clotting factor synthesis would have been observed in these patients. This has not been the case, although fetal bone abnormalities have been reported (Tejani, 1973; Pettifor and Benson, 1975) to be associated with coumarin anticoagulant therapy during the first trimester of pregnancy. There are, however, no data to indicate that these abnormalities were associated with an impaired synthesis of the Gla-containing bone protein, and they may have resulted from pharmacological action of the drug not related to its vitamin K antagonism. A number of factors may have contributed to a failure to observe widespread clinical problems associated with an effect on synthesis of other vitamin K dependent proteins during routine anticoagulant therapy. The vitamin K dependent carboxylase in these other tissues may not be so sensitive to coumarin anticoagulants as is that in the liver, or the normal level of these proteins may be in excess of that needed for their physiological function. It may also be that these proteins do not serve an important physiological function at this time and are merely an evolutionary vestige of a once-important functional system.

4.7.4. Other Effects of Vitamin K Deficiency or Coumarin Treatment

As with any essential nutrient, there have been a number of observations of effects of a deficiency of the nutrient made in the past which do not now seem reasonable in view of the current understanding of its metabolic role. In the case of vitamin K, this is considerably complicated by the fact that the absence of the vitamin and the presence of a coumarin anticoagulant have often been considered to be equivalent conditions. Some of the effects observed from coumarin administration might well be related to a pharmacological action not related to its role as a vitamin K antagonist. Confusing reports suggesting that the role of vitamin K in the control of prothrombin biosynthesis was at the level of DNA transcription, RNA translation, or protein glycosylation have been discussed previously, and there appears to be no support for these views at the present time. The postulation of a generalized effect of vitamin K on glycosylation reactions in both vertebrates and bacteria (Martius *et al.*, 1971) would also appear unfounded. One observation of a specific effect of vitamin K that has not been challenged is the role of vitamin K in the induction of the enzyme 3-ketodihydrosphingosine synthetase in *Bacteroides melaninogenicus*, which was described by Lev and Milford (1971, 1973). The basis for this response has not been determined, but it appears to be a real response to the vitamin. Biezunski (1970) has reported an inhibition of *in vitro* microsomal protein synthesis in preparations from rats previously given an *in vivo* treatment of Warfarin. The significance of this observation is, however, doubtful, as an effect of Warfarin on protein synthesis in rat liver slices was shown not to be due to the vitamin K antagonistic effect of Warfarin (Pool and Borchgrevink, 1964). There is also evidence that hepatic protein synthesis in intact animals (Hill *et al.*, 1968) or isolated perfused livers (Suttie, 1967) and the induction of a specific hepatic protein (Paolucci *et al.*, 1964) are unaffected by vitamin K deficiency. The reports of Bogdanov and Lider (1967, 1968) of alterations in pancreatic enzyme levels by vitamin K deficiency or Warfarin treatment have not been confirmed, nor has the report of von Vogeler *et al.* (1972) been substantiated that coumarin drugs had a specific effect on the synthesis of hepatic enzymes with short half-lives.

The report (Weber *et al.*, 1963) that the C-1 component of complement in chick plasma is vitamin K responsive has not been confirmed. Likewise, the basis for the morphological changes seen in the rough endoplasmic reticulum of rat liver following Warfarin administration (Noonan *et al.*, 1974) has not been explained, nor is there any indication if this condition is of clinical significance. The observations of Olson *et al.* (1968) and Van Buskirk and Kirsch (1973) relative to an alteration in hepatic nucleic acid metabolism in a vitamin K deficiency or following Warfarin treatment have not been adequately explained, but they appear unrelated to the known function of the vitamin. Karpatkin and Karpatkin (1973) have reported that the plasma from coumarin-treated rabbits contains a factor(s) capable of stimulating vitamin K dependent clotting factor synthesis in normal animals. This intriguing response has not been further investigated.

Finally, there is the report of Matthes (1975) that the vitamin K dependent step in prothrombin synthesis is the hydroxylation of an alanine residue in the prothrombin precursor protein to a serine residue in prothrombin and the claim by Lowenthal and Jaeger (1977) that the vitamin K dependent step in prothrombin synthesis is not a carboxylation reaction but is some yet unrecognized reaction. These two reports are in disagreement with the large amount of evidence available to support the vitamin K dependent carboxylation reaction, and further studies are needed to determine if these observations can be rationalized with the data which support such a theory.

4.8. Conclusion

Advances toward a complete understanding of the nutritional and metabolic significance of vitamin K have been made at irregular intervals over the last 35–40 years. Early chemical studies rapidly elucidated the structures of natural compounds which possessed vitamin K activity, and a large number of analogues were synthesized in an effort to understand the structure–function relationships of these compounds. These studies demonstrated that numerous substituted-2-methyl-1,4-naphthoquinones possessed biological activity but that very little could be altered in this basic ring structure without the loss of biological activities.

Early nutritional investigation established that, from a practical nutritional standpoint, poultry species were those in danger of the spontaneous development of a deficiency syndrome, and the supplementation of chick and turkey rations with vitamin K is now a longstanding practice. Clinical nutritionists also established the value of vitamin K supplementation to alleviate the danger of a hemorrhagic disease of the human infant.

Studies of the chemistry of the naturally occurring vitamin K antagonist have led to the synthesis of a wide range of synthetic coumarin derivatives which have been of enormous value, both as rodenticides and as clinical anticoagulants. The widespread use of radioactive isotopes which occurred 15–20 years after the discovery of vitamin K provided the tools to obtain preliminary information on both the biosynthesis and the degradation of the vitamin. Both of these fields are still under active investigation, but the basic metabolic pathways appear to have been established.

Real advances in an understanding of the metabolic role of this vitamin in the regulation of the synthesis of prothrombin and the other vitamin K dependent proteins were not possible until the general process of protein synthesis was established. The last 10–15 years have, however, been a period during which a rapid advance has been made in an understanding of the biochemical function of vitamin K. It has been demonstrated that the vitamin functions as an essential cofactor for a microsomal carboxylase that converts peptide-bound glutamyl residues to γ-carboxyglutamyl residues. This carboxylase appears to be unique in that it requires oxygen and has no demonstrable requirement for ATP. Present evidence would indicate that the energy to drive this carboxylation event comes from the reoxidation of the reduced form of the vitamin, but how this is accomplished has not been determined.

Recent discoveries of γ-carboxyglutamic acid residues in protein other than the long-recognized vitamin K dependent clotting factors have opened new areas of investigation that promise to yield further insight into the physiological and biochemical roles of this vitamin. It would therefore appear that the role of this vitamin at the molecular level might well be established in the next few years.

4.9. References

Aggeler, P. M., White, S. G., Glendening, M. B., Page, E. W., Leake, T. B., and Bates, G., 1952, Plasma thromboplastin component (PTC) deficiency: A new disease resembling hemophilia, *Proc. Soc. Exp. Biol. Med.* **79**:692.

Almquist, H. J., 1941, Vitamin K, *Physiol. Rev.* **21**:194.

Almquist, H. J., 1975, The early history of vitamin K, *Am. J. Clin. Nutr.* **28**:656.

Almquist, H. J., and Klose, A. A., 1939*a*, The anti-hemorrhagic activity of pure synthetic phthiocol, *J. Am. Chem. Soc.* **61**:1611.

Almquist, H. J., and Klose, A. A., 1939*b*, Synthetic and natural antihemorrhagic compounds, *J. Am. Chem. Soc.* **61**:2557.

Almquist, H. J., and Stokstad, E. L. R., 1935*a*, Dietary haemorrhagic disease in chicks, *Nature (London)* **136**:31.

Almquist, H. J., and Stokstad, E. L. R., 1935*b*, Hemorrhagic chick disease of dietary origin, *J. Biol. Chem.* **111**:105.

Anderson, G. F., and Barnhart, M. I., 1964*a*, Intracellular localization of prothrombin, *Proc. Soc. Exp. Biol. Med.* **116**:1.

Anderson, G. F., and Barnhart, M. I., 1964*b*, Prothrombin synthesis in the dog, *Am. J. Physiol.* **206**:929.

Babior, B. M., 1966, The role of vitamin K in clotting factor synthesis. I. Evidence for the participation of vitamin K in the conversion of a polypeptide precursor to factor VII, *Biochim. Biophys. Acta* **123**:606.

Babior, B. M., and Kipnes, R. S., 1970, Vitamin K dependent formation of factor VII by a cell-free system from rat liver, *Biochemistry* **9**:2564.

Baldwin, R. M., Snyder, C. D., and Rapoport, H., 1974, Biosynthesis of bacterial menaquinones: Dissymetry in the naphthalenic intermediate, *Biochemistry* **13**:1523.

Bang, N. U., O'Doherty, G. O. P., and Barton, R. D., 1975, Selective suppression of vitamin K-dependent procoagulant synthesis by compounds structurally unrelated to vitamin K, *Clin. Res.* **23**:521A.

Bell, R. G., and Caldwell, P. T., 1973, The mechanism of warfarin resistance: Warfarin and the metabolism of vitamin K_1, *Biochemistry* **12**:1759.

Bell, R. G., and Matschiner, J. T., 1969*a*, Synthesis and destruction of prothrombin in the rat, *Arch. Biochem. Biophys.* **135**:152.

Bell, R. G., and Matschiner, J. T., 1969*b*, Intracellular distribution of vitamin K in the rat, *Biochim. Biophys. Acta* **184**:597.

Bell, R. G., and Matschiner, J. T., 1970, Vitamin K activity of phylloquinone oxide, *Arch. Biochem. Biophys.* **141**:473.

Bell, R. G., and Matschiner, J. T., 1972, Warfarin and the inhibition of vitamin K activity by an oxide metabolite, *Nature (London)* **237**:32.

Bell, R. G., and Stark, P., 1976, Inhibition of prothrombin synthesis and epoxidation of vitamin K_1 by anticoagulants *in vitro*, *Biochem. Biophys. Res. Commun.* **72**:619.

Bell, R. G., Sadowski, J. A., and Matschiner, J. T., 1972, Mechanism of action of warfarin: Warfarin and metabolism of vitamin K_1, *Biochemistry* **11**:1959.

Bell, R. G., Caldwell, P. T., and Holm, E. E. T., 1976, Coumarins and the vitamin K-K epoxide cycle: Lack of resistance to coumatetralyl in warfarin-resistant rats. *Biochem. Pharmacol.* **25**:1067.

Benson, B. J., and Hanahan, D. J., 1975, Structural studies on bovine prothrombin: Isolation and partial characterization of the Ca^{2+} binding and carbohydrate containing peptides of the *N*-terminus region, *Biochemistry* **14**:3265.

Bentley, R., 1975*a*, Biosynthesis of vitamin K and other natural naphthoquinones, *Pure Appl. Chem.* **41**:47.

Bentley, R., 1975*b*, Biosynthesis of quinones, in: *Biosynthesis*, Vol. 3, Specialists Periodical Reports, Chemical Society, pp. 181–246, Burlington House, London.

Beyer, R. E., and Kennison, R. D., 1959, Relationship between prothrombin time and oxidative phosphorylation in chick liver mitochondria, *Arch. Biochem. Biophys.* **84**:63.

Biezunski, N., 1970, Action of warfarin injected into rats on protein synthesis *in vitro* by liver microsomes as related to its anticoagulating action, *Biochem. Pharmacol.* **19**:2645.

Biggs, R., Douglas, A. S., Macfarlane, R. G., Dacie, J. V., Pitney, W. R., Merskey, C., and O'Brien, J. R., 1952, Christmas disease: A condition previously mistaken for haemophilia, *Br. Med. J.* **2**:1378.

Billeter, M., and Martius, C., 1960, Über die Umwandlung von Phyllochinon (Vitamin K_1) in Vitamin $K_{2(20)}$ im Tierkörper, *Biochem. Z.* **333**:430.

Billeter, M., and Martius, C., 1961, Über die Umwandlung von Vitamin $K_{2(20)}$ und $K_{2(10)}$ in Vitamin $K_{2(20)}$ in Organisms von Vogeln und Saugetieren, *Biochem. Z.* **334**:304.

Billeter, M., Bollinger, W., and Martius, C., 1964, Untersuchungen über die Umwandlung von verfütterten K-Vitaminen durch Austausch der Seitenkette und die Rolle der Darmbakterien hierbei, *Biochem. Z.* **340**:290.

Binkley, S. B., McKee, R. W., Thayer, S. A., and Doisy, E. A., 1940, The constitution of vitamin K_2, *J. Biol. Chem.* **133**:721.

Bjork, I., and Stenflo, J., 1973, A conformational study of normal and dicoumarol-induced prothrombin, *FEBS Lett.* **32**:343.

Bligh, E. G., and Dyer, W. J., 1959, A rapid method of total lipid extraction and purification, *Can. J. Biochem. Physiol.* **37**:911.

Blomstrand, R., and Forsgren, L., 1968, Vitamin K_1-^{3}H in man: Its intestinal absorption and transport in the thoracic duct lymph, *Int. Z. Vit. Forsch.* **38**:46.

Bogdanov, N. G., and Lider, V. A., 1967, The effect of coumarin derivatives on the activity of some external secreted enzymes (in Russian), *Farmakol. Toksikol.* **30**:727.

Bogdanov, N. G., and Lider, V. A., 1968, Role of vitamin K in biosynthesis of some exoenzymes of the digestive organs, *Byul. Exsp. Biol. Med.* **65**:60.

Boyle, C. M., 1960, Case of apparent resistance of *Rattus norvegicus barkenhout* to anticoagulant poisons, *Nature (London)* **188**:517.

Brodie, A. F., and Ballantine, J., 1960a, Oxidative phosphorylation in fractionated bacterial systems. II. The role of vitamin K, *J. Biol. Chem.* **235**:226.

Brodie, A. F., and Ballantine, J., 1960b, Oxidative phosphorylation in fractionated bacterial systems. III. Specificity of vitamin K reactivation, *J. Biol. Chem.* **235**:232.

Brodie, A. F., and Watanabe, T., 1966, Mode of action of vitamin K in microorganisms, *Vitamins Hormones* **24**:447.

Brodie, A. F., Weber, M. M., and Gray, C. T., 1957, The role of vitamin K_1 in coupled oxidative phosphorylation, *Biochim. Biophys. Acta* **25**:448.

Brozovic, M., and Gurd, L. J., 1973, Prothrombin during warfarin treatment, *Br. J. Haematol.* **24**:579.

Bryant, R. W., Jr., and Bentley, R., 1976, Menaquinone biosynthesis: Conversion of *o*-succinylbenzoic acid to 1,4-dihydroxy-2-naphthoic acid and menaquinones by *Escherichia coli* extracts, *Biochemistry* **15**:4792.

Bucher, D., Nebelin, E., Thomsen, J., and Stenflo, J., 1976, Identification of γ-carboxyglutamic acid residues in bovine factors IX and X, and in a new vitamin K-dependent protein, *FEBS Lett.* **68**:293.

Caldwell, P. T., Ren, P., and Bell, R. G., 1974, Warfarin and metabolism of vitamin K, *Biochem. Pharmacol.* **25**:3353.

Campbell, H. A., and Link, K. P., 1941, Studies on the hemorrhagic sweet clover disease. IV. The isolation and crystallization of the hemorrhagic agent, *J. Biol. Chem.* **138**:21.

Campbell, I. M., Robins, D. J., Kelsey, M., and Bentley, R., 1971, Biosynthesis of bacterial menaquinones (vitamins K_2), *Biochemistry* **10**:3069.

Carlisle, T. L., Shah, D. V., Schlegel, R., and Suttie, J. W., 1975, Plasma abnormal prothrombin and microsomal prothrombin precursor in various species, *Proc. Soc. Exp. Biol. Med.* **148**:140.

Cesbron, N., Boyer, C., Guillin, M.-C., and Ménaché, D., 1973, Human coumarin prothrombin: Chromatographic, coagulation and immunologic studies, *Thromb. Diath. Haemorrh.* **30**:437.

Combs, A. B., Porter, T. H., and Folkers, K., 1976, Anticoagulant activity of a naphthoquinone analog of vitamin K and an inhibitor of coenzyme Q_{10}–enzyme systems, *Res. Commun. Chem. Pathol. Pharmacol.* **13**:109.

Cook, S. F., and Scott, K. G., 1935a, Apparent intoxication in poultry due to nitrogenous bases, *Science* **82**:465.

Cook, S. F., and Scott, K. G., 1935b, A bioassay of certain protein supplements when fed to baby chicks, *Proc. Soc. Exp. Biol. Med.* **33**:167.

Couri, D., and Wosilait, W. D., 1966, The effect of coumarin anticoagulants on the adenine nucleotide content and protein synthesis in rat liver, *Biochem. Pharmacol.* **15**:1349.

Dam, H., 1929a, Die Bestimmung von Cholesterin nach der Digitoninmethode, besonders in Hühnereiern und Hühnchen, *Biochem. Z.* **215**:468.

Dam, H., 1929b, Cholesterinstoffwechsel in Hühnereiern und Hühnchen, *Biochem. Z.* **215**:475.

Dam, H., 1930, Über die Cholesterinsynthese in Tierkörper, *Biochem. Z.* **220**:158.

Dam, H., 1934, Haemorrhages in chicks reared on artificial diets: A new deficiency disease, *Nature (London)* **133**:909.

Dam, H., 1935a, The antihaemorrhagic vitamin of the chick: Occurrence and chemical nature, *Nature (London)* **135**:652.

Dam, H., 1935b, The antihaemorrhagic vitamin of the chick, *Biochem. J.* **29**:1273.

Dam, H., 1942, Vitamin K, its chemistry and physiology, *Adv. Encymol.* **2**:285.

Dam, H., 1948, Vitamin K, *Vitamins Hormones* **6**:27.

Dam, H., Schønheyder, F., and Tage-Hansen, E., 1936, Studies on the mode of action of vitamin K, *Biochem. J.* **30**:1075.

Dam, H., Geiger, A., Glavind, J., Karrer, P., Karrer, W., Rothschild, E., and Salomon, H., 1939, Isolierung des Vitamins K in hochgereinigter Form, *Helv. Chim. Acta* **22**:310.

Davie, E. W., and Fujikawa, K., 1975, Basic mechanisms in blood coagulation, *Annu. Rev. Biochem.* **44**:799.

Davie, E. W., and Ratnoff, O. D., 1964, Waterfall sequence for intrinsic blood clotting, *Science* **145**: 1310.

DeLuca, H. F., and Suttie, J. W., 1970, *The Fat-Soluble Vitamins*, University of Wisconsin Press, Madison.

Denson, K. W. E., 1971, The levels of factors, II, VII, IX and X by antibody neutralization techniques in the plasma of patients receiving phenindione therapy, *Br. J. Haematol.* **20**:643.

Dialameh, G. H., Yekundi, K. G., and Olson, R. E., 1970, Enzymatic alkylation of menaquinone-*o* to menaquinones by microsomes from chick liver, *Biochim. Biophys. Acta* **223**:332.

Dialameh, G. H., Taggart, W. V., Matschiner, J. T., and Olson, R. E., 1971, Isolation and characterization of menaquinone-4 as a product of menadione metabolism in chicks and rats, *Int. J. Vit. Nutr. Res.* **41**:391.

DiScipio, R. G., Hermodson, M. A., Yates, S. G., and Davie, E. W., 1977, A comparison of human prothrombin, factor IX (Christmas factor), factor X (Stuart factor), and protein S, *Biochemistry* **16**:698.

Doisy, E. A., 1971, Vitamin K in human nutrition, in: *The Biochemistry, Assay, and Nutritional Value of Vitamin K and Related Compounds*, pp. 79–92, Association of Vitamin Chemists, Chicago.

Doisy, E. A., and Matschiner, J. T., 1970, Biochemistry of Vitamin K, in: *Fat-Soluble Vitamins* (R. A. Morton, ed.), pp. 293–331, Pergamon Press, Oxford.

Downing, M. R., Butkowski, R. J., Clark, M. M., and Mann, K. G., 1975, Human prothrombin activation, *J. Biol. Chem.* **250**:8897.

Drummond, D., 1966, Rats resistant to warfarin, *New Sci.* **30**:771.

Dua, P. N., and Day, E. J., 1966, Vitamin K activity of menadione dimethylpyrimidinol bisulfite in chicks, *Poultry Sci.* **45**:94.

Duello, T. J., and Matschiner, J. T., 1970, Identification of phylloquinone in horse liver, *Arch. Biochem. Biophys.* **138**:640.

Duello, T. J., and Matschiner, J. T., 1971*a*, Characterization of vitamin K from pig liver and dog liver, *Arch. Biochem. Biophys.* **144**:330.

Duello, T. J., and Matschiner, J. T., 1971*b*, The relationship between the storage forms of vitamin K and dietary phylloquinone in the dog, *Int. J. Vit. Nutr. Res.* **41**:180.

Duello, T. J., and Matschiner, J. T., 1972, Characterization of vitamin K from human liver, *J. Nutr.* **102**:331.

Dulock, M. A., and Kolmen, S. N., 1968, Influence of vitamin K on restoration of prothrombin complex proteins and fibrinogen in plasma of depleted dogs, *Thromb. Diath. Haemorrh.* **20**:136.

Elion, J., Benarous, R., and Labie, D., 1976, Isolation and preliminary characterization of a vitamin K dependent peptide from human prothrombin, *Thromb. Haemostas.* **35**:82.

Elliott, G. R., Odam, E. M., and Townsend, M. G., 1976, An assay procedure for the vitamin K_1 2,3-epoxide-reducing system of rat liver involving high-performance liquid chromatography, *Biochem. Soc. Trans.* **4**:615.

Elliott, M. C., Isaacs, B., and Ivy, A. C., 1940, Production of "prothrombin deficiency" and response to vitamin A, D and K, *Proc. Soc. Exp. Biol. Med.* **43**:240.

Enfield, D. L., Ericsson, L. H., Walsh, K. A., Neurath, H., and Titani, K., 1975, Bovine factor X_1 (Stuart factor): Primary structure of the light chain, *Proc. Natl. Acad. Sci. USA* **72**:16.

Ernster, L., Hall, J. M., Lind, C., Rase, B., and Golvano, M. P., 1973, DT diaphorase: Reaction mechanism and possible metabolic function, *Atti Sem. Stud. Biol.* **5**:217.

Esmon, C. T., and Suttie, J. W., 1976, Vitamin K-dependent carboxylase: Solubilization and properties, *J. Biol. Chem.* **251**:6238.

Esmon, C. T., Grant, G. A., and Suttie, J. W., 1975*a*, Purification of an apparent rat liver prothrombin precursor: Characterization and comparison to normal rat prothrombin, *Biochemistry* **14**:1595.

Esmon, C. T., Sadowski, J. A., and Suttie, J. W., 1975*b*, A new carboxylation reaction: The vitamin K-dependent incorporation of $H^{14}CO_3^-$ into prothrombin, *J. Biol. Chem.* **250**:4744.

Esmon, C. T., Suttie, J. W., and Jackson, C. M., 1975*c*, The functional significance of vitamin K action: Difference in phospholipid binding between normal and abnormal prothrombin, *J. Biol. Chem.* **250**:4095.

Esmon, C. T., Stenflo, J., Suttie, J. W., and Jackson, C. M., 1976, A new vitamin K-dependent protein: A phospholipid-binding zymogen of a serine esterase, *J. Biol. Chem.* **251**:3052.

Esnouf, M. P., 1976, The blood clotting mechanism, in: *Progress in Clinical and Biological Research*, Vol. 5 (G. A. Jamieson and T. J. Greenwalt, eds.), pp. 69–84, Alan R. Liss, Inc., New York.

Esnouf, M. P., and Prowse, C. V., 1977, The gamma-carboxy glutamic acid content of human and bovine prothrombin following warfarin treatment, *Biochim. Biophys. Acta* **490**:471–476.

Esnouf, M. P., Lloyd, P. H., and Jesty, J., 1973, A method for the simultaneous isolation of factor X and prothrombin from bovine plasma, *Biochemistry* **131**:781.

Farrand, S. K., and Taber, H. W., 1974, Changes in menaquinone concentration during growth and early sporulation in *Bacillus subtilis, J. Bacteriol.* **117**:324.

Fernlund, P., 1976, γ-Carboxyglutamic acid in human urine, *Clin. Chim. Acta* **72**:147.

Fernlund, P., Stenflo, J., Roepstorff, R., and Thomsen, J., 1975, Vitamin K and the biosynthesis of prothrombin. V. γ-Carboxyglutamic acids, the vitamin K-dependent structures in prothrombin, *J. Biol. Chem.* **250**:6125.

Fieser, L. F., 1939a, Synthesis of 2-methyl-3-phytyl-1,4-naphthoquinone, *J. Amer. Chem. Soc.* **61**:2559.

Fieser, L. F., 1939b, Identity of synthetic 2-methyl-3-phytyl-1,4-naphthoquinone and vitamin K_1, *J. Am. Chem. Soc.* **61**:2561.

Fieser, L. F., Tishler, M., and Sampson, W. L., 1941, Vitamin K activity and structure, *J. Biol. Chem.* **137**:659.

Frick, P. G., Riedler, G., and Brogli, H., 1967, Dose response and minimal daily requirement for vitamin K in man, *J. Appl. Physiol.* **23**:387.

Friedman, P. A., and Shia, M., 1976, Some characteristics of a vitamin K-dependent carboxylating system from rat liver microsomes, *Biochem. Biophys. Res. Commun.* **70**:647.

Ganrot, P. O., and Nilehn, J. E., 1968, Plasma prothrombin during treatment with dicumarol. II. Demonstration of an abnormal prothrombin fraction, *Scand. J. Clin. Lab. Invest.* **22**:23.

Gardner, E. J., 1977, The effect of menadione (MK-o) and its derivatives on the carboxylation of pre-prothrombin in a detergent solubilized rat liver microsomal system, *Fed. Proc.* **36**:307 (abstr.).

Gel'man, N. S., Lukoyanova, M. A., and Ostrovskii, D. N., 1967, *Respiration and Phosphorylation of Bacteria*, Plenum, New York.

Girardot, J.-M., Delaney, R., and Johnson, B. C., 1974, Carboxylation, the completion step in prothrombin biosynthesis, *Biochem. Biophys. Res. Commun.* **59**:1197.

Girardot, J.-M., Mack, D. O., Floyd, R. A., and Johnson, B. C., 1976, Evidence for vitamin K semiquinone as the functional form of vitamin K in the liver vitamin K-dependent protein carboxylation reaction, *Biochem. Biophys. Res. Commun.* **70**:655.

Glover, J., 1970, Biosynthesis of the fat-soluble vitamins, in: *Fat-Soluble Vitamins* (R. A. Morton, ed.), pp. 161–221, Pergamon Press, Oxford.

Goodman, S. R., Houser, R. M., and Olson, R. E., 1974, Ineffectiveness of phylloquinone epoxide as an inhibitor of prothrombin synthesis in the rat, *Biochem. Biophys. Res. Commun.* **61**:250.

Grant, G. A., and Suttie, J. W., 1976a, Rat liver prothrombin precursors: Purification of a second, more basic form, *Biochemistry* **15**:5387.

Grant, G. A., and Suttie, J. W., 1976b, Rat prothrombin: Purification characterization, and activation, *Arch. Biochem. Biophys.* **176**:650.

Green, J. P., Søndergaard, E., and Dam, H., 1956, Intracellular distribution of vitamin K in beef liver, *Biochim. Biophys. Acta* **19**:182.

Griminger, P., 1965, Relative vitamin K potency of two water-soluble menadione analogues, *Poultry Sci.* **44**:211.

Griminger, P., 1966, Biological activity of the various vitamin K forms, *Vitamins Hormones* **24**:605.

Griminger, P., 1971, Nutritional requirements for vitamin K-animal studies, in: *The Biochemistry, Assay, and Nutritional Value of Vitamin K and Related Compounds*, pp. 39–59, Association of Vitamin Chemists, Chicago.

Guillin, M.-C., Aronson, D. L., Bezeaud, A., Ménaché, D., Schlegel, N., and Amar, M., 1977, The purification of human acarboxy prothrombin: Characterization of its derivatives after thrombin cleavage, *Thromb. Res.* **1**:223.

Hadler, M. R., and Shadbolt, R. S., 1975, Novel 4-hydroxycoumarin anticoagulants active against resistant rats, *Nature (London)* **253**:275.

Hadler, M. R., Redfern, R., and Rowe, F. P., 1975, Laboratory evaluation of difenacoum as a rodenticide, *J. Hyg. Camb.* **74**:441.

Hamilton, J. W., and Dallam, R. D., 1968, Isolation of vitamin K from animal tissue, *Arch. Biochem. Biophys.* **123**:514.

Harris, R. S., Wool, I. G., Loraine, J. A., Marrian, G. F., and Thiman, K. V., 1966, *Vitamins and Hormones*, Vol. 24, Academic Press, New York.

Harris, R. S., Munson, P. L., Diczfalusy, E., and Glover, J., 1974, *Vitamins and Hormones*, Vol. 32, Academic Press, New York.

Hart, K. T., 1958, Study of hydrolysis of urinary metabolites of 2-methyl-1,4-naphthoquinone, *Proc. Soc. Exp. Biol. Med.* **97**:848.

Hauschka, P. V., Lian, J. B., and Gallop, P. M., 1975, Direct identification of the calcium-binding amino acid γ-carboxyglutamate, in mineralized tissue, *Proc. Natl. Acad. Sci. USA* **72**:3925.

Hauschka, P. V., Reid, M. L., Lian, J. B., Friedman, P. A., and Gallop, P. M., 1976a, Probable vitamin K-dependence of γ-carboxyglutamate formation in bone, *Fed. Proc.* **35**:1354 (abstr.).

Hauschka, P. V., Friedman, P. A., Traverso, H. P., and Gallop, P. M., 1976b, Vitamin K-dependent γ-carboxyglutamic acid formation by kidney microsomes *in vitro*, *Biochem. Biophys. Res. Commun.* **71**:1207.

Hazell, K., and Baloch, K. H., 1970, Vitamin K deficiency in the elderly, *Gerontol. Clin.* **12**:10.

Hemker, H. C., and Reekers, P. P. M., 1974, Isolation and purification of proteins induced by vitamin K absence, *Thromb. Diath. Haemorrh. Suppl.* **57**:83.

Hemker, H. C., Veltkamp, J. J., Hensen, A., and Loeliger, E. A., 1963, Nature of prothrombin biosynthesis: Preprothrombinaemia in vitamin K-deficiency, *Nature (London)* **200**:589.

Hemker, H. C., Muller, A. D., and Loeliger, E. A., 1970, Two types of prothrombin in vitamin K deficiency, *Thromb. Diath. Haemorrh.* **23**:633.

Henriksen, A., Christensen, T. B., and Helgeland, L., 1976, On the significance of the carbohydrate moieties of bovine prothrombin for clotting activity, *Biochim. Biophys. Acta* **421**:348.

Hermodson, M. A., Suttie, J. W., and Link, K. P., 1969, Warfarin metabolism and vitamin K requirement in the warfarin-resistant rat, *Am. J. Physiol.* **217**:1316.

Hill, R. B., Paul, F., and Johnson, B. C., 1966, Oxygen consumption and NADPH oxidation in microsomes from vitamin K-deficient, warfarin- and dicumarol-treated rats, *Proc. Soc. Exp. Biol. Med.* **121**:1287.

Hill, R. B., Gaetani, S., Paolucci, A. M., RamaRao, P. B., Alden, R., Ranhotra, G. S., Shah, D. V., Shah, V. K., and Johnson, B. C., 1968, Vitamin K and biosynthesis of protein and prothrombin, *J. Biol. Chem.* **243**:3930.

Hollander, D., 1973, Vitamin K_1 absorption by everted intestinal sacs of the rat, *Am. J. Physiol.* **225**:360.

Hollander, D., and Rim, E., 1976, Vitamin K_2 absorption by rat everted small intestinal sacs, *Am. J. Physiol.* **231**:415.

Hollander, D., and Truscott, T. C., 1974a, Colonic absorption of vitamin K-3, *J. Lab. Clin. Med.* **83**:648.

Hollander, D., and Truscott, T. C., 1974b, Mechanism and site of vitamin K-3 small intestinal transport, *Am. J. Physiol.* **226**:1516.

Hollander, D., Muralidhara, K. S., and Rim, E., 1976, Colonic absorption of bacterially synthesized vitamin K_2 in the rat, *Am. J. Physiol.* **230**:251.

Holst, W. F., and Halbrook, E. R., 1933, A "scurvy-like" disease in chicks, *Science* **77**:354.

Horth, C. E., McHale, D., Jeffries, L. R., Price, S. A., Diplock, A. T., and Green, J., 1966, Vitamin K and oxidative phosphorylation, *Biochem. J.* **100**:424.

Hoskin, F. C. G., Spinks, J. W. T., and Jaques L. B., 1954, Urinary excretion products of menadione (vitamin K_3), *Can. J. Biochem. Physiol.* **32**:240.

Hougie, C., Barrow, E. M., and Graham, J. B., 1957, Stuart clotting defect. I. Segregation of an hereditary hemorrhagic state from the heterogeneous group heretofore called "stable factor" (SPCA, proconvertin, factor VII) deficiency, *J. Clin. Invest.* **36**:485.

Howard, J. B., and Nelsestuen, G. L., 1975, Isolation and characterization of vitamin K-dependent region of bovine blood clotting factor X, *Proc. Natl. Acad. Sci. USA* **72**:1281.

Isler, O., and Wiss, O., 1959, Chemistry and biochemistry of the K vitamins, *Vitamins Hormones* **17**:53.

Isler, O., Rüegg, R., Chapard-dit-Jean, L. H., Winterstein, A., and Wiss, O., 1958, Synthese und Isolierung von Vitamin K_2 und isoprenologen Verbindungen, *Helv. Chim. Acta* **41**:786.

IUPAC-IUB, 1966, IUPAC-IUB—Tentative rules: Nomenclature of quinones with isoprenoid side chains, *J. Biol. Chem.* **241**:2989.

Jackson, W. B., and Kaukeinen, D., 1972, Resistance of wild Norway rats in North Carolina to warfarin rodenticide, *Science* **176**:1343.

Jackson, W. B., Brooks, J. E., Bowerman, A. M., and Kaukeinen, D. E., 1975, Anticoagulant resistance in Norway rats as found in U.S. cities, *Pest Control* **43**:12.

Johnson, H. V., Martinovic, J., and Johnson, B. C., 1971, Vitamin K and the biosynthesis of the glycoprotein prothrombin, *Biochem. Biophys. Res. Commun.* **43**:1040.

Johnson, H. V., Boyd, C., Martinovic, J., Valkovich, G., and Johnson, B. C., 1972, A new blood protein which increases with vitamin K deficiency, *Arch. Biochem. Biophys.* **148**:431.

Johnston, M. F. M., and Olson, R. E., 1972a, Studies of prothrombin biosynthesis in cell-free systems. II. Incorporation of L-[U-^{14}C]leucine into prothrombin by rat liver microsomes and ribosomes, *J. Biol. Chem.* **247**:3994.

Johnston, M. F. M., and Olson, R. E., 1972b, Studies of prothrombin biosynthesis in cell-free systems. Regulation by vitamin K and warfarin of prothrombin biosynthesis in rat liver microsomes, *J. Biol. Chem.* **247**:4001.

Jones, J. P., Fausto, A., Houser, R. M., Gardner, E. J., and Olson, R. E., 1976, Effect of vitamin K homologues on the conversion of preprothrombin to prothrombin in rat liver microsomes, *Biochem. Biophys. Res. Commun.* **72**:589.

Jones, J. P., Gardner, E. J., Cooper, T. G., and Olson, R. E., 1977, The species of carbon dioxide utilized in the vitamin K dependent carboxylation of preprothrombin, *Fed. Proc.* **36**:307 (abstr.).

Josso, F., Lavergne, J. M., Gouault, M., Prou-Wartelle, O., and Soulier, J. P., 1968, Differents états moleculaires du facteur II (Prothrombine): Leur etude à l'aide de la staphylocoagulase et d'anticorps anti-facteur II, *Thromb. Diath. Haemorrh.* **20**:88.

Josso, F., Lavergne, J. M., and Soulier, J. P., 1970, Les dysprothrombinemies constitutionelles et acquises, *Nouv. Rev. Fr. Hematol.* **10**:633.

Kabat, H., Stohlman, E. F., and Smith, M. I., 1944, Hypoprothrombinemia induced by administration of indandione derivatives, *J. Pharmacol. Exp. Ther.* **80**:160.

Karpatkin, M., and Karpatkin, S., 1973, Evidence for a humoral agent capable of raising vitamin K-dependent coagulation factors in rabbits, *Br. J. Haematol.* **24**:553.

Karrer, P., and Geiger, A., 1939, Vitamin K aus Alfalfa, *Helv. Chim. Acta* **22**:945.

Karrer, P., Geiger, A., Legler, R., Rüegger, A., and Salomon, H., 1939, Über die Isolierung des α-Phyllochinones (Vitamin K aus Alfalfa) sowie uber dessen Entdeckungsgeschechter, *Helv. Chim. Acta* **22**:1464.

Keenan, W. J., Jewett, T., and Glueck, H. I., 1971, Role of feeding and vitamin K in hypoprothrombinemia of the newborn, *Am. J. Dis. Child.* **121**:271.

Kipfer, R. K., and Olson, R. E., 1970, Reversal by vitamin K of cycloheximide inhibited biosynthesis of prothrombin in the isolated perfused rat liver, *Biochem. Biophys. Res. Commun.* **38**:1041.

Kisiel, W., Ericsson, L. H., and Davie, E. W., 1976, Proteolytic activation of protein C from bovine plasma, *Biochemistry* **15**:4893.

Knauer, T. E., Siegfried, C., Willingham, A. K., and Matschiner, J. T., 1975, Metabolism and biological activity of *cis*- and *trans*-phylloquinone in the rat, *J. Nutr.* **105**:1519.

Knauer, T. E., Siegfried, C. M., and Matschiner, J. T., 1976, Vitamin K requirement and the concentration of vitamin K in rat liver, *J. Nutr.* **106**:1747.

Koller, F., Loeliger, A., and Duckert, F., 1951, Experiments on a new clotting factor (factor VII), *Acta Haematol.* **6**:1.

Konishi, T., Baba, S., and Sone, H., 1973, Whole-body autoradiographic study of vitamin K distribution in rat, *Chem. Pharm. Bull.* **21**:220.

Koser, S. A., 1968, *Vitamin Requirements of Bacteria and Yeast*, Thomas, Springfield, Ill.

Lev, M., and Milford, A. F., 1971, Vitamin K stimulation of sphingolipid synthesis, *Biochem. Biophys. Res. Commun.* **45**:358.

Lev, M., and Milford, A. F., 1973, The 3-ketodihydrosphingosine synthetase of *Bacteroides melaninogenicus:* induction by vitamin K, *Arch. Biochem. Biophys.* **157**:500.

Lian, J. B., and Prien, E. L., Jr., 1976, γ-Carboxyglutamic acid in the calcium binding matrix of certain kidney stones, *Fed. Proc.* **35**:1763 (abstr.).

Link, K. P., 1959, The discovery of dicumarol and its sequels, *Circulation* **19**:97.

Lorusso, D. J., and Suttie, J. W., 1972, Warfarin binding to microsomes isolated from normal and warfarin-resistant rat liver, *Mol. Pharmacol.* **8**:197.

Losito, R., Owen, C. A., Jr., and Flock, E. V., 1967, Metabolism of [C^{14}]menadione, *Biochemistry* **6**:62.

Losito, R., Owen, C. A., Jr., and Flock, E. V., 1968, Metabolic studies of vitamin K$_1$-^{14}C and menadione-^{14}C in the normal and hepatectomized rats, *Thromb. Diath. Haemorrh.* **19**:383.

Lowenthal, J., 1970, Vitamin K analogs and mechanisms of action of vitamin K, in: *The Fat-Soluble Vitamins* (H. F. DeLuca and J. W. Suttie, eds.), pp. 431–436, University of Wisconsin Press, Madison.

Lowenthal, J., and Birnbaum, H., 1969, Vitamin K and coumarin anticoagulants: Dependence of anticoagulant effect on inhibition of vitamin K transport, *Science* **164**:181.

Lowenthal, J., and Chowdhury, M. N. R., 1970, Synthesis of vitamin K₁ analogs: A new class of vitamin K₁ antagonists, *Can. J. Chem.* **48**:3957.

Lowenthal, J., and Jaeger, V., 1977, Synthesis of clotting factors by a cell-free system from rat liver in response to the addition of vitamin K₁ *in vitro, Biochem. Biophys. Res. Commun.* **74**:25.

Lowenthal, J., and MacFarlane, J. A., 1964, The nature of the antagonism between vitamin K and indirect anticoagulants, *J. Pharmacol. Exp. Ther.* **143**:273.

Lowenthal, J., and MacFarlane, J. A., 1965, Vitamin K-like and antivitamin K activity of substituted para-benzoquinones, *J. Pharmacol. Exp. Ther.* **147**:130.

Lowenthal, J., and MacFarlane, J. A., 1967, Use of a competitive vitamin K antagonist, 2-chloro-3-phytyl-1,4-naphthoquinone, for the study of the mechanism of action of vitamin K and coumarin anticoagulants, *J. Pharmacol. Exp. Ther.* **157**:672.

Lowenthal, J., and Simmons, E. L., 1967, Failure of actinomycin D to inhibit appearance of clotting activity by vitamin K *in vitro, Experientia* **23**:421.

Lowenthal, J. MacFarlane, J. A., and McDonald, K. M., 1960, The inhibition of the antidotal activity of vitamin K₁ against coumrin anticoagulant drugs by its chloro analogue, *Experientia* **16**:428.

Lund, M., 1964, Resistance to warfarin in the common rat, *Nature (London)* **203**:778.

MacCorquodale, D. W., Cheney, L. C., Binkley, S. B., Holcomb, W. F., McKee, R. W., Thayer, S. A., and Doisy, E. A., 1939, The constitution and synthesis of vitamin K₁, *J. Biol. Chem.* **131**:357.

Macfarlane, R. G., 1964, Haematology an enzyme cascade in the blood clotting mechanism, and its function as a biochemical amplifier, *Nature (London)* **202**:498.

Mack, D. O., Suen, E. T., Girardot, J.-M., Miller, J. A., Delaney, R., and Johnson, B. C., 1976, Soluble enzyme system for vitamin K-dependent carboxylation, *J. Biol. Chem.* **251**:3269.

Macy, L. R., and Winkler, V. W., 1971, Chemistry of vitamin K and its relationship to biological activity, in: *The Biochemistry, Assay, and Nutritional Value of Vitamin K and Related Compounds*, pp. 61–77, Association of Vitamin Chemists, Chicago.

Magnusson, S., 1973, Primary structure studies on thrombin and prothrombin, 1973, *Thromb. Diath. Haemorrh. Suppl.* **54**:31.

Magnusson, S., Sottrup-Jensen, L., Petersen, T. E., Morris, H. R., and Dell, A., 1974, Primary structure of the vitamin K-dependent part of prothrombin, *FEBS Lett.* **44**:189.

Magnusson, S., Petersen, T. E., Sottrup-Jensen, L., and Claeys, H., 1975, Complete primary structure of prothrombin: Isolation, structure and reactivity of ten carboxylated glutamic acid residues and regulation of prothrombin activation by thrombin, in: *Proteases and Biological Control*, Vol. 2 (E. Reich, D. B. Rifkin, and E. Shaw, eds.), pp. 123–149, Cold Spring Harbor Conference on Cell Proliferation, Cold Spring Harbor, N.Y.

Malhotra, O. P., 1972a, Atypical prothrombins induced by dicoumarol, *Nature (London) New Biol.* **239**:59.

Malhotra, O. P., 1972b, Terminal amino acids of normal and dicoumarol-treated prothrombin, *Life Sci.* **11**:455.

Malhotra, O. P., and Carter, J. R., 1971, Isolation and purification of prothrombin from dicumarolized steers, *J. Biol. Chem.* **246**:2665.

Malhotra, O. P., and Carter, J. R., 1972, Biological and nonbiological activation of normal and dicoumarol-treated prothrombin, 1972, *Life Sci.* **11**:445.

Marshall, F. N., 1972a, 2,3,5,6-Tetrachloro-4-pyridinol: A new chemical structure for anticoagulant activity, *Proc. Soc. Exp. Biol. Med.* **139**:223.

Marshall, F. N., 1972b, Potency and coagulation factor effects of 2,3,5,6-tetrachloropyridinol compared to warfarin and its antagonism by vitamin K, *Proc. Soc. Exp. Biol. Med.* **139**:806.

Martius, C., 1961, Recent investigations on the chemistry and function of vitamin K, in: *Ciba Foundation Symposium on Quinones in Electron Transport* (G. E. W. Wolstenholme and C. M. O'Connor, eds.), pp. 312–326, Little, Brown, Boston.

Martius, C., 1967, Chemistry and function of vitamin K, in: *Blood Clotting Enzymology* (W. H. Seegers, ed.), pp. 551–575, Academic Press, New York.

Martius, C., and Esser, H. O., 1958, Über die Konstitution des im Tierkörper aus Methyl Naphthochinon gebildeten K-Vitamines, *Biochem. Z.* **331**:1.

Martius, C., and Nitz-Litzow, D., 1953, Über den Wirkungsmechanismus des Dicumarols und verwandter Verbindungen, *Biochim. Biophys. Acta* **12**:134.

Martius, C., and Nitz-Litzow, D., 1954, Oxydative Phosphorylierung und Vitamin K Mangel, *Biochim. Biophys. Acta* **13**:152.

Martius, C., Burkart, W., and Stalder, R., 1971, On the mechanism of action of vitamin K in vertebrates and bacteria, *FEBS Lett.* **18**:257.

Matschiner, J. T., 1970, Occurrence and biopotency of various forms of vitamin K, in: *The Fat-Soluble Vitamins* (H. F. DeLuca and J. W. Suttie, eds.), pp. 377–397, University of Wisconsin Press, Madison.

Matschiner, J. T., 1971, Isolation and identification of vitamin K from animal tissue, in: *The Biochemistry, Assay, and Nutritional Value of Vitamin K and Related Compounds*, pp. 21–37, Association of Vitamin Chemists, Chicago.

Matschiner, J. T., and Amelotti, J. M., 1968, Characterization of vitamin K from bovine liver, *J. Lipid Res.* **9**:176.

Matschiner, J. T., and Bell, R. G., 1972, Metabolism and vitamin K activity of *cis* phylloquinone in rats, *J. Nutr.* **102**:625.

Matschiner, J. T., and Bell, R. G., 1973, Effect of sex and sex hormones on plasma prothrombin and vitamin K deficiency, *Proc. Soc. Exp. Biol. Med.* **144**:316.

Matschiner, J. T., and Taggart, W. V., 1967, Separation of vitamin K and associated lipids by reversed-phase partition column chromatography, *Anal. Biochem.* **18**:88.

Matschiner, J. T., and Taggart, W. V., 1968, Bioassay of vitamin K by intracardial injection in deficient adult male rats, *J. Nutr.* **94**:57.

Matschiner, J. T., and Willingham, A. K., 1974, Influence of sex hormones on vitamin K deficiency and epoxidation of vitamin K in the rat, *J. Nutr.* **104**:660.

Matschiner, J. T., Taggart, W. V., and Amelotti, J. M., 1967, The vitamin K content of beef liver, detection of a new form of vitmain K, *Biochemistry* **6**:1243.

Matschiner, J. T., Bell, R. G., Amelotti, J. M., and Knauer, T. E., 1970, Isolation and characterization of a new metabolite of phylloquinone in the rat, *Biochim. Biophys. Acta* **201**:309.

Matschiner, J. T., Zimmerman, A., and Bell, R. G., 1974, The influence of Warfarin on vitamin K epoxide reductase, *Thromb. Diath. Haemorrh. Suppl.* **57**:45.

Matthes, K. J., 1975, Prothrombin—Relationship between structure, function and vitamin K dependence, *Med. Welt* **26**:1777.

Mayer, H., and Isler, O., 1971, Vitamin K group—Chemistry, in: *The Vitamins*, Vol. III (W. H. Sebrell, Jr., and R. S. Harris, eds.), pp. 418–443, Academic Press, New York.

McFarlane, W. D., Graham, W. R., and Richardson, F., 1931, The fat-soluble vitamin requirements of the chick. I. The vitamin A and vitamin D content of fish meal and meat meal, *Biochem. J.* **25**:358.

McKee, R. W., Binkley, S. B., MacCorquodale, D. W., Thayer, S. A., and Doisy, E. A., 1939, The isolation of vitamins K_1 and K_2, *J. Am. Chem. Soc.* **61**:1295.

Morris, H. R., Dell, A., Petersen, T. E., Sottrup-Jensen, L., and Magnusson, S., 1976, Mass-spectrometric identification and sequence location of the ten residues of the new amino acid (γ-carboxyglutamic acid) in the *N*-terminal region of prothrombin, *Biochem. J.* **153**:663.

Morrison, S. A., and Esnouf, M. P., 1973, The nature of the heterogeneity of prothrombin during dicoumarol therapy, *Nature (London) New Biol.* **242**:92.

Morrissey, J. J., Jones, J. P., and Olson, R. E., 1973, Isolation and characterization of isoprothrombin in the rat, *Biochem. Biophys. Res. Commun.* **54**:1075.

Munns, T. W., Johnston, M. F. M., Liszewski, M. K., and Olson, R. E., 1976, Vitamin K-dependent synthesis and modification of precursor prothrombin in cultured H-35 hepatoma cells, *Proc. Natl. Acad. Sci. USA* **73**:2803.

Nardacci, N. J., Jones, J. P., Hall, A. L., and Olson, R. E., 1975, Synthesis of nascent prothrombin and albumin in a heterologous system using rat liver messenger RNA purified on oligo (dT)-cellulose, *Biochem. Biophys. Res. Commun.* **64**:51.

Nelsestuen, G. L., and Suttie, J. W., 1971, Properties of asialo and aglycoprothrombin, *Biochem. Biophys. Res. Commun.* **45**:198.

Nelsestuen, G. L., and Suttie, J. W., 1972a, The purification and properties of an abnormal prothrombin protein produced by dicoumarol-treated cows: A comparison to normal prothrombin, *J. Biol. Chem.* **247**:8176.

Nelsestuen, G. L., and Suttie, J. W., 1972*b*, Mode of action of vitamin K: Calcium binding properties of bovine prothrombin, *Biochemistry* **11**:4961.

Nelsestuen, G. L., and Suttie, J. W., 1972*c*, The carbohydrate of bovine prothrombin: Partial structural determination demonstrating the presence of α-galactose residues, *J. Biol. Chem.* **247**:6096.

Nelsestuen, G. L., and Suttie, J. W., 1973, The mode of action of vitamin K: Isolation of a peptide containing the vitamin K-dependent portion of prothrombin, *Proc. Natl. Acad. Sci. USA* **70**:3366.

Nelsestuen, G. L., Zytkovicz, T. H., and Howard, J. B., 1974, The mode of action of vitamin K: Identification of γ-carboxyglutamic acid as a component of prothrombin, *J. Biol. Chem.* **249**:6347.

Noonan, S. M., Lorusso, D. J., Barnhart, M. I., and Suttie, J. W., 1974, Warfarin effect on rat liver ultrastructure, RNA content and polysome distribution, *Thromb. Diath. Haemorrh.* **32**:366.

Nyquist, S. E., Matschiner, J. T., and Morré, D. J. J., 1971, Distribution of vitamin K among rat liver cell fractions, *Biochim. Biophys. Acta* **244**:645.

Olson, J. P., Miller, L. L., and Troup, S. B., 1966, Synthesis of clotting factors by the isolated perfused rat liver, *J. Clin. Invest.* **45**:690.

Olson, R. E., 1964, Vitamin K induced prothrombin formation: Antagonism by actinomycin D, *Science* **145**:926.

Olson, R. E., 1974, New concepts relating to the mode of action of vitamin K, *Vitamins Hormones* **32**:483.

Olson, R. E., Philipps, G., and Wang, N., 1968, The regulatory action of vitamin K, *Adv. Enzyme Regulation* **6**:213.

O'Reilly, R. A., 1971, Vitamin K in hereditary resistance to oral anticoagulant drugs, *Am. J. Physiol.* **221**:1327.

O'Reilly, R. A., 1976, Vitamin K and the oral anticoagulant drugs, *Annu. Rev. Med.* **27**:245.

Owen, C. A., 1971, Vitamin K group: Deficiency effects in animals and human beings, in: *The Vitamins*, Vol. III (W. H. Sebrell, Jr., and R. S. Harris, eds.), pp. 470–491, Academic Press, New York.

Owen, C. A., Jr., Magath, T. B., and Bollman, J. L., 1951, Prothrombin conversion factors in blood coagulation, *Am. J. Physiol.* **166**:1.

Paolucci, A. M., Rao, P. B. R., and Johnson, B. C., 1963, Vitamin K deficiency and oxidative phosphorylation, *J. Nutr.* **81**:17.

Paolucci, A. M., Gaetani, S., and Johnson, B. C., 1964, Vitamina K e sintesi proteica. I. Sintesi indotta di triptofano pirrolasi in ratti carenti di vitamina K, *Quad. Nutr.* **24**:275.

Parmar, S. S., and Lowenthal, J., 1962, Oxidative phosphorylation in mitochondria from animals treated with 2-chloro-3-phytyl-l,4-naphthoquinone, an antagonist of vitamin K_1, *Biochem. Biophys. Res. Commun.* **8**:107.

Pereira, M., and Couri, D., 1971, Studies on the site of action of dicumarol on prothrombin biosynthesis, *Biochim. Biophys. Acta* **237**:348.

Pereira, M. A., and Couri, D., 1972, Site of inhibition by dicoumarol of prothrombin biosynthesis: Carbohydrate content of prothrombin from dicoumarol-treated rats, *Biochim. Biophys. Acta* **261**:375.

Pettifor, J. M., and Benson, R., 1975, Congenital malformations associated with the administration of oral anticoagulants during pregnancy, *J. Pediatr.* **86**:459.

Polson, J. B., and Wosilait, W. D., 1969, Effect of selected antibiotics on prothrombin time of rats and the relationship to coumarin anticoagulants, *Proc. Soc. Exp. Biol. Med.* **132**:963.

Pool, J. G., and Borchgrevink, C. F., 1964, Comparison of rat liver response to coumarin administered *in vivo* versus *in vitro*, *Am. J. Physiol.* **206**:229.

Pool, J. G., O'Reilly, R. A., Schneiderman, L. J., and Alexander, M., 1968, Warfarin resistance in the rat, *Am. J. Physiol.* **215**:627.

Price, P. A., Otsuka, A. S., Poser, J. W., Kristaponis, J., and Raman, N., 1976*a*, Characterization of a γ-carboxyglutamic acid-containing protein from bone, *Proc. Natl. Acad. Sci. USA* **73**:1447.

Price, P. A., Poser, J. W., and Raman, N., 1976*b*, Primary structure of the γ-carboxyglutamic acid-containing protein from bovine bone, *Proc. Natl. Acad. Sci. USA* **73**:3374.

Prowse, C. V., Mattock, P., Esnouf, M. P., and Russell, A. M., 1976, A variant of prothrombin induced in cattle by prolonged administration of warfarin, *Biochim. Biophys. Acta* **432**:265.

Prydz, H., 1964, Studies on proconvertin (factor VII). V. Biosynthesis in suspension cultures of rat liver cells, *Scan. J. Clin. Lab. Invest.* **16**:540.

Pyörälä, K., 1965, Determinants of the clotting factor response to warfarin in the rat, *Ann. Med. Exp. Biol. Fenn.* **43**:Suppl. 3.

Quick, A. J., 1937, The coagulation defect in sweet clover disease and in the hemorrhagic chick disease of dietary origin: A consideration of the source of prothrombin, *Am. J. Physiol.* **118:**260.

Ramasarma, T., 1968, Lipid quinones, *Adv. Lipid Res.* **6:**107.

Ranhotra, G. S., and Johnson, B. C., 1969, Vitamin K and the synthesis of factors VII–X by isolated rat liver cells, *Proc. Soc. Exp. Biol. Med.* **132:**509.

Reekers, P. P. M., Lindhout, M. J., Kop-Klaassen, B. H. M., and Hemker, H. C., 1973, Demonstration of three anomalous plasma proteins induced by vitamin K antagonist, *Biochim. Biophys. Acta* **317:**559.

Ren, P., Laliberte, R. E., and Bell, R. G., 1974, Effects of warfarin, phenylindanedione, tetrachloro-pyridinol, and chloro-vitamin K_1 on prothrombin synthesis and vitamin K metabolism in normal and warfarin-resistant rats, *Mol. Pharmacol.* **10:**373.

Renk, E., and Stoll, W. G., 1968, Orale antikoagulantien, *Progr. Drug Res.* **11:**226.

Rennison, B. D., and Hadler, M. R., 1975, Field trials of difenacoum against Warfarin-resistant infestations of *Rattus norvegicus*, *J. Hyg. Camb.* **74:**449.

Rez, G., and Prydz, H., 1971, The mode of action of warfarin: An intracellular precursor for factor VII in rat liver, *Biochim. Biophys. Acta* **244:**495.

Rietz, P., Gloor, U., and Wiss, O., 1970, Menachinone aus menschlicher Leber und Faulschlamma, *Int. Z. Vitaminforsch.* **40:**351.

Sadowski, J. A., 1976, The *in vivo* and *in vitro* metabolism of vitamin K_1 during the synthesis of prothrombin: The significance of the cyclic interconversion of vitamin K_1 and its 2,3-epoxide, Ph. D. thesis, University of Wisconsin-Madison.

Sadowski, J. A., and Suttie, J. W., 1974, Mechanism of action of coumarins: Significance of vitamin K epoxide, *Biochemistry* **13:**3696.

Sadowski, J. A., Esmon, C. T., and Suttie, J. W., 1976, Vitamin K-dependent carboxylase: Requirements of the rat liver microsomal enzyme system, *J. Biol. Chem.* **251:**2770.

Sadowski, J. A., Schnoes, H. K., and Suttie, J. W., 1977, Vitamin K epoxidase: Properties and relationship to prothrombin synthesis, *Biochemistry* **16:**3856.

Sasarman, A., Purvis, P., and Portelance, V., 1974, Role of menaquinone in nitrate respiration in *Staphylococcus aureus*, *J. Bacteriol.* **117:**911.

Scott, M. L., 1966, Vitamin K in animal nutrition, *Vitamins Hormones* **24:**633.

Sebrell, W. H., Jr., and Harris, R. S., 1954, *The Vitamins*, Vol. II, Academic Press, New York.

Sebrell, W. H., Jr., and Harris, R. S., 1971, *The Vitamins*, Vol. III, Academic Press, New York.

Seegers, W. H., Novoa, E., Henry, R. L., and Hassouna, H. I., 1976, Relationship of "new" vitamin K-dependent protein C and "old" autoprothrombin II-A, *Thromb. Res.* **8:**543.

Seerley, R. W., Charles, O. W., McCampbell, H. C., and Bertsch, S. P., 1976, Efficacy of menadione dimethylpyrimidinol bisulfite as a source of vitamin K in swine diets, *J. Anim. Sci.* **42:**599.

Shah, D. V., and Suttie, J. W., 1971, Mechanism of action of vitamin K: Evidence for the conversion of a precursor protein to prothrombin in the rat, *Proc. Natl. Acad. Sci. USA* **68:**1653.

Shah, D. V., and Suttie, J. W., 1972, The effect of vitamin K and warfarin on rat liver prothrombin concentrations, *Arch. Biochem. Biophys.* **150:**91.

Shah, D. V., and Suttie, J. W., 1973, The chloro analog of vitamin K: Antagonism of vitamin K action in normal and Warfarin-resistant rats, *Proc. Soc. Exp. Biol. Med.* **143:**775.

Shah, D. V., and Suttie, J. W., 1974, The vitamin K-dependent, *in vitro* production of prothrombin, *Biochem. Biophys. Res. Commun.* **60:**1397.

Shah, D. V., Tews, J. K., Harper, A. E., and Suttie, J. W., 1978, Metabolism and transport of γ-carboxyglutamic acid, *Biochim. Biophys. Acta* (in press).

Shah, D. V., Suttie, J. W., and Grant, G. A., 1973, A rat liver protein with potential thrombin activity: Properties and partial purification, *Arch. Biochem. Biophys.* **159:**483.

Shearer, M. J., and Barkhan, P., 1973, Studies on the metabolites of phylloquinone (vitamin K_1) in the urine of man, *Biochim. Biophys. Acta* **297:**300.

Shearer, M. J., Barkhan, P., and Webster, G. R., 1970, Absorption and excretion of an oral dose of tritiated vitamin K_1 in man, *Br. J. Haematol.* **18:**297.

Shearer, M. J., Mallinson, C. N., Webster, G. R., and Barkhan, P., 1972, Clearance from plasma and excretion in urine, faeces and bile of an intravenous dose of tritiated vitamin K_1 in man, *Br. J. Haematol.* **22:**579.

Shearer, M. J., McBurney, A., and Barkhan, P., 1973, Effect of warfarin anticoagulation on vitamin-K_1 metabolism in man, *Br. J. Haematol.* **24:**471.

Shearer, M. J., McBurney, A., and Barkan, P., 1974, Studies on the absorption and metabolism of phylloquinone (vitamin K_1) in man, *Vitamins Hormones* **32**:513.

Shineberg, B., and Young, I. G., 1976, Biosynthesis of bacterial menaquinones: The membrane-associated 1,4-dihydroxy-2-naphthoate octaprenyltransferase of *Escherichia coli*, *Biochemistry* **15**:2754.

Skotland, T., Holm, T., Østerud, B., Flengsrud, R., and Prydz, H., 1974, The localization of a vitamin K-induced modification in an *N*-terminal fragment of human prothrombin, *Biochem. J.* **143**:29.

Sommer, P., and Kofler, M., 1966, Physicochemical properties and methods of analysis of phyllo-quinones, menaquinones, ubiquinones, phosphoquinones, menadione and related compounds, *Vitamins Hormones* **24**:349.

Stahmann, M. A., Huebner, C. F., and Link, K. P., 1941, Studies on the hemorrhagic sweet clover disease. V. Identification and synthesis of the hemorrhagic agent, *J. Biol. Chem.* **138**:513.

Stenflo, J., 1970, Dicumarol-induced prothrombin in bovine plasma, *Acta Chem. Scand.* **24**:3762.

Stenflo, J., 1972, Vitamin K and the biosynthesis of prothrombin. II. Structural comparison of normal and dicoumarol-induced bovine prothrombin, *J. Biol. Chem.* **247**:8167.

Stenflo, J., 1973, Vitamin K and the biosynthesis of prothrombin. III. Structural comparison of an NH_2 terminal fragment from normal and from dicoumarol-induced bovine prothrombin, *J. Biol. Chem.* **248**:6325.

Stenflo, J., 1974, Vitamin K and the biosynthesis of prothrombin. IV. Isolation of peptides containing prosthetic groups from normal prothrombin and the corresponding peptides from dicoumarol-induced prothrombin, *J. Biol. Chem.* **249**:5527.

Stenflo, J., 1976, A new vitamin K-dependent protein: Purification from bovine plasma and preliminary characterization, *J. Biol. Chem.* **251**:355.

Stenflo, J., and Ganrot, P. O., 1972, Vitamin K and the biosynthesis of prothrombin. I. Identification and purification of a dicoumarol-induced abnormal prothrombin from bovine plasma, *J. Biol. Chem.* **247**:8160.

Stenflo, J., and Ganrot, P. O., 1973, Binding of Ca^{2+} to normal and dicoumarol-induced prothrombin, *Biochem. Biophys. Res. Commun.* **50**:98.

Stenflo, J., and Suttie, J. W., 1977, Vitamin K-dependent formation of γ-carboxyglutamic acid, *Annu. Rev. Biochem.* **46**:157.

Stenflo, J., Fernlund, P., Egan, W., and Roepstorff, P., 1974, Vitamin K dependent modifications of glutamic acid residues in prothrombin, *Proc. Natl. Acad. Sci. USA* **71**:2730.

Suttie, J. W., 1967, Control of prothrombin and factor VII biosynthesis by vitamin K, *Arch. Biochem. Biophys.* **118**:166.

Suttie, J. W., 1970, The effect of cycloheximide administration on vitamin K-stimulated prothrombin formation, *Arch. Biochem. Biophys.* **141**:571.

Suttie, J. W., 1973*a*, Mechanism of action of vitamin K: Demonstration of a liver precursor of prothrombin, *Science* **179**:192.

Suttie, J. W., 1973*b*, Anticoagulant-resistant rats: Possible control by the use of the chloro analog of vitamin K, *Science* **180**:741.

Suttie, J. W., 1974, Metabolism and properties of a liver precursor to prothrombin, *Vitamins Hormones* **32**:463.

Suttie, J. W., and Jackson, C. M., 1977, Prothrombin structure, activation, and biosynthesis, *Physiol. Rev.* **57**:1.

Suttie, J. W., Grant, G. A., Esmon, C. T., and Shah, D. V., 1974, Postribosomal function of vitamin K in prothrombin synthesis, *Mayo Clin. Proc.* **49**:933.

Suttie, J. W., Hageman, J. M., Lehrman, S. R., and Rich, D. H., 1976, Vitamin K-dependent car-boxylase: Development of a peptide substrate, *J. Biol. Chem.* **251**:5827.

Taggart, W. V., and Matschiner, J. T., 1969, Metabolism of menadione-6,7-^{3}H in the rat, *Biochemistry* **8**:1141.

Tai, J. Y., and Liu, T.-Y., 1976, Isolation of a coagulating enzyme from *Limulus polyphemus*, *Fed. Proc.* **35**:1486.

Taylor, J. D., Millar, G. J., Jaques, L. B., and Spinks, J. W. T., 1956, The distribution of administered vitamin K_1-^{14}C in rats, *Can. J. Biochem. Physiol.* **34**:1143.

Tejani, N., 1973, Anticoagulant therapy with cardiac valve prosthesis during pregnancy, *Obstet. Gynecol.* **42**:785.

Thierry, M. J., and Suttie, J. W., 1969, Distribution and metabolism of menadiol diphosphate in the rat, *J. Nutr.* **97**:512.

Thierry, M. J., and Suttie, J. W., 1971, Effect of warfarin and the chloro analog of vitamin K on phylloquinone metabolism, *Arch. Biochem. Biophys.* **147**:430.

Thierry, M. J., Hermodson, M. A., and Suttie, J. W., 1970, Vitamin K and warfarin distribution and metabolism in the warfarin-resistant rat, *Am. J. Physiol.* **219**:854.

Van Buskirk, J. J., and Kirsch, W. M., 1973, Loss of hepatoma ribosomal RNA during warfarin therapy, *Biochem. Biophys. Res. Commun.* **52**:562.

Vermeer, C., Soute, B. A. M., Govers-Riemslag, J., and Hemker, H. C., 1976, *In vitro* prothrombin synthesis from a purified precursor protein. I. Development of a bovine liver cell-free system, *Biochim. Biophys. Acta* **444**:926.

von Vogeler, E. F., Reinauer, H., and Hollmann, S., 1972, The influence of vitamin K on clotting factors and enzymes with short half lives in rat liver, *Z. Klin. Chem. Klin. Biochem.* **10**:207.

Wagner, G. C., Kassner, R. J., and Kamen, M. D., 1974, Redox potentials of certain vitamins K: Implications for a role in sulfite reduction by obligately anaerobic bacteria, *Proc. Natl. Acad. Sci. USA* **71**:253.

Wallin, R., and Prydz, H., 1975, Purification from bovine blood of the warfarin-induced precursor of prothrombin, *Biochem. Biophys. Res. Commun.* **62**:398.

Weber, F., and Wiss, O., 1971, Vitamin K group: Active compounds and antagonists, in: *The Vitamins*, Vol. III (W. H. Sebrell, Jr., and R. S. Harris, eds.), pp. 457–466, Academic Press, New York.

Weber, F., Wiss, O., and Isliker, H., 1963, Über die Beeinflussung des Serumkomplementes durch Vitamin-K-mangel beim Küken, *Experientia* **19**:142.

Willingham, A. K., and Matschiner, J. T., 1974, Changes in phylloquinone epoxidase activity related to prothrombin synthesis and microsomal clotting activity in the rat, *Biochem. J.* **140**:435.

Willingham, A. K., Laliberte, R. E., Bell, R. G., and Matschiner, J. T., 1976, Inhibition of vitamin K epoxidase by two non-coumarin anticoagulants, *Biochem. Pharmacol.* **25**:1063.

Wiss, O., and Gloor, H., 1966, Absorption, distribution, storage and metabolites of vitamin K and related quinones, *Vitamins Hormones* **24**:575.

Woolley, D. W., 1947, Recent advances in the study of biological competition between structurally related compounds, *Physiol. Rev.* **27**:308.

Woolley, D. W., and McCarter, J. R., 1940, Antihemorrhagic compounds as growth factors for the Johne's bacillus, *Proc. Soc. Exp. Biol. Med.* **45**:357.

Wosilait, W. D., 1961, Role of vitamin K in electron transport, *Fed. Proc.* **20**:1005.

Wosilait, W. D., 1966, Effect of vitamin K deficiency on the adenosine nucleotide content of chicken liver, *Biochem. Pharmacol.* **15**:204.

Zimmerman, A., and Matschiner, J. T., 1974, Biochemical basis of hereditary resistance to warfarin in the rat, *Biochem. Pharmacol.* **23**:1033.

Zytkovicz, T. H., and Nelsestuen, G. L., 1976, γ-Carboxyglutamic acid distribution, *Biochim. Biophys. Acta* **444**:344.

Index